COMMUNICATION SYSTEMS ANALYSIS AND DESIGN
a systems approach

Richard A. Williams, Ph.D., P.E.
University of Akron

PRENTICE-HALL, INC., Englewood Cliffs, New Jersey 07632

Library of Congress Cataloging-in-Publication Data

WILLIAMS, RICHARD A.
  Communication systems analysis and design.

  Bibliography: p.
  Includes index.
  1. Telecommunication systems.   I. Title.
TK5101.W46   1987   621.38′0413   86-17015
ISBN 0-13-153777-6

Editorial/production supervision: *Edith Riker*
Cover design: *Lundgren Graphics, Ltd.*
Manufacturing buyer: *Rhett Conklin*

Printed in the United States of America

10   9   8   7   6   5   4   3   2   1

ISBN   0-13-153777-6   01

Prentice-Hall International (UK) Limited, *London*
Prentice-Hall of Australia Pty. Limited, *Sydney*
Prentice-Hall of Canada Inc., *Toronto*
Prentice-Hall Hispanoamericana, S.A., *Mexico*
Prentice-Hall of India Private Limited, *New Delhi*
Prentice-Hall of Japan, Inc., *Tokyo*
Prentice-Hall of Southeast Asia Pte. Ltd., *Singapore*
Editora Prentice-Hall do Brasil, Ltda., *Rio de Janeiro*

CONTENTS

This book is dedicated to my wife, Dawn, and to my four boys—Mark, Matthew, Dylan, and Collin—with apologies for the many hours they had to spend without me while the manuscript was in preparation.

PREFACE

This book is the outgrowth of a set of notes that was developed over a period of several years for what was first a one-quarter (10-week) course that grew into a full-year course sequence (two 15-week semesters), and later expanded again with the addition of another one-semester course in data communications.

The objective of this set of courses is to prepare the senior-year electrical engineering student with enough of a theoretical and technical background in the broad field of communications systems to allow the student to move into any of the many specialty areas which are involved with modern communication systems with enough of an understanding of the field as a whole to see how and where his or her specialty area interfaces with the entire system. The theoretical and technical background of the student is assumed to include the following: mathematics through an understanding of basic integral calculus and basic solid geometry; modern physics (solid-state theory and devices, atomic-line emission and absorption phenomena, thermal radiation theory, etc.); electrical network analysis techniques; basic analog and digital electronic devices and circuits; basic electromagnetics; and the fundamentals of electrical transmission-line theory.

Supplemental topic areas that might prove helpful include random-signal theory and analysis techniques; communication and noise-signal theory; advanced electromagnetics (including T-line fields and antenna, propagation, and scattering theory); quantum electronics (laser and maser theory); advanced electronic circuitry (including integrated-circuit theory and techniques, etc.); digital computer hardware, architecture, systems, and programming; linear programming and scheduling techniques; optics; and physiological responses to audio and visual stimuli. Which of this long list of supplementary topics of study a student might select to complement a course sequence in basic communications systems would depend greatly upon the student's own interest in one or more of these speciality areas.

The general outline of this book is as follows:

Part I: Chapters 1 through 6 An introduction to a number of the different background areas which are basic to an understanding of many different types of communications systems.

Part II: Chapters 7 through 12 An examination of some of the typical important examples of present-day communications system technology. Rather than trying to cover all the possible types of systems which currently exist, the objective in these chapters is to illustrate the application of various techniques and methods to some of the more common types of systems.

As was mentioned earlier, no one book has covered in sufficient depth all the topics that were felt to be important to a communication *systems* course sequence, although many good texts are available that cover bits and pieces of the material (albeit often in more detail than was felt to be desirable for the intent of the courses which I was establishing). In this book, an attempt has been made to cover most of the major topic areas which are felt to be important to a depth that is sufficient to allow an engineering understanding of the theory and practice of each topic area, without getting the student "lost" in all the underlying details, while at the same time still trying to avoid "handwaving" types of explanations. Obviously, since many different topic areas are covered, and since some of these could rightly justify a complete textbook in and of themselves, there may be some feeling of incompleteness as the reader finishes each chapter. To some degree this is intentional. It should serve to remind the reader that behind the general technical area of communications systems, there are a multitude of theoretical and practical topic areas, any one of which could provide months, years, or even a lifetime of fascinating study and work opportunities. It is hoped that this book provides an introduction to at least a few of these areas and encourages the student to pursue actively his or her special interests further into those fields.

I would like to acknowledge the help and encouragement that I received from my colleagues and students at the University of Akron. Special thanks are due my students for their many comments regarding the content and organization of the notes which formed the basis for this textbook and for suffering through the various homework and exam problems, many of which appear as end-of-chapter problems in this book. I would also like to thank those persons associated with Prentice-Hall and its subsidiary, Reston Publishing Company, and the various reviewers who assisted in getting the book into its finished form.

1

INTRODUCTION TO THE SYSTEM CONCEPT

In this book we will be discussing electronic communication systems in which information is transported from one person or place (the sender) to another (the receiver). To begin, we will have to define what is meant by the word *information* and try to specify it as to *type*, *quality*, and *quantity*. Then we shall examine how information originates and is processed at the originating end of the system; how it is converted into an electrical signal to be sent over the communication channel; what the communication channel does to the signal and how this affects the information content in the signal; and, finally, how the signal is processed at the receiving end to recover in part the information that was sent and to present that information to the person for whom the message is intended. The criterion of how well the system transmits information will be determined by how well the information presented to the intended receiver reproduces the information originated by the sender—this means that we will be looking for errors (noise) introduced into the information as it is transmitted and processed by the system.

It is assumed that anyone reading this book already has a background in the basic areas of electrical engineering and electronics and is familiar with the fundamentals of circuit analysis, basic electromagnetic fields, and the circuit details of electronic circuits such as amplifiers, oscillators, basic digital circuits, and the like. Therefore, there will be no discussion in detail of the operation of electronic circuits as such (except where special-purpose circuits are being considered); instead the amplifiers, oscillators, detectors, and so on of a communication system will be treated as "black boxes," with only the terminal characteristics of each black box comprising the total system being specified. Terms such as gain, loss, impedance level, frequency of operation, frequency stability, passband characteristics, signal spectral characteristics, noise sources, and noise figure will be defined and used in the following text.

1.1 MESSAGES AND THEIR INFORMATION CONTENT

When we think of the means by which a thought or message may be conveyed from one person to another, we immediately think of things like printing (plain language text), computer data, speech, music, photographs, paintings and drawings, and TV or motion pictures, for instance. A computer printout, for example, may convey a very large amount of useful (needed) information to the reader in a brief, cryptic format, while an English language text message may convey the same amount of information to the reader, but in a more *redundant* form. For instance, one might be able to cross out about half the letters in the English language text and the reader would still be able to figure out the message with a good probability of being correct. Similarly, a person may talk in a terse manner and convey much information in very few words, or that person may be verbose and do a lot of talking while conveying little actual information.

If a voice message is sent over a high-information-rate (high-bandwidth, low-noise) channel, the receiver will be able to recognize the person who is speaking and will be able to hear the tonal inflections and other characteristics of the speaker's voice, but if the same information (voice message) is sent over a low-information-rate channel (low-bandwidth and high-noise), the receiving person may have difficulty in recognizing the speaker but may still be able to understand all the words that are spoken. In the latter case, no message content will be lost unless the tonal and inflection qualities of the speaker's voice are important to conveying the content of the message.

One often hears the statement that "one picture is worth a thousand words." Actually, the information content of a detailed photograph may well be equal to that of many thousands of words of English text, but much of the information content in the picture may not be necessary to convey the intended message from the sender to the receiver. When we start considering systems like motion pictures, TV, and color TV, we find systems in which the actual information *rate* (information content per unit of time) is quite high, but in which the needed information rate may be quite small. In other words, much more information is being sent than is needed, or can even be assimilated, by the receiver.

We can see from this discussion that information and message content are nebulous quantities to work with in general, so to get a good "handle" on the situation from the viewpoint of the engineer interested in a communications system design, we will be working with the *maximum* possible information content of a given electrical signal and will leave the question of how much of this information content is really necessary up to the psychologists and the philosophers for the time being, although we later will be briefly considering the amount of information that is actually needed to convey a voice message (the subject of vocoder techniques) and the amount of detail necessary in a color TV picture to satisfy the average viewer.

1.2 THE MAXIMUM INFORMATION CONTENT OF A SIGNAL

Suppose that we are given an analog signal such as the one shown in Figure 1-1 where the maximum Fourier frequency component is f_H[1] and where the receiving system is capable of discerning n possible amplitude levels (either due to limitations in the receiver or due to noise on the signal or both).   From the sampling theorem of communication and information theory, we are told that we have enough information to reconstruct the signal if we take a sample of the signal at a rate of $R_s = 2 \times f_H$ samples per second.[2] For each sample, we could determine in which of the n discernible levels the signal falls by answering m yes/no questions, where $m = \log_2(n)$, or where $n = 2^m$. If, for example, we have eight discernible levels, then we would need three yes/no questions to determine in which of the eight levels the signal fell at that particular sample time, or three yes/no answers for each sample taken.   Thus, to specify the signal completely, we would need to answer $(3 \times 2 \times f_H)$ yes/no questions per second.   The answer to a yes/no question is considered to be the smallest amount of information that one person can convey to another: it is called one *bit* of information.   Thus, in the example, the *information rate* would be $6 \times f_H$ bits per second (with 16 discernible levels, it would be $8 \times f_H$, with 32 levels, $10 \times f_H$, etc.).   In general, the information rate in a signal is given by

$$R_I = (m_i)(R_s) \text{ bits per second} \tag{1-1}$$

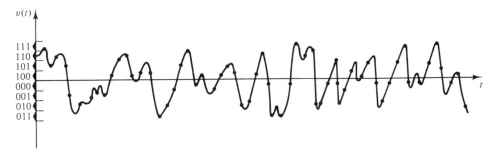

Figure 1-1   A communications signal with eight discernable levels (the dots show sample values equally spaced in time).

[1] See the discussion of Fourier components in Chapter 2.

[2] This applies only to signals having components throughout the entire frequency range from 0 to f_H Hz. A more comprehensive discussion is presented in Section 2.4.

Continuous-Time Analog	Discrete-Time Analog
Continuous-Time Digital	Discrete-Time Digital

Figure 1-2 A general classification of electrical signals.

where $m_i = \log_2 (n_{dis})$, $R_s = 2\,BW_I$, n_{dis} is the number of discernible signal levels, and BW_I is the *information bandwidth* of the signal. Thus, we see that the information content does not vary in direct proportion to the number of discernible signal levels. In Section 3.5, we will be relating the number of discernible levels in a signal to its signal-to-noise ratio (making the assumption that this ratio is the limiting factor in determining the number of levels).

In most of the cases we will be discussing, the information-bearing signals will be electrical voltages and currents. One manner in which we might categorize signals of this type is shown in Figure 1-2. All signals are, of course, defined at all points of time within the time region of interest and at all voltage or current levels within the range covered by the signal. Thus, in one sense, all signals are *continuous-time analog* in nature. However, in many applications we may only be interested in the value of the signal at certain discrete points on the time axis (usually, but not always, separated by a fixed time interval Δt). We might then categorize these signals as *discrete-time* signals since we are interested in their defined values or states only at these certain points on the time axis. Similarly, we might use electrical voltage or current levels to define a finite number of possibilities instead of the infinite number of possible values that a signal may theoretically have within any given voltage or current range. When we do this, we have what is called a *digital* signal. Digital signals may have any finite integer number of defined possibilities (2, 3, 4, etc.), but the binary (two-state) digital signal is the most popular, although digital signals with 4, 8, 12, 16, 24, and 32 possible states are also somewhat prevalent.

Thus far we have discussed electrical signals in terms of defining them as voltage or current time functions (the *time-domain* definition of a signal). A signal may also be defined in other ways. For deterministic signals (those that are a known function of time), a complete definition of the signal may also be given as a function of complex frequency (the *frequency-domain* definition of the signal). In other cases (nondeterministic or random signals), a complete definition of the signal may not be possible in either the time or the frequency domain, and the best description of the signal may be given by combining a time-domainlike partial definition (such as the voltage or current probability-distribution-density function) with a frequency-domainlike partial description (such as the signal's power-spectral-density function). The time-domain and frequency-domain definitions of a signal and the relationships and transformations between them will be discussed more fully in Chapter 2 in the sections on Fourier series and the Fourier and LaPlace transforms.

1.3 AN EXAMPLE OF A COMMUNICATION SYSTEM

Before going any further, let us take a look at a rather simple communication system and discuss the various factors that we would want to consider in analyzing this system. We will use a standard AM radio link as our example: the block diagram of this is shown in Figure 1-3. We assume that we start with voice or music going into the studio microphone and finish up at the loudspeaker of a home receiver. The block diagram shows all the various pieces of equipment and paths through which the signal must travel when passing from the microphone to the loudspeaker. Let's look at the signal and the path at each of these points along the way.

We begin at the studio microphone. At this point we may not only have the desired voice or music signal to be transmitted going into the microphone, but also perhaps some noise (the hiss of the air-conditioning system, a door slamming, papers rustling, etc.). The microphone itself is an imperfect device, not only adding noise of its own, but also introducing some degree of distortion by limiting the frequency range of the signal or by distorting the linearity of the signal as it converts it from an acoustical signal to an electrical signal. However, at the terminals of the microphone we do have an electrical signal that is more or less the electrical analog of the acoustical signal into the microphone. This we call the baseband electrical signal.

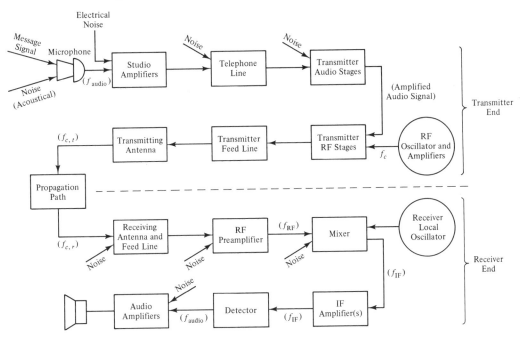

Figure 1-3 A typical communications system (a broadcast-band transmitter/receiver).

This electrical signal is then amplified (and further distorted) by the studio amplifiers, the studio-to-transmitter telephone line, and the transmitter audio amplifiers (including the high-power modulation amplifier). At this point the amplified audio baseband signal (along with the introduced noise and distortion) is used in an RF (radio-frequency) stage of the transmitter in a process called *modulation* to produce the amplitude-modulated RF signal. Essentially, what the modulation process does is to multiply the audio signal with a signal from an RF oscillator to produce a signal consisting of a number of RF components (the *carrier* and the *sidebands*). If we would look at one (the upper) of the AM sidebands, we would see that it is similar to the audio signal, except that each frequency component is increased in frequency by the amount f_c, where f_c is the frequency of the signal from the RF oscillator (the carrier frequency). The composite modulated RF signal now goes through a transmission line to the transmitting antenna, through space to the receiving antenna, and from the receiving antenna through the receiving feed line to the input of the radio receiver. During this travel, the RF signal undergoes attenuation and may also experience distortion of various types (selective time delay of different frequency components, bandwidth limiting, nonlinear distortion, etc.). Also, a great amount of noise is introduced in most cases, especially at the receiving antenna and in the receiving feed line.

At the input terminals of the radio receiver, we are now working with an electrical signal whose amplitude may well be less than a microvolt. To use this signal, it is necessary to amplify it. Perhaps this is first done in an RF amplifier that amplifies the signal at the frequency at which it is received. More noise is always introduced onto the signal at this point. If the receiver is of the superheterodyne type, the signal is then translated in frequency to another RF band of frequencies called the intermediate frequency (IF). This is done by mixing the signal with the RF signal from an oscillator in the receiver called the *local oscillator*. The process of mixing is just another type of modulating or multiplying of two signals together. Additional noise is introduced at this point because of noise generated by the mixer stage and noise on the signal from the local oscillator stage. Some of this latter noise may be in the form of frequency instability of the local oscillator signal.

The IF signal produced by the mixer is then further amplified by the IF amplifier (really just another RF amplifier operating at the intermediate frequency). The signal from the output of the IF amplifier contains additional noise introduced by the IF amplifier itself. This signal is then fed into the AM detector stage that converts the RF-type IF signal back into a baseband audio signal. Much of the noise and distortion that was introduced onto the signal while it was at RF or IF frequencies will now be converted by the detector into noise and distortion on the resulting audio signal.

The electrical signal out of the detector is then amplified by an audio amplifier or amplifiers to a power level sufficient to drive a loudspeaker that converts it into an equivalent acoustical signal. The audio amplifier(s) and the loudspeaker will introduce additional noise and distortion onto the signal, so that what comes out of the loudspeaker, while resembling the acoustical signal that went into the microphone at the sending end of the system, nevertheless may differ considerably from it. It is our job, in analyzing a communication system, to determine just what happens to the signal as it passes through a channel such as the one just described, and to determine just how good a replica of the original message signal our received message signal is.

In the foregoing we have described one of the simplest radio communication systems. But, as you can see, even it is quite involved. In many cases, however, we find that some of the factors that produce noise and/or distortion are negligible and for all practical purposes can be neglected. On the other hand, more sophisticated communication systems, such as video systems, telemetry and data communication systems, are much more complicated, involving other types of devices such as physical sensors, data processors, camera tubes, analog and digital computers, sophisticated modulation techniques, digital-to-analog and analog-to-digital converters, and so on.

PROBLEMS

1-1. Assume an analog signal has n discernible levels. Find the bits of information in each sample of the signal if
    a. $n = 64$ levels.
    b. $n = 512$ levels.
    c. $n = 800$ levels.
    d. $n = 1000$ levels.
    e. $n = 1600$ levels.
    Note: It is customary when discussing bits of information to retain the fractional portion of the number.
    Hint:

$$\log_a x = \frac{\log_b x}{\log_b a}$$

1-2. Find the information rate for each of the five signals of Problem 1-1 if the signal has an information bandwidth of
    a. $BW_I = 4$ KHz   (a telephone signal).
    b. $BW_I = 15$ KHz   (a "hi-fi" music signal).
    c. $BW_I = 4.2$ MHz   (a TV video signal).

1-3. A signal has 200 discernible levels and an information rate of 2000 bits per second.
    a. What is the information bandwidth of the signal?
    b. At what minimum rate must it be sampled?
    c. How many bits of information are present in each sample?

1-4. Discuss the differences between digital and analog signals. Where might each type be used? For what purposes?

1-5. To convert a continuous-time analog signal to a discrete-time digital signal, two processes are necessary. What are those two processes? How might you reverse the process (i.e., convert the discrete-time digital signal back into a continuous-time analog signal)?

1-6. Consider the various noise inputs shown in Figure 1-3. What might cause these noises? Which of the noise inputs might be the most troublesome? Why?

1-7. Figure 1-2 is based upon the time-domain definition of the four signal types. How might you characterize these four signal types in the frequency domain?

2

BASIC CONCEPTS AND ANALYSIS TECHNIQUES

In this chapter we will be considering some of the basic methods used for analyzing and characterizing communication signals and for describing what happens to these signals as they pass along through the various parts of the communication system. For the most part we will be considering signals that fall into the *continuous-time analog* category described in Chapter 1. This type of signals may be further subcategorized as follows:

1. *Deterministic versus random*—Those signals that have a defined time function are called *deterministic* signals whereas *random* signals can only be defined statistically.

2. *Periodic versus aperiodic*—A deterministic signal is periodic if it repeats itself every T seconds (T is called the *period* of the signal and $f_o = 1/T$ is called its *fundamental frequency* (in Hertz).

3. *Defined over all time versus defined from a finite point in time*—If a signal is defined all the way back to $t = -\infty$, it is said to be defined over all time. Practically, we apply this definition to any signal whose starting point is long enough past that all transient effects associated with the start-up of the signal have died out. Signals that have begun recently enough so that all the

transient effects have not died out are treated as signals defined from a definite point in time.

4. *Correlated versus uncorrelated signals*—Signals that are an exact copy of each other except for an amplitude scaling factor are considered to be completely correlated. If the amplitude scaling factor is negative, then the signals are completely negatively correlated (or anticorrelated). The time average of the product of the time functions of two such signals is plus or minus the product of their root-mean-square (rms) values. Two signals that are completely uncorrelated will have a time-function product that averages out to zero.

2.1 SIGNAL MODIFICATION BY THE SYSTEM

As a signal passes through a communications system, it is adulterated by a number of different mechanisms:

1. *Noise*—Noise is an unwanted signal (usually random in nature) that is added onto the signal at various points in its passage through the communications system. Thus, at any point in the system, the signal that is present at that point consists of the original information signal (perhaps somewhat distorted by the mechanisms to be discussed shortly) plus a noise signal representing the noise that has already been added. We usually characterize each of these signal components by their respective power levels (using S for the power of the information signal and N for the power of the noise signal). The ratio of the signal power to the noise power is called the *signal-to-noise ratio* (S/N) and is useful as a figure of merit when discussing the system's noise performance.

2. *Other unwanted signals*—These may be other communication signals, signals from other equipment, or spurious signals from devices within the communications system itself. These signals go by various names (image signals, parasitic oscillations, oscillator leakage, harmonic outputs, impulse noise, etc.) and will be discussed or mentioned at various points in the following chapters.

3. *Frequency-dependent distortion*—This is a distortion of the information-signal time function due to frequency-dependent variations in the attenuation or amplification or frequency-dependent time delay of the various frequency components (Fourier components) present in the information signal as it passes through the system.

4. *Time-dependent distortion*—This is an effect that occurs primarily in the propagation of RF (radio-frequency) or optical signals from the transmitting antenna to the receiving antenna and involves attenuation, phase-shift, or field-polarization variations that change as a function of time.

5. *Nonlinear distortion*—This is due to portions of the system being nonlinear in amplitude response. It causes harmonics to be generated from sinusoidal

(single-frequency) signals (harmonic distortion) and also causes the generation of sum-frequency and difference-frequency signals whenever two or more Fourier components are present in the input signal to the nonlinear portion of the system (intermodulation distortion).

All the foregoing mechanisms (if severe enough) will cause the system's output waveform to be a poor reproduction of its input waveform and are to be avoided or minimized as much as possible.   One of the design criteria for a *hi-fidelity* music system, for example, is to minimize the effects of such mechanisms in the recording, transmission, and acoustical reproduction of voice and music signals.   In some cases one type of distortion may be more noticeable than another.   For example, small amounts of frequency-dependent time-delay variation are not detectable on voice signals and perhaps not even noticeable on music signals, but will produce disastrous consequences when they occur on a TV video signal or a high-data-rate digital data signal.

2.2 *FOURIER SIGNAL ANALYSIS*

If we look at a particular communications signal, we may classify it either as periodic or nonperiodic.   If it is a periodic signal, we may analyze it into its *Fourier series* components, where the frequency of the nth component or *harmonic* is given by n times f_o, where f_o is the basic *fundamental* frequency of the waveform, and where $f_o = 1/T$, where T is the period of the waveform in seconds.   This gives f_o the units of Hertz (or cycles per second).   In all practical waveforms there will be a limit to the number of components present (i.e., there is some upper limit to n, which we call n_{max}).   Thus the maximum frequency component present in the waveform is $n_{max} \times f_o = f_{max}$.

If we pass a periodic signal made of Fourier components as just described through a practical communications channel, we will, in general, find that some of the components are attenuated or amplified more than others, with the result that the waveform at the output of the system, although still periodic with a period T, may be quite different in appearance than the waveform at the input.   We call this type of distortion of a signal passing through a system *amplitude-frequency distortion*.   In addition to the amplitude-frequency distortion, if the various components are delayed by different amounts of times as they pass through the system, we will also have *phase-frequency distortion* present on the signal.   This type of distortion occurs if the phase shift that the system introduces onto each component is not proportional to the frequency of that component by the same proportionality factor.

In the case of nonperiodic signals, we may analyze the signal by means of *Fourier integral analysis* where the Fourier series summation is replaced by the continuum of the Fourier integral.   Essentially, what we are doing here is to let the period of the waveform approach infinity so that the fundamental frequency approaches zero.   There then become an infinite *number* of frequency components spaced infinitesimally close in frequency.   However, there is still, in the practical

Figure 2-1 A system component processes the input signal or signals to produce an output signal.

case, an upper limit to the frequency of the components contained in the waveform, and this is still called f_{max}. Again, each of this infinite number of components may be thought to have a phase and amplitude—and passing the signal through a channel where the amplification or attenuation is not constant over all the frequencies up to f_{max}, or where the phase shift introduced is not proportional to the frequency of the component being considered, will produce amplitude or phase distortion of the output waveform just as happens in the case of the periodic signal's Fourier series components.

An ideal communication channel with no distortion should have a gain or an attenuation that is constant over all the frequencies involved in the signal and a phase shift that is proportional to the signal frequency for all frequencies involved in the signal to be transmitted. Note that, in the case of the phase shift, a phase shift that is proportional to the frequency of the signal passing through the system means that the system will introduce the *same* time delay on all signals passing through, independent of their frequency.

Putting this into mathematical terms, the input and output of a system or a system component (black box) such as is shown in Figures 2-1 and 2-2 become, for a periodic signal of fundamental angular frequency ω_0,

$$v_{in}(t) = V_0 + V_1 \cos(\omega_0 t + \theta_1) + V_2 \cos(2\omega_0 t + \theta_2)$$
$$+ \cdots + V_{n_{max}} \cos(n_{max}\omega_0 t + \theta_{n_{max}}) \tag{2-1}$$

and

$$v_{out}(t) = V'_0 + V'_1 \cos(\omega_0 t + \theta'_1) + V'_2 \cos(2\omega_0 t + \theta'_2)$$
$$+ \cdots + V'_{n_{max}} \cos(n_{max}\omega_0 t + \theta'_{n_{max}}) \tag{2-2}$$

There will be no amplitude distortion of the signal if

$$V'_n = A_{Vn} \times V_n \tag{2-3a}$$

where A_{Vn} is independent of ω (i.e., the voltage gain does not vary with the frequency).

There will be no phase distortion if

$$\theta'_n = \theta_n - \omega_n \tau_D = \theta_n - (n \times \omega_0)\tau_D \tag{2-3b}$$

where τ_D (the time delay of the component) is a constant that does not depend upon frequency.

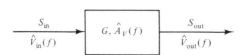

Figure 2-2 An amplifier (or other device) amplifies an input voltage phasor (at frequency f) to give the output phasor $\hat{V}_{out} = \hat{A}_v \hat{V}_{in}$. Also $S_{out} = G S_{in}$.

The Fourier transform pair in terms of frequency f in hertz and time t in seconds is

$$v(t) = \int_{-\infty}^{\infty} V(f)\, \varepsilon^{j2\pi ft}\, df \quad \text{and} \quad V(f) = \int_{-\infty}^{\infty} v(t)\, \varepsilon^{-j2\pi ft}\, dt$$

Thus, for a nonperiodic signal extending over the complete time axis but limited in frequency, the equivalents of Equations 2-1 and 2-2 will be

$$v_{in}(t) = \int_{f=0}^{f_{max}} V_1(f) \cos(2\pi ft + \theta(f))\, df \tag{2-4}$$

and

$$v_{out}(t) = \int_{f=0}^{f_{max}} V_1'(f) \cos(2\pi ft + \theta'(f))\, df \tag{2-5}$$

(using the one-sided Fourier integral representation), where $V_1(f)$, $V_1'(f)$, $\theta(f)$, and $\theta'(f)$ are all functions of f and when plotted are called the amplitude and phase spectrums of the signals.   For distortionless transmission through the system, we must have

$$V_1'(f) = A_V(f) \times V_1(f) \tag{2-6a}$$

where $A_V(f)$ is a constant for $0 < f < f_{max}$ and

$$\theta'(f) = \theta(f) - 2\pi f \tau_D(f) \tag{2-6b}$$

where τ_D is a constant for $0 < f < f_{max}$.

2.3 THE DECIBEL SYSTEM OF AMPLITUDE RATIO AND POWER RATIO SPECIFICATION

Consider a part of a communication channel as being represented by the box shown in Figure 2-2.   A signal goes in at the left with a voltage amplitude (rms) of V_{in} and a power S_{in} and comes out at the right with a voltage V_{out} and a power level S_{out}.   The *power gain ratio* is given by

$$G = \frac{S_{out}}{S_{in}} = \left(\frac{V_{out}}{V_{in}}\right)^2 \times \frac{Z_{in}}{Z_{out}} \tag{2-7}$$

where Z_{in} and Z_{out} are the impedance levels (resistances) at the input and the output, respectively (both assumed to be real).   The voltage gain of the amplifier is given by $A_V = V_{out}/V_{in}$, and if Z_{in} is equal to Z_{out}, the power gain is just equal to the square of A_V. G is called the *power*-gain *ratio* and A_V is called the *voltage*-gain *ratio*.   As will be seen later, it is quite often convenient[1] to use a measure

---

[1] One reason for using the decibel system is that the response of many physical sensors is logarithmic.   For example, the human ear can detect about 1 dB of change in the power level of acoustic sound reaching the ear at any sound level.

for these gains that is proportional to the logarithm of these ratios than to use the ratios themselves. This is the system called the decibel system (where one decibel is one-tenth of a "bel," the "bel" being named in honor of Alexander Graham Bell).

To express a power gain G in terms of the decibel (dB) system, we use the relationship

$$(G)_{dB} = 10 \log_{10} G = 10 \log_{10} \frac{S_{out}}{S_{in}} \qquad (2\text{-}8a)$$

But also notice that

$$(G)_{dB} = 10 \log_{10} \left(A_V^2 \times \frac{Z_{in}}{Z_{out}} \right)$$

$$= 20 \log_{10} A_V + 10 \log_{10} \frac{Z_{in}}{Z_{out}} \qquad (2\text{-}8b)$$

which allows us to obtain the gain expressed in decibels from either the power-gain ratio or from the voltage-gain ratio and the ratio of the input and output impedance levels. In working with the decibel system, there are some common relations between G as a ratio and G in decibels that should be memorized:

3 dB is approximately equivalent to $G = 2$.
10 dB is equivalent to $G = 10$.
20 dB is equivalent to $G = 100$.
30 dB is equivalent to $G = 1000$, and so on.

Note that the decibel unit is always used to indicate a *ratio* between the power of *two* signals, *never to indicate an absolute level of power or voltage*. As such, the decibel system is often used to indicate the gain or loss of a device through which a signal passes.

However, there is a closely related technique that is used to describe absolute signal power levels. This is the decibel Watt (dBW) or decibel milliwatt (dBm) system. What this consists of is referring the power level of the signal in question to some reference power level (1 Watt for the decibel Watt system and 1 milliwatt for the decibel milliwatt system). We then specify the level of the signal by giving how many decibels above or below the reference power level its power is. Let's take a look at a couple of examples.

Example:

Let S_i be the power level (in watts) of the signal of interest; if $S_i = 1$ milliwatt, then we say that $(S_i)_{dBm} = 0.0$ dBm or $(S_i)_{dBW} = -30.0$ dBW. Similarly, if $S_i = 10$ watts, $(S_i)_{dBm} = +40.0$ dBm and $(S_i)_{dBW} = +10$ dBW. Also, if $S_i = 1$ microwatt, $(S_i)_{dBm} = -30$ dBm and $(S_i)_{dBW} = -60$ dBW.

The decibel system and its companion decibel milliwatt system are very useful in communication systems analysis where the signal must pass through many different parts of the system in cascade. For example, suppose that there are several stages having different gain ratios (G greater than unity for the gain ratios of active devices and less than unity for devices that attenuate the signal). If one calculates the overall gain using the individual gain ratios of each device in the cascade, there is much multiplication involved. However, if one uses the decibel system, the gains (expressed as positive and negative decibel quantities) are simply added algebraically. The advantage becomes especially apparent when one begins making modifications to the system, such as changing the gain of one of the devices in the chain. It is much easier to calculate and see the effects of such a change when the decibel system is employed.

Example:

> Suppose that a signal passes through four devices, the first of which has a power gain of 100, the second of which has an attenuation that reduces the signal power by a factor of 4, the third has a power gain of 20, and the fourth has an attenuation factor of 8. A signal with a power level of 4 microwatts goes into the system. What is the power of the signal coming out of the system?

> **Solution:** Power into system $(+6 - 30)$ = -24 dBm
> First device   = $+20$ dB
> Second device $(-3 + -3)$ = -6 dB
> Third device $(+3 + 10)$ = $+13$ dB
> Fourth device $(-3 - 3 - 3)$ = $\underline{-9\text{ dB}}$
> Power out of the device   = -6 dBm

In terms of milliwatts, -6 dBm is 6 dB below 1 milliwatt or $\frac{1}{4}$ milliwatt output power. (Note that the relationship that the decibel equivalent of a ratio of 2 is approximately 3 dB has been used in these calculations).

2.4 SYSTEM IMPEDANCE LEVELS

Figure 2-3 shows a simple system consisting of an encoder, three system devices, and a decoder. Consider that the signal between each of the boxes is an electrical signal (the same considerations as we are about to discuss here would also apply to acoustical or other signal forms). One of the characteristics of any two-port device is its input and output impedances. In communication systems the normal procedure is to match the output impedance of a device (or at least match its magnitude) to the input impedance of the next device in the cascade. When this is done, we then talk about the *impedance level* between the two devices. For instance, in Figure 2-3, Z_1 would be the output impedance of the first transducer (encoder) as well as the input impedance of system device 1, Z_2 is the output impedance of device 1 and the input impedance of device 2, and so on.

Quite often it is found that the input impedance of a device is dependent upon the impedance attached to its output (and to a lesser degree in some cases, its output impedance may depend upon the impedance attached to its input). Manufacturers of communication system devices usually specify the input and output impedance *levels* of the devices, which means that when the output of the

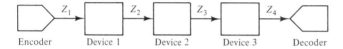

Figure 2-3 System impedance levels.

device is terminated in an impedance equal to its output impedance, the device's input impedance will be equal to the specified input impedance value, and if the impedance of the generator driving the input is equal to the specified input impedance, then the impedance seen looking back into the device output terminals will be equal to the specified output impedance. In many devices (line filters, attenuators, etc.), the specified input and output impedance levels may be equal.

One of the places where impedance matching is of prime importance is when the system device is a length of transmission line (audio, video, or radio frequency), because mismatches at the receiving end[2] of the line can cause standing waves on the line, resulting in excess power loss as well as various types of echo effects. The matching of impedances between amplifier stages and other devices is of importance to maximize the transfer of signal power from one stage to the next and to reduce the introduction of excess noise into the system (i.e., for the maintenance of good signal-to-noise ratios).

2.5 BANDWIDTH CONSIDERATIONS, BODE PLOTS, AND AMPLIFIERS

As discussed in Section 2.1, in passing a signal through a communications channel, we will have to consider how some of the frequency components of the signal may be attenuated or phase shifted by the devices in the system that constitutes the channel. For the information signal itself, we are usually talking about signals that contain components that range from nearly zero frequency to no more than a few megahertz (for example, a good-quality audio signal may range from 15 Hz to about 17,000 Hz, a voice communications signal from 50 Hz to about 3500 Hz, and a TV signal from about 30 Hz to about 4.5 MHz). These band regions are what are called the *information-signal bandwidths* or *baseband bandwidths* of the signals involved. After the signal is used to modulate some type of carrier, we may have a much wider bandwidth (especially when angle modulation such as FM or PM is used) for the resulting modulated signal than we had for the information signal. These latter bandwidths are usually referred to as RF (or IF) bandwidths. An information-signal bandwidth is usually characterized by extending over a large frequency range compared to the median frequency, while a modulated signal usually has a fairly small bandwidth-to-median frequency ratio.

For example, in a standard AM broadcast radio signal, the baseband bandwidth of the audio modulating signal may extend from 20 to 10,000 Hz, but the

---

[2] Note that although mismatches at the sending end do not affect the standing-wave ratio on the line, they can cause repetitive echoing if there is also a mismatch at the receiving end of the line. Also, an impedance mismatch at the sending end of a transmission line prohibits transferring maximum power into the transmission line.

modulated RF signal might extend from 990 KHz to 1010 KHz and the equivalent signal in the receiver IF amplifier from 445 to 465 KHz (in the IF case the bandwidth is about 4.4% of the IF center frequency).

From our earlier discussions in this chapter, so as not to have distortion on the signal, it was necessary for any device in the system to have a constant attenuation or amplification factor over all the frequencies contained in the signal and also to have a constant time delay for the passage of all these signal components (a phase delay that is proportional to the signal frequency). This is illustrated in Figure 2-4a for a signal with frequency components between f_1 and f_2 where $f_2 \geq f_L$ and $f_2 \leq f_H$. However, the situation of Figure 2-4a is ideal condition, which is almost never achieved in practice.

The normal situation is more like that shown in Figure 2-4b. Here we see that the amplitude part of the complex gain function is not a constant from f_1 to f_2 but that, rather, the gain function amplitude rises from near zero at a frequency below f_1 to the nominal value at a frequency somewhat above f_1. The dropoff to zero above f_2 is somewhat similar, with some of the signal components just below f_2 receiving less than the desired amplification. The passband of such a "black box" is usually specified as the difference between the "3 dB downpoints," which are designated f_L and f_H, respectively (or in some cases f_L and f_U). These are points at which the power gain of the device or amplifier has dropped to one-half

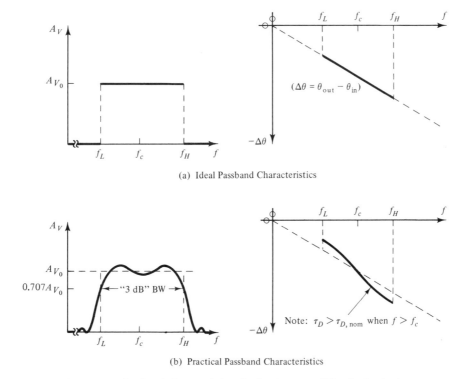

(a) Ideal Passband Characteristics

(b) Practical Passband Characteristics

Figure 2-4   Passband characteristics of a bandpass amplifier (bode plots).

of its value at midband. This corresponds to where the voltage gain has dropped to 0.707 of its midband value. These points (f_L and f_U) are therefore called the *half-power* points as well as the -3 dB points. The general region about each of the half-power points is referred to as a *skirt* of the amplitude characteristic curve (i.e., the region where the gain is falling from its nominal midband value to zero). Within the actual passband itself, the gain may not be uniform but may vary up and down slightly with frequency, with the number of "humps" present on the top of the characteristic being some indication of the complexity of the device's circuitry.

When the phase-shift characteristics are considered, the ideal case of a phase shift that is exactly proportional to the frequency is usually not obtainable (see Figure 2-4a). Instead, the phase characteristics usually resemble more what is shown in Figure 2-4b, and this results in some amount of phase distortion (also called time-lag distortion) being present on the output signal. As mentioned earlier, this phase distortion may be of little importance for an audio signal, but of great importance in the transmission of a TV signal or a digital data signal.

The preceding discussion concentrated on passband characteristics where the bandwidth was small compared to the median frequency. In these cases, the gain amplitude is usually plotted versus the frequency on linear scales. However, when we go to situations involving audio and video amplifiers and baseband communication channels, the range of frequencies is so large that it becomes more convenient to plot the logarithm of the gain versus the logarithm of the frequency. Plots of this type are shown in Figure 2-5 and are called *Bode plots*. Usually $\log_{10} G$ is plotted against $\log_{10} f$ for the amplitude plot and θ versus $\log_{10} f$ for the phase plot. Very often, since the gain in decibel units is proportional to $\log_{10} G$ the function $G_{\mathrm{dB}}(f)$ is plotted versus $\log_{10} f$. In these baseband characteristic plots, no center frequency is designated, but there still occur lower and upper half-power (or -3 dB) frequency points. For a single R-C- or R-L-type circuit (such as occurs in the input or output coupling circuits of an audio amplifier), the gain outside of the passband region changes by 20 dB for every factor-of-10 change in the frequency. Since a change in frequency by a factor of 10 is called a *decade* change in frequency (a change by a factor of 2 is called an *octave*), we say that the gain of a simple R-C-coupled audio amplifier stage "rolls off" or decreases in gain at a rate of 20 dB per decade. If this asymptotic rolloff characteristic line is plotted on the Bode plot and extended until it meets the constant (horizontal) asymptote for the midband gain, the two *asymptotes* meet at the point f_L. The actual gain curve at this point is 3 dB different from the point where the two asymptotes meet.

The plot shown in Figure 2-5 is that of a typical R-C-coupled amplifier with an emitter bypass capacitor. The low-frequency characteristics are due to the presence of the input and output coupling capacitors and the emitter bypass capacitor, and the high-frequency rolloff characteristics are due to the effective internal capacitance of the transistor itself. The amplitude of the gain rolls off at high frequencies at the -20-db-per-decade rate, but the low-frequency rolloff is at the rate of 40 dB per decade because there are two R-C circuits involved (due to the input coupling capacitor and the output coupling capacitor) and each circuit contributes 20 dB per decade to the rolloff rate. For the common emitter type of amplifier from which the data presented in Figure 2-5 are taken, there is an

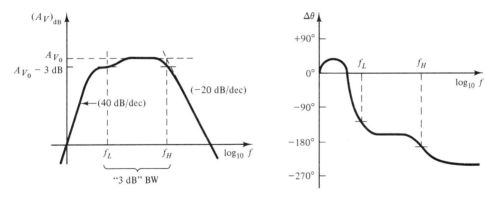

Figure 2-5   Passband characteristics of an audio or video amplifier (bode plots).

inherent -180-degree phase shift at midband. In addition to this there is an additional phase shift of from 0 to -90 degrees in the high-frequency region or additional phase shift of from 0 to nearly 270 degrees at the lower frequencies (this additional low-frequency phase shift reaches a maximum at some frequency greater than 0 Hz and then drops back to $+180$ degrees at 0 Hz for the amplifier just described).

2.6 NONLINEAR DISTORTION

Sections 2.2 and 2.5 discussed two types of distortion that may be present in a communication system: amplitude-frequency and phase-frequency distortion. These two types of distortion are present even in systems that are perfectly linear. However, perfectly linear systems exist in theory only, and all practical systems contain some degree of nonlinearity. This nonlinearity can lead to serious *nonlinear distortion* of a signal passing through the system.

Consider a sinusoidal signal that is applied to the input of a system whose output can be described as

$$v_{\text{out}}(t) = Av_{\text{in}}(t) + B[v_{\text{in}}(t)]^2 \tag{2-9}$$

If $v_{\text{in}}(t)$ is given by $V_{m_1} \cos \omega_1 t$, then

$$v_{\text{out}}(t) = AV_{m_1} \cos \omega_1 t + B(V_{m_1} \times 2) \cos^2 \omega_1 t$$

$$= (AV_{m_1} \cos \omega_1 t) + \frac{(BV_{m_1}^2)}{2}$$

$$+ \frac{(BV_{m_1}^2)}{2} \cos 2 \omega_1 t \tag{2-10}$$

As can be seen, the output not only contains a component at the original signal frequency ω_1 but also components at d.c. and at the second harmonic ($2 \omega_1$) of the signal frequency. Quite often, the d.c. term disappears from the output (unless d.c. coupling is used in the system), but the second harmonic term passes through, giving rise to what is called *harmonic distortion* on the output signal. This is one of the consequences of the nonlinear distortion mechanism. In the example given, we have considered a type of nonlinear distortion that gives rise only to second harmonic distortion terms in the output, but in many cases higher-order nonlinear distortion will be present and the output will contain many different harmonics of the input signal. If V_{o_1} is the root-mean-square (rms) voltage of the fundamental signal in the output, V_{o_2} the rms voltage of the second harmonic, V_{o_3} the rms voltage of the third harmonic, and so on, then the percentage of harmonic distortion on the output signal is given as

$$D_H = \frac{\sqrt{V_{o_2}^2 + V_{o_3}^2 + V_{o_4}^2 + \cdots}}{V_{o_1}} \times 100\% \qquad (2\text{-}11)$$

If the input signal contains two or more sinusoidal components, the system nonlinearities will produce (in addition to harmonic distortion) output signal components at the sums and differences of all of the frequencies involved in the input signal. This type of distortion is called *intermodulation distortion*. For example, if V_{out} is again described by Equation 2-9 and V_{in} is given by the relation $v_{\text{in}}(t) = V_{m_1} \cos \omega_1 t + V_{m_2} \cos \omega_2 t + V_{m_3} \cos \omega_3 t$, then the output becomes

$$\begin{aligned}
v_{\text{out}}(t) &= A[V_{m_1} \cos \omega_1 t + V_{m_2} \cos \omega_2 t + V_{m_3} \cos \omega_3 t] \\
&\quad + B[V_{m_1}^2 \cos^2 \omega_1 t + V_{m_2}^2 \cos \omega_2 t + V_{m_3}^2 \cos \omega_3 t] \\
&\quad + 2B[V_{m_1} V_{m_2} \cos \omega_1 t \cos \omega_2 t + V_{m_1} V_{m_3} \cos \omega_1 t \cos \omega_3 t \\
&\quad + V_{m_2} V_{m_3} \cos \omega_2 t \cos \omega_3 t] \\
&= A[V_{m_1} \cos \omega_1 t + V_{m_2} \cos \omega_2 t + V_{m_3} \cos \omega_3 t] \\
&\quad + \frac{B}{2}[V_{m_1}^2 + V_{m_2}^2 + V_{m_3}^2] \\
&\quad + \frac{B}{2}[V_{m_1}^2 \cos 2\omega_1 t + V_{m_2}^2 \cos 2\omega_2 t + V_{m_3}^2 \cos 2\omega_3 t] \\
&\quad + B[V_{m_1} V_{m_2} \{\cos (\omega_2 + \omega_1)t + \cos (\omega_2 - \omega_1)t\} \\
&\quad + V_{m_1} V_{m_3} \{\cos (\omega_3 + \omega_1)t + \cos (\omega_3 - \omega_1)t\} \\
&\quad + V_{m_2} V_{m_3} \{\cos (\omega_3 + \omega_2)t + \cos (\omega_3 - \omega_2)t\}]
\end{aligned} \qquad (2\text{-}12)$$

Referring to the second part of Equation 2-12, the first pair of square brackets contains amplified signal-frequency terms at ω_1, ω_2, and ω_3. The second pair of square brackets contains the d.c. terms, and the third pair contains the harmonic-distortion terms at the frequencies $2\omega_1$, $2\omega_2$, and $2\omega_3$ and the fourth pair contains the intermodulation-distortion terms at $(\omega_3 \pm \omega_1)$, $(\omega_3 \pm \omega_2)$, and $(\omega_2 \pm \omega_1)$ (assuming that $\omega_3 > \omega_2 > \omega_1$). The percentage of intermodulation distortion can be calculated in a manner similar to that for the harmonic-distortion:

$$D_I = \frac{\sqrt{V_{31}^2 + V_{21}^2 + V_{32}^2 + \text{etc.}}}{V_M} \times 100\% \qquad (2\text{-}13)$$

where

$$V_{31} = \sqrt{V_{3+1}^2 + V_{3-1}^2} \quad \text{and} \quad V_{21} = \sqrt{V_{2+1}^2 + V_{2-1}^2}, \text{ etc.}$$

where V_{3+1} = the voltage amplitude of the component at $\omega = (\omega_3 + \omega_1)$, which has a value of $[2 \times B \times \frac{1}{2} \times V_{m_1} \times V_{m_3}]$, and so on. V_{3-1} similarly is the amplitude of the voltage at $\omega = (\omega_3 - \omega_1)$, and so on. $V_M = A\sqrt{V_{m_1}^2 + V_{m_2}^2 + V_{m_3}^2 + \text{etc.}}$ is equal to $\sqrt{2}$ times the root-mean-square value of the combined *signal* output components.

Since nonlinear distortion in a system produces both harmonic and intermodulation distortion, we usually consider both of the latter at one time when making distortion measurements and define the total distortion as

$$D = \sqrt{D_{H_1}^2 + D_{H_2}^2 + \cdots + D_I^2} \% \qquad (2\text{-}14)$$

where D_{H_1} is the percentage of *harmonic* distortion due to the first signal component, D_{H_2} is the percentage of *harmonic* distortion due to the second signal component, while D_I is the percentage of *intermodulation* distortion.

Quite often in giving the performance specifications for audio amplifiers and for receivers that produce an audio output, the output noise is included along with the distortion products and a quantity called the SINAD ratio is generated. It is a power ratio and is usually expressed in terms of the decibel system. The equation for the *ratio* itself is

$$\text{SINAD}_{(\text{ratio})} = \left[\frac{S_o + N_o + P_D}{N_o + P_D}\right] = \frac{\text{total audio output power}}{\text{noise power + distortion power}} \qquad (2\text{-}15)$$

where S_o = the output signal power, N_o = the output noise power, and P_D = the power of the distortion products in the output. (P_D has been used here, but quite often this equation is given with D used in place of P_D. When this is done, it is *not* the same D as we have been using in the preceding discussion.) The SINAD is usually specified with a given input signal level and frequency or frequencies applied to the input of the amplifier or receiver under test.[3] P_D depends upon the level of S_o and is usually small until S_o becomes quite large, at which point it begins to increase quite rapidly. The SINAD starts at 1 (0 dB) when $S_o = 0$, becomes 2 (3 dB) when S_o is approximately equal to N_o, goes to a maximum (usually around 100 to 120 dB) when S_o is much greater than N_o and P_D is still small, and then decreases as P_D becomes significant (see Figure 2-6). If some minimum (SINAD)$_{\text{dB}}$ ratio is taken to be an acceptable minimum level of performance for the device or system, then the range of S_o over which the (SINAD)$_{\text{dB}}$

---

[3] See, Bernard M. Oliver and John M. Cage, *Electronic Measurements and Instrumentation* (New York: McGraw-Hill, 1971).

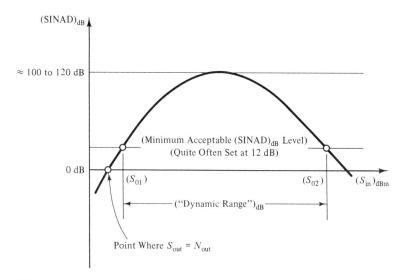

Figure 2-6 A plot of the SINAD as a function of the input signal level, showing the definition of system in terms of the SINAD.

is greater than this minimum is said to be the dynamic range attainable with that specific device or system (this is usually given in terms of (S_{o2}/S_{o1}) expressed in terms of decibels.

2.7 THE SAMPLING THEOREM AND INFORMATION RATES

In the introduction, we discussed very briefly the maximum information content in a signal and showed that it depended upon the highest frequency present in the signal and upon the number of discernible signal levels. We said that the smallest quantity of information was the *bit*, which we defined as the answer to a yes/no type of question. One bit is also one digit in the binary number system. Thus a five-digit binary number can contain five bits of information. A digital communication channel that can transmit 25 binary pulses per second has a data-rate capacity of 25 data bits per second (bps), but a system that can transmit 25 four-level pulses per second has a data-rate capacity of 50 bps, since each four-level pulse contains two data bits. Similarly, transmitting 25 eight-level pulses per second is equivalent to transmitting at a data rate of 75 bps.

In the preceding discussion we have used both the term *information bits* and the term *data bits*. We shall use the former when discussing the information content or the information rate of an analog signal and the latter when talking about the number of bits or bits per second in a digital data signal. Another related term that is often confused with the "bit rate" of a digital data signal is its *Baud rate*. The latter is defined as *the number of electrical transitions (changes in state) per second* of the electrical form of the digital data signal. This is sometimes also called the "modulation rate." Each electrical transition can be a change in the voltage or current level, a change in the frequency, a change in the phase (or some combination of these) of the signal. Depending upon the digital coding and modulation techniques employed, the Baud rate may be equal to, less than, or greater than the digital data rate. These concepts will be considered in more detail in the sections on digital modulation techniques (Sections 4.5.5 and 4.6) and in the Chapter on data communications (Chapter 11).

Sometimes it is more convenient to work with the common or natural logarithm systems than with the $\log_2$ system as was done in Equation 1-1. Thus if we let n_{dis} = the number of discernible signal levels and m_i = the number of bits of information, the number of bits in a single n-level pulse is given by

$$m_i = \log_2 n_{\mathrm{dis}} = 1.44 \log_e n_{\mathrm{dis}} = 3.32 \log_{10} n_{\mathrm{dis}} \qquad (2\text{-}16)$$

The *information rate* for a baseband analog signal is

$$R_I = m_i \times R_s \text{ bps} \qquad (2\text{-}17)$$

where

R_I = the information rate in bits per second

m_i = the bits per sample

R_s = the sample rate = $2(f_H - f_L) = 2\,\mathrm{BW}_I$

f_H = the highest-frequency component of the signal

f_L = the lowest-frequency component of the signal

The *maximum information capacity* of a communications *channel* is also given by Equation 2-17 if we let n = the number of discernible levels possible on the channel (usually determined by the channel noise and the encoder/decoder capabilities) and let $(f_H - f_L)$ be the *information* bandwidth BW_I of the channel (also sometimes referred to as the video bandwidth). The sampling theorem gives the minimum sample rate R_s as being equal to $2\,\mathrm{BW}_I$. However, this is a theoretical limit, and usually if we want to reproduce a signal adequately with an information bandwidth of BW_I, we will want to sample that signal at something more than the R_s rate given. If we actually sample the signal at a rate $R_s' > R_s = 2\,\mathrm{BW}_I$ and then later re-create a continuous signal from the samples, the maximum bandwidth of valid information contained in the regenerated signal is still the BW_I of the original signal and *not* $R_s'/2$.

An interesting point is posed when the sampled signal is a sinusoid that varies in amplitude or phase very slowly. Then the information bandwidth f_I becomes very small, even though the actual approximate frequency is very high. For example, consider a 5-MHz signal whose amplitude *varies* at a frequency not to exceed 10 Hz. Then the total bandwidth of this signal (see Section 4.2 on AM modulation) would extend from $f_L = 4.999990$ to $f_H = 5.000010$ MHz or $\Delta f = 20$ Hz. However, all the information is contained in *one* of the "sidebands" of this signal (4.999990 to 5.0 or 5.0 to 5.000010) or $\Delta f = 10$ Hz. Thus, this 5-MHz signal could be sampled at a rate of only 20 times per second, and the amplified samples sent to the receiving end where they are used to excite a very narrow (10-Hz) bandwidth filter whose passband is centered about 5 MHz to produce a replica of the original AM-modulated 5-MHz signal. The BW_I of the channel needed to transmit the information needed to reconstruct this 5-MHz signal would be only 20 Hz.

Note that increasing the information bandwidth BW_I of a signal by some factor K will also increase its maximum information rate by a factor K; but increasing the number of discernible signal levels in the signal by a factor K only increases the maximum information rate by a factor K', where $K' < K$ and is given by

$$K' = \frac{\log_2 (K * n_{dis})}{\log_2 n_{dis}} = \log_{n_{dis}}(K * n_{dis}) \tag{2-18}$$

where n_{dis} = the original number of discernible levels.

As mentioned earlier, in many cases the signal-to-noise ratio of a signal is the limiting factor in determining the number of discernible signal levels. For the case where a signal and the noise on the signal both have Gaussian properties, the number of discernible levels is given by (see Section 3.5)

$$n_{dis} = \left(1 + \frac{S}{N}\right)^{1/2} \tag{2-19}$$

so that the maximum information rate of the signal is

$$R_I = 2(BW_I) \log_2\left[\left(1 + \frac{S}{N}\right)^{1/2}\right] \text{ bps} \tag{2-20}$$

Looking at this another way, since S and N are powers, n_{dis} as given by Equation 2-19 is the *ratio* of the rms value of the voltage of the signal with the noise to the rms value of the noise alone, where σ_N is taken as the rms value of the Gaussian noise signal.

2.8 DOPPLER FREQUENCY SHIFT

Whenever a signal is transmitted from a moving source to a moving receiver (see Figure 2-7), we may encounter the problem of the receiver receiving the signal at a different frequency than that at which it was transmitted. This phenomenon is

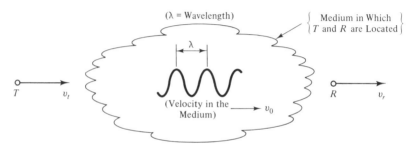

Figure 2-7 Propagation through a medium.

familiar to anyone who has ever been alongside a railroad track when a train passes by sounding its whistle. As the train passes by a distinct decrease in the pitch (frequency) of the whistle or horn sound is noticed. As the train was approaching, the frequency of the sound heard was higher than that actually being produced by the horn or whistle, and as the train is going away from the observer, the observed frequency is lower than that produced. This is the type of effect called the *Doppler effect*.

2.8.1 Doppler in a Medium

First, consider the case where the signal is being propagated through a medium (such as sound waves in air or water, etc.) and refer to Figure 2-7. Here T represents the transmitter (sound producer) and R represents the receiver (the hearer of the sound). If both are standing still with respect to the medium and if the transmitter produces a signal at frequency f_t, as this signal travels through the medium the wavelength of the signal will be given by

$$\lambda_o = \frac{v_o}{f_t} \tag{2-21}$$

where v_o is the velocity of propagation (travel) of the signal in the medium. The receiver will intercept cycles of the signal waveform at a rate of $v_o/\lambda_o = f_t$ cycles per second (Hertz). Thus, the receiver hears the same frequency as was sent by the transmitter.

However, if the transmitter is moving toward the receiver with a velocity v_t, then the wavelength of the signal that travels through the medium is shorter because the transmitter has moved a distance v_t/f_t since sending the same point in the previous cycle of the wave. Thus

$$\lambda = \lambda_o - \left(\frac{v_t}{f_t}\right) = \frac{(v_o - v_t)}{f_t} \tag{2-22}$$

If $v_r = 0$ (receiver not moving), then the received frequency is

$$f_r = f_t \left(\frac{v_o}{v_o - v_t}\right) \tag{2-23}$$

If v_r is negative, the receiver will intercept the incoming wave at a more rapid rate than when it is standing still, giving a received frequency of

$$f_r = \frac{(v_o - v_r)}{\lambda} = f_t \left(\frac{v_o - v_r}{v_o - v_t}\right) \tag{2-24}$$

The equations are algebraically valid—that is, if the transmitter and receiver are moving in directions opposite to that shown in Figure 2-7, it is only necessary to use negative values for the velocities in order to obtain the correct results.

It is interesting to note that for waves in a medium, the Doppler shift that occurs is *not* strictly a function of the *relative* velocity between the transmitter and the receiver, but instead depends upon which one is moving. For example, consider the case where only one is moving at a time and let $v_o = 1100$ feet per second:

First, let the transmitter be moving toward the receiver with $v_t = 110$ feet per second, and let $v_r = 0$. Then

$$f_r = f_t \left(\frac{1100}{1100 - 110} \right) = 1.1111 f_t \quad \text{Hz} \tag{2-25}$$

But now suppose that $v_t = 0$ and $v_r = -110$ feet per second. Then

$$f_r = f_t \left(\frac{1100 + 110}{1100} \right) = 1.1000 f_t \quad \text{Hz} \tag{2-26}$$

Thus, the movement of the transmitter toward the receiver gives a greater Doppler shift than does the movement of the receiver toward the transmitter!

Keeping this in mind, it should be pointed out also that for the case of sound waves through an air mass, the possible movement of the air mass itself must be considered and v_t and v_r should actually be the velocity of the transmitter and the receiver *relative to the air mass* (which as just indicated, may itself be in motion). Thus, a transmitter or a receiver that is actually stationary relative to the earth's surface must be considered to be in motion if the wind is blowing! An interesting result of this phenomenon occurs when the receiver and the transmitter are both fixed relative to the earth but the wind is blowing from one to the other (see Equation 2-24 with v_r equal to v_t). Here no Doppler shift occurs at the receiver although the wavelength λ of the signal as it is propagating is either larger or smaller than the λ given by Equation 2-21.

2.8.2 Doppler Shift of an EM Wave in Space

For an electromagnetic (EM) wave in space, Einstein's theory of relativity says that the wave moves away from the transmitter at the speed of light, c, no matter how fast the transmitter may be moving with respect to any arbitrary reference system. Thus, consider for the moment the reference system to be centered on the transmitter, T (which is to say that the transmitter is standing still). Then the wavelength of the transmitted wave will be $\lambda = \lambda_o = c/f_t$ and the frequency seen at the receiver will be given by

$$f_{r1} = (c/\lambda) \left(\frac{c - v_r}{c} \right) = f_t \left(\frac{c - v_r}{c} \right) \tag{2-27a}$$

However, both R and T may be in motion in another coordinate system, and thus it may be more general to replace v_r by the *relative* velocity between R and T. Let us call this relative velocity v_A. Then

$$f_{r1} = f_t \left(\frac{c + v_A}{c} \right) = f_t \left(\frac{c + |v_r - v_t|}{c} \right) \tag{2-27b}$$

where $v_A = |v_r - v_t|$ is called the relative approach velocity. v_A is positive when R and T are getting closer together.

Now, if we put the reference system on R instead of on T we find that T is moving and R is standing still in *this* reference system. The velocity of the wave coming into R is still given by c. This is one of the paradoxes (to our normal way of thinking) of the theory of relativity. Now, since T is moving, we find that the wavelength of the radiation arriving at R is given by $\lambda = \lambda_o(c - v_t)/c = (c - v_t)/f_t$, and therefore the received frequency is

$$f_{r2} = c/\lambda = f_t\left(\frac{c}{c - v_t}\right) \tag{2-28a}$$

But again both T and R may be moving with respect to some other coordinate system, so we replace v_t by the relative velocity of T with respect to R, which is to say that we replace v_t with v_A so that we now have

$$f_{r2} = f_t\left(\frac{c}{c - v_A}\right) = f_t\left(\frac{c}{c - |v_t - v_r|}\right) \tag{2-28b}$$

We now have two equations for the received frequency, and both cannot be correct because for a given value of v_A they will give two different results. Our problem is that we have used an assumption of the relativity theory and mixed it with a classical method. However, we do find that for cases where $v_A <<< c$ Equations 2-27b and 2-28b both give answers that are approximately correct. Without getting any more deeply involved in the theory of relativity, we will merely state that the correct solution of the problem is given by taking the geometric mean of the two results f_{r1} and f_{r2}. This gives us the relativistic Doppler equation for an EM wave:

$$f_r = \sqrt{(f_{r1})(f_{r2})} = \left(\sqrt{\frac{c + v_A}{c - v_A}}\right)(f_t) \tag{2-29}$$

From the preceding discussion, we see that there are two basic types of Doppler phenomenon—that in which we have propagation through a medium (i.e., where the medium is actually deformed in some manner by the wave as it passes through) and that in which an EM wave is passing through empty space.[4]

2.9 FREQUENCY STABILITY

In the previous section the Doppler effect was discussed. This is but one of the many effects that may cause the frequency of the signal arriving at the receiver to be different from that sent by the sender. Especially where radio links are con-

---

[4] *Historical note*: In the early years of EM wave theory, people could not imagine a radiation or wave that did not need a medium for propagation, so they postulated a theoretical medium that nobody could see or detect and called it "ether." As you read early books on radio you will find reference to "ether waves" and other such descriptions of EM radiation. Later work in the field showed that as far as we know now, there is no medium involved in the propagation of EM waves. EM waves include infrared, visible light, ultraviolet, X-ray, and gamma radiation as well as radio-wave radiation.

sidered where the information signal is modulated upon a carrier waveform and perhaps shifted (mixed) with other oscillator signals to higher or lower frequencies as it passes through the various parts of the channel, frequency-variation effects may be introduced. Some of these will be due to the instability of the signals from the carrier oscillators as well as from the local mixer oscillators in transmitters and receivers.   In an amplitude-modulated system, some of these variations in frequency may produce little or no variation in the frequency of the detected output heard by the receiver (i.e., the re-created information signal), while in other systems (like narrow-band FM, single-sideband modulation, frequency-shift keyed digital systems, etc.), even a small amount of such oscillator frequency instability may have very severe effects upon the detected output.   Some of these frequency variations may be due to factors that allow the variation to be described in a definitive manner (as definite functions of time, etc.), but other types of variations may be caused by mechanisms that are random in nature and must be described in a statistical manner.   This is done by giving the *statistics* of the frequency variation such as the *mean* (average frequency), the *variance* about the mean, or the *standard deviation* of the frequency (the square root of the variance).

The standard deviation is denoted by σ and the variance by σ^2. For example, a frequency variation that is Gaussian in nature may be described by sketching its frequency-probability density function as is shown in Figure 2-8.   In this case the mean frequency, f_m, is slightly offset from the desired or design frequency, f_d, by some amount, and in addition, the frequency varies in a symmetrical manner about f_m, with a 68.25% probability that the actual frequency will lie between $f_m - \sigma$ and $f_m + \sigma$.   There is a 99.75% probability that the instantaneous frequency will lie between $(f_m - 3\sigma)$ and $(f_m + 3\sigma)$.   The value of σ will depend upon how *unstable* the source is.   For a stable, well-designed oscillator, the value of σ (or, more properly, the value of σ/f_m) will be quite small (10^{-5} to 10^{-8} or smaller), while for a poorer, less costly oscillator, the σ/f_m ratio will be larger.

Sources of frequency instability are changes in the oscillator's component parts due to aging; motion of the wires, components, or other parts of the oscillator; variations in the d.c. supply voltage to the oscillator; changes in the loading on the oscillator output; changes due to temperature variations; radiation from nuclear, IR, or optical sources; and the like.   For systems that employ some type of elec-

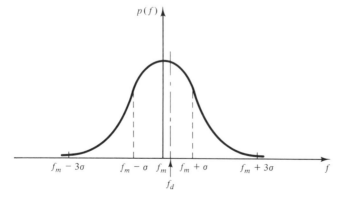

Figure 2-8 Variation of a frequency from the design value (normal distribution).

tromechanical system (such as tape recorders, etc.), the mechanical speed variations that occur may also produce frequency instability of the electronic signals.

PROBLEMS

2-1. Why do we use Fourier series or Fourier transform analysis techniques in analyzing communication systems? (Give some reasons.)

2-2. How are *harmonic* signals related to the *fundamental* signal component insofar as frequency and amplitude are concerned?

2-3. Give the equations for the expansion coefficients in the *complex* and *trigonometric* forms of the Fourier series (see a mathematics handbook or a text on transform calculus).

2-4. A baseband communication signal has frequency components that range from 1.5 KHz to 4.7 KHz and has 800 discernible amplitude levels. What is the maximum rate of information that this channel can carry?

2-5. A signal with a 3.3-KHz information bandwidth has an S/N ratio of 7 dB. How many discernible levels are there in this signal and what is its maximum information data rate?

2-6. A certain digital coding system produces an output of 500 12-level digital pulses per second. What is the digital data rate of this signal?

2-7. Assume a communications channel with 4 discernible levels. By what factor do you increase the channel capacity by doubling the number of levels to 8? By what factor (additional) by doubling again to 16? By doubling again to 32? What total factor (ratio) in channel capacity increase is obtained in going from 4 to 32 levels (8 times)? Suppose that the increase was from 64 to 8 times 64 (or 512) levels; what capacity increase (ratio) would occur?

2-8. If an analog channel has a bandwidth of 1500 Hz and a rated channel capacity of 12,000 bits per second, what is (a) the S/N ratio on the channel and (b) the number of distinguishable signal levels?

2-9. What is the maximum information rate of a signal whose information bandwidth is 4.5 MHz and whose signal-to-noise ratio is 23 dB (a typical TV signal)?

2-10. In an amplifier which is impedance matched into a system the input impedance of the amplifier is 70 ohms and the output impedance is 300 ohms. The input voltage to the amplifier is 15 microvolts, and the output signal voltage is 0.65 volts (both rms values). What is the power gain in decibels of the amplifier?

2-11. An amplifier's input resistance is 600 ohms, its output resistance is 8 ohms, and its power gain is 45 dB when operating with a matched load. If the input voltage is 10 mV, what is the power delivered to the load? What is the current through the load?

2-12. A "hi-fi" amplifier has an input impedance of 600 ohms and an output impedance of 8 ohms. When the amplifier is tested with matched source and load on both the input and the output the rms, input voltage is 1 mV and the rms output voltage is 8 volts. (a) What is the gain (in decibels) of the amplifier? (b) What is the voltage-gain ratio, A_v? (c) What are the input and output signals levels expressed in terms of dBm?

2-13. In a communications system a signal starts out at a level of 0.1 milliwatt, is amplified by an amplifier having a voltage gain ratio of 100 at an impedance level of 300 ohms, is transmitted through 5 miles of 300-ohm cable having an attenuation of 1 dB per 1000 feet, is amplified by an amplifier having a power gain ratio of 1000, and is delivered to a decoding device. Work this problem using the dB/dBm system and find the output power level into the decoding device in dBm and convert it to Watts.

2-14.

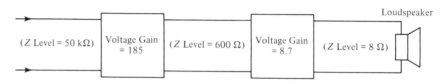

If $V_1 = 500$ mV rms, find P_{out}. What is the gain of each amplifier and the total gain expressed in decibels?

2-15. List and briefly describe the various types of distortion that may be present in a baseband amplifier (an audio amplifier for example).

2-16. What are meant by the terms *frequency distortion* and *phase distortion* as they are used in some electronic textbooks?

2-17. If a system is to transmit a signal with a baseband bandwidth of 100 KHz without distortion, what can be said about the requirements concerning the amplification or attenuation in the system and the phase shift or time delay in the system?

2-18. a. What causes *harmonic* distortion (discuss and explain)?

b. What causes *intermodulation* distortion (discuss and explain)?

2-19. An aircraft flying at 30,000 feet is above a point 20 miles distant from a radio navigation station toward which the aircraft is flying. If the aircraft speed is 750 mph, and the navigation station transmits on 112.4 MHz, what is the received frequency at the aircraft? Give the answer correct to the nearest 0.1 Hz.

2-20. Boat A is traveling north at 15 mph and boat B is traveling south at 20 mph. Boat A is 3 miles northwest of boat B when boat A blows its horn ($f_{horn} = 80$ Hz). What is the frequency of the horn signal received by boat B?

2-21. An aircraft moving at 600 mph at 30,000 feet is transmitting on a frequency of 130 MHz to a ground station directly in front of it which is 12 miles away (ground distance) from the aircraft's location. What frequency will the ground station receive? What frequency will a station 12 ground-miles distant and off the right wing of the aircraft receive?

2-22. A hypersonic aircraft traveling at 8000 miles per hour at 30,000 feet transmits to a ground station 10 miles in front of it (ground distance) on 135.955 KHz. What will be the received frequency at the ground station? (Carry the result out to the nearest 0.1 Hz.)

3

THE PROPERTIES OF SYSTEM NOISE

In this chapter we will look at the aspects of noise theory as applied to electronic communication systems. Noise theory has been a widely discussed and written-upon topic for many years, but the only aspects of this vast quantity of theory and knowledge to be covered here are those aspects necessary to do intelligent analysis and design work on basic communication systems.

There are many types of noise from many different possible sources that may interfere with a communications channel. For example, we are all familiar with the types of noise that occur in radio and TV reception in the form of disrupted sound or pictures due to interference (noise) from automotive ignition systems, thunderstorm lightning, defective light bulbs, Silicon-controlled-rectifier (SCR) light dimmers, fluorescent light fixtures, arcing switches, or loose electrical connections. This is what is termed *impulse noise*, and we usually take great pains to eliminate it at the source or to shield our communication system from it as much as possible. We will be discussing it later.

There is another form of noise that we cannot reduce or eliminate and which is much more fundamental in its nature. This is the noise that is produced naturally in nature due to the discrete (quantum) characteristics of everything in the universe: the fact that there is no charge of electricity smaller than one unit of electronic charge (i.e., the charge on one electron). One type of noise in this category is what is called *thermal noise*, and this is what we will be discussing in most of the remainder of this chapter.

Before beginning the discussion of noise itself, we first introduce the concept of *power-distribution functions*. If we consider nonperiodic signals, we cannot say that there are so many Watts of power *at* a certain frequency f; instead, we have to say that there are so many Watts of power *in* the frequency range from f to $f + \Delta f$. If we let this amount of power in that frequency range be denoted by $P_{\Delta f}$, we can determine $P_{\Delta f}$ from the integration of a function $p(f)$ called the *power-density function* over the region from f to Δf, or

$$P_{\Delta f} = \int_{f}^{f + \Delta f} p(f)\, df \quad \text{Watts} \tag{3-1}$$

The power-density function $p(f)$ has the units of Watts per unit of frequency, or Watts per Hertz. But since the physical units of the Hertz are actually seconds^{-1}, we can say that $p(f)$ has the units of Watt-sec. We shall be talking about the power-density function of both noise signals and information signals. The power-density function is often called the *power-spectral-density function*. For noise, the power in the bandwidth $\Delta f = (f_2 - f_1)$ is given by

$$N = \int_{f_1}^{f_2} p_n(f)\, df \quad \text{Watts} \tag{3-2}$$

Returning to types of noise signals, thermal noise is often classed as *Gaussian white* noise, which means that the amplitude-time statistics of the noise are Gaussian in nature and that the power-density function of the noise is constant over the range of frequencies on which the noise is defined to exist. One of the characteristics of most Gaussian noise is that it is completely random in nature: the value of the noise signal at any one given instant of time has no dependence upon the value at any other point in time. The amplitude-probability function follows the Gaussian or normal distribution curve. We will not go into great detail discussing the characteristics of noise signals at this point, but will merely make use of the results of noise theory to determine how communication systems are affected by such noise.

In the following sections we will be using the concept of the *noise bandwidth* of a block in a communications system. In this case we will assume that the input noise power spectral density $p_n(f) = p_n$ is constant over a wide range of frequency and that the system block has a voltage or current transfer function $H(f)$ with a maximum value of 1.0. The noise power out of the system block is then given by

$$N = \int |H(f)|^2 p_N\, df = B_N p_N \tag{3-3}$$

from which we have that the noise bandwidth is

$$B_N = \int |H(f)|^2\, df \tag{3-4}$$

where the limits of the integration are such as to include all the frequencies where $|H(f)|$ has any appreciable value. The noise bandwidth B_N is not necessarily equal to the signal bandwidth (usually the latter is taken to be the -3 dB or half-power bandwidth) of the system block. For example, if the system block's response function is that of a singly tuned resonant circuit, it will be found that B_N is equal to $(\pi/2)\,BW_{-3\,dB}$.

3.1 THERMAL NOISE SOURCE IN A RESISTANCE

Consider now the case of a resistance R at a temperature of T degrees Kelvin. This resistance will be a generator of thermal noise power (plus perhaps other types of noise in the case of some types of resistors). At the lower frequencies (frequencies less than the infrared for R at room temperature), the noise generated by R will be very nearly Gaussian and white in its nature. One way of showing R as a generator of noise is to represent it as a current generator $\overline{i_n^2}$ in parallel with a perfect noiseless resistor of the value of R (see Figure 3-1). Under these conditions, the current source has a mean-square value of

$$\overline{i_n^2} = \frac{4\,kTB_N}{R}\quad\text{Amperes}^2 \tag{3-5}$$

where B_N is the range of frequencies being considered, R is the ohmic value of the resistor, T is the Kelvin temperature, and k is Boltzman's constant (1.381×10^{-23} watt-sec/°K). Let us attach to R another resistor R' that has the same value as R. Then we find that of the power delivered by the generator *in R* the resistor R' will absorb one-half, or, referring to Figure 3-1,

$$P_{NR'} = (\overline{i_{R'}^2})R' = \left(\frac{1}{2}\right)^2\left(\frac{4\,kTB_N}{R}\right)R'$$
$$= kTB_n \quad\text{Watts} \tag{3-6a}$$

if $R = R'$.

Equations 3-5 and 3-6a come from the Rayleigh-Jeans approximation to Planck's law of thermal radiation (see Appendix A). Equation 3-6a will be very important to us, since many of the cases that we will be considering will be those where the value of the noisy source resistance will be equal to the value of the load resistance. If this is not the case, Equation 3-6 will include the square of a multiplying constant

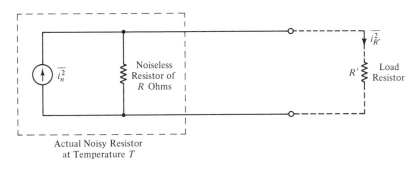

Actual Noisy Resistor
at Temperature T

Figure 3-1 Representation of a resistor generating thermal noise.

other than $\frac{1}{2}$, which depends upon the relative values of R and R':

$$P_{NR'} = \left(\frac{R}{R + R'}\right)^2 (4kTB_N)\left(\frac{R'}{R}\right)$$

$$= \left[\frac{R \times R'}{(R + R')^2}\right](4kTB_N) \quad \text{Watts}$$

(3-6b)

Equations 3-6a and 3-6b may be stated in words as follows:

> The Gaussian noise available from a source at temperature T is directly proportional to the value of T and the bandwidth being considered so long as the ratio hf/kT is small, where f is the frequency being considered, h is Planck's constant, k is Boltzman's constant, and T is the temperature in degrees Kelvin.

The foregoing statement and Equations 3-6a and 3-6b would seem to indicate that noise power would flow from resistor R to the resistor R', and indeed this is so if R' is a perfect noiseless resistance. However, if R' itself is a practical (noisy) resistor that generates thermal noise of its own, we find that it generates and delivers an equal amount of thermal noise power that goes back to the original resistor R (assuming both R and R' to be at the same temperature). However, if one resistor is at a lower temperature than the other, there *will* be a net flow of electrical power from the hotter resistor to the cooler one. For example, consider the case in Figure 3-2 where one resistor is in a can of boiling water and the other is in a pumped liquid helium bath at 2°K. Resistor R_1 will deliver 373 kB_N Watts of power to R_2, and R_2 will deliver 2 kB_N Watts of power to R_1, or there will be a net flow of 371 kB_N Watts of power from R_1 to R_2. Using some numerical values for an example by letting $B_N = 10.0$ MHz $= 10^7$Hz, we find that the net flow of power is 5.11×10^{-14} Watt, which is a small power level, but it may nevertheless be comparable to the received signal power from an interplanetary space probe vehicle.

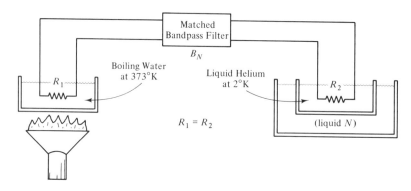

Figure 3-2   Resistors of equal value not in thermal equilibrium.

3.2 ADDITION OF SIGNALS AND NOISE IN A NOISY SIGNAL

Earlier we had defined the signal-to-noise ratio as the ratio of the signal *power* to the noise *power* (see Section 2.1).   We can express this ratio in decibel form as

$$(S/N)_{i,\,\mathrm{dB}} = 10 \log_{10} \left(\frac{S_i}{N_i} \right) \mathrm{dB} \tag{3-7}$$

where the subscript *i* represents some particular point in the system under analysis. Let us now consider the case where there may be more than one information signal present at a given place in the system as well as more than one noise signal.

Information signals are usually coherent to some degree.   If two information signals are *completely* coherent, we may add their rms *voltage* values, or

$$V_{1+2} = V_1 + V_2 \quad \mathrm{Volts} \tag{3-8}$$

This means that we cannot add their powers

$$P_{S_1 + S_2} \neq P_{S_1} + P_{S_2} \quad \mathrm{Watts} \tag{3-9}$$

In general, we find that to get the power of the two signals, we must have an equation of the form

$$
\begin{aligned}
P_{S_1 S_2} &= \frac{V_1^2 + V_2^2 + (2\,Q V_1 V_2)}{R_o} \\
&= P_{S_1} + P_{S_2} + 2Q \sqrt{P_{S_1} + P_{S_2}}
\end{aligned} \tag{3-10}
$$

where R_o is the impedance level at which the signals are being considered and Q is the amount of correlation between the two signals ($Q = 1$ for perfect correlation). If, however, there is less than perfect correlation between the two information signals, we will find that Q is less than unity, and if the two signals are completely uncorrelated, then $Q = 0$, and in *that* case *only* may we simply add the two signal *powers* together (but now we cannot add the signal voltages—in fact, if Q is *anything* less than unity, we cannot add the signal voltages).   Note that this prohibition applies only to adding the *rms* voltages of the signal—the voltage time functions, if known, can always be added together at any value of time, or the expression

$$v_{1+2}(t) = v_1(t) + v_2(t) \tag{3-11}$$

is always true, whether the two signals are correlated or not.

In the case of noise signals, there is often little or no correlation between the two noise signals.   Thus,

$$N_{1+2} = N_1 + N_2 + 2Q_N \sqrt{N_1 + N_2} \cong N_1 + N_2 \tag{3-12}$$

since Q_N is usually zero or nearly zero.   Thus, we can normally add the noise *powers* from all the noise sources in a communication system.   Since there is usually only one information signal source in a system but perhaps dozens of noise sources, we usually do our system calculations in terms of power levels rather than voltage levels. By doing this, we can combine the effects of the noise sources by simple addition of their noise powers.

At this point some discussion of the words "coherent" and "correlation" is probably in order. As pointed out in Chapter 2, two signals are perfectly correlated or coherent if one is some multiple of the other, that is, if $v_2(t) = A_{21}v_1(t)$, where A_{21} is a constant. Two sinusoidal signals of different frequencies have zero correlation, and two signals of the same frequency have perfect correlation if they are exactly in phase but zero correlation if they are 90 degrees out of phase. If they are 180 degrees out of phase, they are said to have perfect negative correlation (i.e., $Q = -1$). The correlation factor Q of two sinusoidal signals of the same frequency is given by $Q = \cos(\delta)$, where δ is the phase difference between the two signals. Two completely random signals (such as Gaussian noise signals) have zero correlation.

3.3 SYSTEM NOISE CONSIDERATIONS

In Figure 3-3, we have shown a radio receiving system where some sort of system device has been inserted into the transmission line ahead of the receiver (or, alternatively, we could have considered an amplifier at either RF or audio frequencies instead of the receiver—the analysis is the same). Let the receiver have a noise bandwidth of B_N Hz, and let the device inserted have a loss factor of L (i.e., $100 * L\%$ of the signal into the device is lost in the device and only $100 * (1 - L)\%$ comes out of the right-hand port). Let the ambient temperature of the lossy device be given by $T_a°K$. We will first calculate the signal-to-noise ratio of the signal presented to the receiver under the condition that the input signal to the lossy device has an infinite S/N ratio (i.e., no noise on the signal into the device).

Whenever we have a lossy device in a system, the noise power per unit of bandwidth that is generated by the device itself is given by[1]

$$p_n(f) = kT_aL \quad \text{Watts/Hz} \tag{3-13}$$

where L is the loss factor. Thus, the noise power in the bandwidth of interest (the BW of the following part(s) of the system as denoted by B_N) is given by

$$N = kT_aLB_N \quad \text{Watts} \tag{3-14}$$

Since the loss factor is L, the signal power out of the device and into the receiver will be

$$S_r = S_i(1 - L) \quad \text{Watts} \tag{3-15}$$

Figure 3-3 A lossy device in a communication system.

[1] See Appendix B.

or the S/N ratio at the receiver input will be

$$(S/N)_r = \frac{S_i(1 - L)}{kT_aLB_N} \tag{3-16}$$

Equation 3-16 gives the S/N ratio at the receiver input under the conditions that the incoming signal presented to the lossy device has no noise on it. This is very seldom the case, so let us proceed to modify Equation 3-16 to include the more general case of a signal arriving at the input of the device already containing noise of power N_i, or where the input signal-to-noise ratio $(S/N)_i = S_i/N_i$. Now the total noise power presented to the receiver will be given by

$$N_r = [N_i(1 - L)] + (kT_aLB_N) \tag{3-17}$$

or, in terms of the equivalent noise temperature (by dividing by kB_N as explained shortly),

$$T_{N_r} = [T_{N_i}(1 - L)] + [L \quad T_a] \tag{3-18}$$

Note that N_r (or T_{N_r}) may be either larger or smaller than N_i (or T_{N_i}) depending upon the relative values of T_{N_i}, L, and T_a. Also, note that in the case of Equation 3-17, N_i is the device input noise power in the bandwidth B_N. The signal power is the same as before (Equation 3-15), so now the signal-to-noise ratio at the receiver input will be given by

$$(S/N)_r = \frac{S_i(1 - L)}{[N_i(1 - L)] + (kT_aLB_N)} \tag{3-19}$$

Sometimes in considering noise in a communications system, it is convenient to work with the *noise temperature* rather than the noise power. The noise temperature at any given point j in the system is defined as

$$T_{N_j} = \left[\frac{N_j}{kB_N}\right] \tag{3-20}$$

Using this definition and applying it to the input noise signal, Equation 3-19 becomes

$$(S/N)_r = \frac{S_i(1 - L)}{(LT_a) + [(1 - L)T_{N_i}]} \tag{3-21}$$

where the quantity $(LT_a) + (1 - L)T_{N_i}$ can be called the *equivalent noise temperature* T_{N_r} at the receiver input terminals. Thus, Equation 3-19 becomes

$$(S/N)_r = \frac{S_r}{kT_{N_r}B_N} \tag{3-22}$$

where T_{N_r} is the equivalent noise temperature at the *receiver* input. The value of $(S/N)_r$ will always be smaller (worse) than $(S/N)_i$. Also, note that the output S/N ratio depends upon knowing the input signal and noise *levels*. Knowing *just* the input S/N *ratio* is *not* sufficient to find the output S/N ratio for this situation.

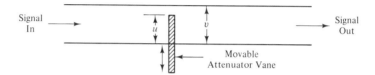

Figure 3-4 A "signal pipe" with an attenuating vane inserted.

Sometimes it is hard to visualize that the noise power added to the output of a lossy device by the losses in the device itself is actually given by Equation 3-14 and the following equations. One way to visualize this relationship physically would be to postulate a "signal pipe" such as is shown in Figure 3-4 with a sliding absorptive vane that can be positioned into the pipe to absorb a desired fraction of the input signal $L = u/v$, where v is the distance across the signal pipe and u is the distance which the vane is inserted into the pipe. The remaining open portion of the pipe passes a fraction $G = 1 - L$ of the input signal through to the right-hand section of the pipe. The attenuator vane itself radiates a noise signal of LT_akB_N Watts from its right-hand surface, and this noise signal adds onto the portion of the input signal that gets past the vane to form the complete output signal.

In addition to the rather nonrigorous explanation just given for the validity of Equation 3-14 and following, Appendix B presents a rigorous proof of these relationships for the case of the T-pad attenuator. Since any passive two-port network can be put into the form of an equivalent-T network, Appendix B is a general proof of these relationships for any passive two-port network.

Consider now several cascaded attenuators or devices as shown in Figure 3-5, where for each device we have $L_j =$ loss factor and $1 - L_j = G_j =$ the transmission (gain) factor for the jth device, where G_j is less than unity. Now starting from the right-hand port of the system and applying the relationships just developed, we get the output signal-to-noise ratio equation in the forms

$$\left(\frac{S}{N}\right)_{\text{out}} = \frac{(1 - L_3)S_3}{k\,B_N[L_3 T_{L_3} + (1 - L_3)T_3]}$$

$$= \frac{(1 - L_3)(1 - L_2)S_2}{k\,B_N\{L_3 T_{L_3} + [1 - L_3][L_2 T_{L_2} + (1 - L_2)T_2]\}} \qquad (3\text{-}23)$$

$$= \frac{(1 - L_3)(1 - L_2)(1 - L_1)S_1}{k\,B_N\{L_3 T_{L_3} + [1 - L_3][L_2 T_{L_2} + (1 - L_2)(L_1 T_{L_1} + (1 - L_1)T_1)]\}}$$

If the bandwidths of the devices involved are not the same, then the bandwidth value that should be used for B_N is the minimum bandwidth of the cascade (i.e.,

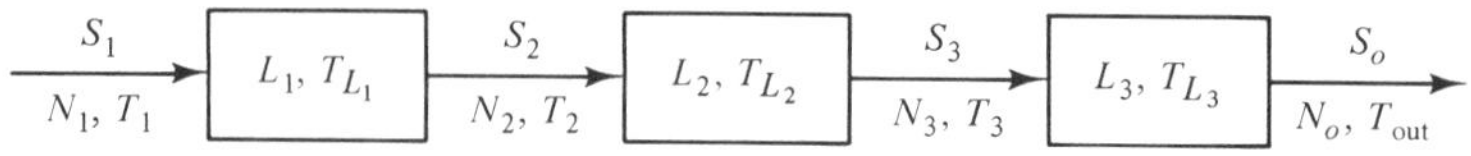

$$S_1 \quad \boxed{L_1, T_{L_1}} \quad S_2 \quad \boxed{L_2, T_{L_2}} \quad S_3 \quad \boxed{L_3, T_{L_3}} \quad S_o$$
$$N_1, T_1 \qquad\qquad N_2, T_2 \qquad\qquad N_3, T_3 \qquad\qquad N_o, T_{\text{out}}$$

Figure 3-5 A cascade of attenuators or attenuating devices.

the difference in frequency between the upper and lower 3-dB rolloff points of the passband curve of the *cascade*).   From Equation 3-23 we see that the calculation of the *S/N* ratio after passing through several lossy devices is a combination of alternate addition and multiplication processes.   Thus, it is not possible to work directly with the decibel system of gain and loss units in this situation.   The only exception to this is the calculation of the numerator factor (the signal level); the denominator factor must be worked out piece by piece to determine the noise power level at the output end of the system.   This again brings out the point that knowing the *S/N* ratio at the input or some other point of the system is not adequate information to calculate the output *S/N* ratio.   The input signal and noise *power levels* must be known, not just their ratio.

Many times the loss in a device is given in decibels rather than as the loss factor *L*.   We cannot convert from decibels to loss factor *L* directly using the standard logarithmic relationship, but must instead work through the gain of the device, *G*, where *G* is less than unity.   For example, if we are told that the loss of a certain device is *X* dB, what this actually means is that the device has a gain of $-X$ dB.   Thus, the gain *ratio*, *G*, is given by

$$G = (10)^{-X/10} < 1 \tag{3-24}$$

and the loss factor *L* is

$$L = (1 - G) = 1 - 10^{-X/10} \tag{3-25}$$

where *X* is the device *loss* expressed in decibel units (i.e., $X > 0$).

Example:

If a device has a loss of 6 dB, its loss factor will be given by

$$L = (1 - 10^{-0.6}) \doteq (1 - \tfrac{1}{4}) = \tfrac{3}{4}$$

3.4 *AMPLIFIER AND RECEIVER NOISE FIGURES*

The noise factor of an amplifier, radio receiver,[2] or other active device is an expression of how much *extra* noise the device itself adds onto the noise already appearing at its output due to amplified input noise.   This extra added noise is often called the *excess noise power* of the device.   We shall now look at the determination of this excess noise power and the definition and determination of the noise factor and noise figure of the device (the *noise figure* is 10 times the common log of the noise factor and is given in decibels).

As we talk about noise in a system, we need to specify the noise bandwidth that we are considering.   Usually this is the noise bandwidth of the smallest-effective-bandwidth block of the system.   In the case of an audio or video amplifier, this usually means the bandwidth of one of the amplifiers or filters at the end of

---

[2] The noise factor of a radio receiver is usually specified in terms of the excess noise power appearing at the input to the demodulation detector in the receiver.

the cascade of system devices. In radio receivers it usually refers to the bandwidth of the narrowest IF amplifier or IF filter.

Now let us define the noise factor, F, of some device or some cascade of devices in a system as being the ratio of the input S/N ratio to the output S/N ratio (i.e., the ratio of two ratios):

$$F = \frac{S_{in}/N_{in}}{S_{out}/N_{out}} > 1 \qquad (3\text{-}26)$$

or

$$\frac{S_{out}}{N_{out}} = \left(\frac{1}{F}\right)\left(\frac{S_{in}}{N_{in}}\right) \qquad (3\text{-}27)$$

F is always greater than unity, and thus we always find that the output S/N ratio is *smaller* than the input S/N ratio. This is due to the excess noise that is introduced by the device itself.

Consider now an amplifier of power gain G as shown in Figure 3-6. The output signal power is equal to the input signal power multiplied by the gain while the output noise power is made up of the sum of the amplified input noise plus the excess noise power added by the amplifier, or

$$N_{out} = \text{excess noise} + GN_{in} \quad \text{Watts} \qquad (3\text{-}28)$$

The output S/N ratio then becomes

$$\frac{S_{out}}{N_{out}} = \left(\frac{1}{F}\right)\left(\frac{S_{in}}{N_{in}}\right) = \frac{GS_{in}}{(\text{excess noise}) + (GN_{in})} \qquad (3\text{-}29)$$

from which we see that

$$FN_{in} = (N_{eq} + N_{in}) = N_{out}/G \qquad (3\text{-}30)$$

where we have defined $N_{eq} = \text{excess noise}/G$. This leads to

$$F = \frac{N_{eq} + N_{in}}{N_{in}} = \frac{N_{out}}{GN_{in}} \qquad (3\text{-}31)$$

From this it is obvious that the noise factor, F, is a function of the input noise signal as well as of the gain and the excess noise (which are characteristic of the amplifier). The term N_{eq} is called the excess noise of the amplifier referred to its input. It is the equivalent input noise power, which, if amplified by a noiseless amplifier, would produce the *excess* noise of a noisy amplifier of the same gain.

The noise factor, F, as defined has a disadvantage in that it is not a very useful parameter to use to specify the performance level of a piece of equipment, since it depends upon the noise on the signal fed into the input of the piece of equipment. To get around this difficulty, we shall define what is called the *standard*

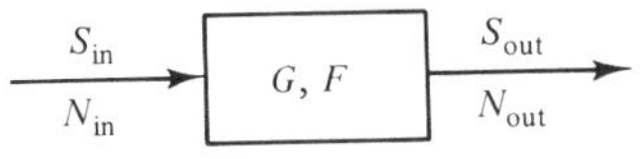

Figure 3-6 An amplifier with a noise figure of F.

noise factor, F_o, which is what the noise factor, F, would be if the input noise came from a resistor of temperature 290°K attached to the input terminals of the amplifier or receiver in question.   Thus,

$$N_{\text{in}} = kB_N 290 = N_o \tag{3-32}$$

which gives

$$F_o = \frac{N_{\text{eq}} + N_o}{N_o} = \frac{N_{\text{out}}}{GN_o} = \frac{N_{\text{out}}}{GkB_N 290} \tag{3-33}$$

Here N_o has been defined as N_{in} from a resistor at 290°K and some bandwidth B_N.[3] N_o is called the *standard noise power* for the noise bandwidth B_N, and we shall define T_o as the standard noise temperature of 290°K.   The noise out of the amplifier can now be written as

$$N_{\text{out}} = \text{excess noise} + GN_o = F_o(GN_o) \tag{3-34}$$

under the conditions that the amplifier input is attached to a resistor R at a temperature of 290°K.   The excess noise then is

$$\text{excess noise} = (F_o - 1)(GN_o) \tag{3-35}$$

for the case where the input noise is N_o.   Now, for *any* input noise level, N_{in}, we get

$$N_{\text{out}} = (F_o - 1)(290kB_N G) + GN_{\text{in}} \tag{3-36}$$

or

$$\begin{aligned} T_{\text{out}} &= [(F_o - 1)(290G)] + T_{\text{in}}G \\ &= G[(F_o - 1)290 + T_{\text{in}}] \end{aligned} \tag{3-37}$$

and since the signal out is GS_{in}, we get

$$\begin{aligned} \frac{S_{\text{out}}}{N_{\text{out}}} &= \frac{GS_{\text{in}}}{(F_o - 1)(290kBG) + GN_{\text{in}}} \\ &= \frac{S_{\text{in}}}{N_{\text{eq}} + N_{\text{in}}} \end{aligned} \tag{3-38}$$

where the G factors cancel in the numerator and the denominator.   Comparing this with Equations 3-29 and 3-30, we see that

$$N_{\text{eq}} = (F_o - 1)(290 \, kB_N) \tag{3-39}$$

which is the noise generated by the amplifier itself referred back to its input terminals.

---

[3] Note that where we don't have the actual noise bandwidth specified, we may wish to use $\int G(f)df$ in place of the product $G \times B_N$. This may also be the better procedure in cases where, although the -3 dB bandwidth is known, the shape of the passband curve is irregular or nonstandard in nature.

The measurement of the noise factor (usually the standard noise factor) of an amplifier may be done with equipment such as is shown in Figure 3-7. Here the desire is to make this measurement without having to determine *absolutely* such things as the noise power into an amplifier or the output noise power of the same amplifier. Such absolute measurements are difficult to make accurately without specialized equipment. However, it is easier to measure and compare two noise powers on a relative basis, especially at the amplifier output. The method about to be discussed uses this second approach.

In the circuit shown in Figure 3-7 two resistors of equal resistance, R_o, are maintained at two different known temperatures. Usually these are noninductive wirewound resistors of some type, since quite often carbon composition resistors introduce additional noise other than strictly thermal noise because of the varying contact resistance between the carbon granules within the carbon composition mixture. The amplifier under test has its input alternately connected to one of these resistors and then to the other. A noise power meter connected to the output of the amplifier records the relative noise power output of the amplifier in each case. Writing the expression for the output noise power for the switch in each position and taking the ratio gives us a factor that we shall call K_p, where

$$K_p = \frac{[(F_o - 1)(290kB_n) + (kB_NT_1)] \times G}{[(F_o - 1)(290kB_N) + (kB_nT_2)] \times G}$$

$$= \frac{[(F_o - 1)290 + T_1]}{[(F_o - 1)290 + T_2]} = \frac{N_{out,1}}{N_{out,2}} \tag{3-40}$$

which, when solved for F_o, gives

$$F_o = 1 + \frac{[(K_pT_2) - T_1]}{(1 - K_p)\,290} \tag{3-41}$$

which gives the standard noise factor in terms of the *ratio* of the two noise powers measured by the output noise meter and the two temperatures of the resistances. The resistor temperatures are easily determined in degrees Kelvin with conventional temperature-measuring equipment (quite often the two resistors are in temperature-controlled ovens), and since only the *ratio* of the output noise powers is needed, the meter used for this has only the requirement of being linear (with respect to power) and does not need to be especially accurate on an absolute basis. Note

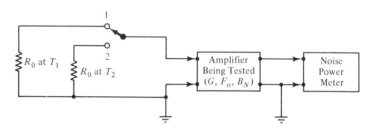

Figure 3-7 The measurement of amplifier noise figure.

that it is *not* necessary to know the gain or noise bandwidth of the amplifier under test.   Some commercial noise-figure measuring sets work on the principle just described by switching rapidly back and forth between two resistors.   The change in the output noise power of the amplifier or receiver being tested then appears as a square wave that is processed by the circuitry of the test set with the noise factor being displayed directly on a meter dial for the convenience of the person making the test.

When several stages of amplification are cascaded as is shown in Figure 3-8, we may determine an overall noise figure for the cascade if we know the noise figures of the individual stages.   Here we assume that the bandwidth of the cascade is the smallest bandwidth in the cascade (or the overall noise bandwidth of the cascade).   Let N_o again be defined as $(kB_N\,290)$.   The signal and noise out of the first stage are

$$S_2 = G_1 S_1 \tag{3-42}$$

$$N_2 = [(F_{o_1} - 1)N_o G_1] + (G_1 N_1) \tag{3-43}$$

and the signal-to-noise ratio therefore is

$$\frac{S_2}{N_2} = \frac{S_1 G_1}{[(F_{o_1} - 1)N_o + N_1]G_1} \tag{3-44}$$

while that out of the second stage is

$$
\begin{aligned}
\frac{S_3}{N_3} &= \frac{S_2 G_2}{[(F_{o_2} - 1)N_o + N_2]G_2} \\[2mm]
&= \frac{(G_1 G_2)S_1}{G_2\{[(F_{o_2} - 1)N_o] + G_1[(F_{o_1} - 1)N_o + N_1]\}} \\[2mm]
&= \frac{(G_1 G_2)S_1}{(G_1 G_2)\left[\left(\dfrac{(F_{o_2} - 1)N_o}{G_1}\right) + (F_{o_1} - 1)N_o + N_1\right]} \\[2mm]
&= \frac{(G_1 G_2)S_1}{\{[(F_o' - 1)N_o] + N_1\}(G_1 G_2)}
\end{aligned}
\tag{3-45}
$$

where F_o' is the equivalent standard noise figure of the cascade of the two stages. We may now solve Equation 3-45 for $F_o' - 1$ and then for F_o'.

$$F_o' - 1 = \left[\frac{(F_{o_2} - 1)}{G_1}\right] + (F_{o_1} - 1) \tag{3-46}$$

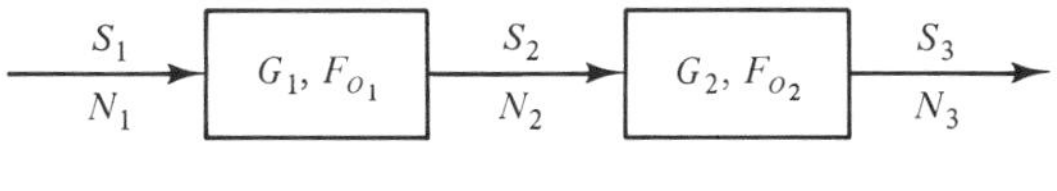

Figure 3-8 Cascaded noisy amplifier stages.

and

$$F'_o = \frac{(F_{o2} - 1)}{G_1} + F_{o1} \tag{3-47}$$

Similarly, for three stages in cascade, the overall noise factor would be

$$F'_o = F_{o1} + \frac{(F_{o2} - 1)}{G_1} + \frac{(F_{o3} - 1)}{G_1 G_2} \tag{3-48}$$

From Equations 3-47 and 3-48, we see that the standard noise factor of a cascade is primarily determined by the noise factor of the first stage, with the contributions by following stages being reduced by the total gain factor of the stages that precede each of them. Thus, in Equation 3-48, there would probably be little contribution to the cascade noise factor due to the third stage if the first and second stages had high gain factors. It is because of this major influence of the first stage upon the cascade's noise factor that we usually find system designers paying the most attention to (and spending the most money upon) the design of the first stage, trying to achieve simultaneously the goal of high gain and low noise factor within this stage.

Sometimes the noise performance of an amplifier or receiver is given in terms of its *equivalent noise temperature*, T_{Ne}, which is N_{eq} as defined in Equation 3-39 divided by kB_N. Table 3-1 tabulates $(F_o)_{dB}$ and also N_{eq} (the latter for a 30-MHz B_N) versus T_{Ne}.

Another figure of merit that may be used for an amplifier, receiver or even a complete receiving system (including the antenna) is the ratio of the system gain prior to some point in the system divided by the equivalent noise temperature at that same point (the "G/T ratio") expressed in decibels per degree Kelvin with $G = 1$ and a 1°K noise temperature as the 0-dB reference point. When using this figure of merit, it is important to know what gains are included in G and what noise powers are contributing to the calculation of the noise temperature T. If this is used as the figure of merit for a satellite-TV receiving system, for example, G would be the net gain of the dish antenna and feedline to the preamplifier input, and T would be the total system noise temperature referenced to that same point.

TABLE 3-1 AMPLIFIER NOISE FIGURE AND
EQUIVALENT EXCESS NOISE POWER FOR A 30-MHz
NOISE BANDWIDTH VERSUS THE AMPLIFIER
STANDARD NOISE FIGURE

T_{Ne}	$(F_o)_{dB}$	N_{eq} for $B_N = 30$ MHz
50° K	0.691 dB	$2.07 * 10^{-14}$ W
90° K	1.174 dB	$3.73 * 10^{-14}$ W
120° K	1.504 dB	$4.97 * 10^{-14}$ W
200° K	2.278 dB	$8.28 * 10^{-14}$ W
300° K	3.085 dB	$12.43 * 10^{-14}$ W
600° K	4.870 dB	$24.84 * 10^{-14}$ W

Note that although the gain of the preamplifier is not involved in calculating G, it *is* involved in determining the equivalent noise temperature since it divides the equivalent-excess-noise-temperature contribution of the receiver that follows the preamplifier to determine its contribution to the equivalent noise temperature at the preamplifier input.

Example:

Let a satellite-TV receiving system have an antenna gain of 34 dB, a feed line loss of 2 dB, a preamplifier equivalent excess noise temperature of 150°K and gain of 18 dB, and a receiver noise figure of 5 dB. Then,

$$G_{\mathrm{dB}} \text{ is given by } 34 \text{ dB } - 2 \text{ dB } = 32 \text{ dB } \quad \text{and}$$

$$T = 30 + 150 + \frac{(10^{5/10} - 1)\ 290}{10^{18/10}}$$

$$= 30 + 150 + 9.94 = 189.94°\mathrm{K}$$

This gives

$$(T)_{\mathrm{dBK}} = 10 \log_{10} \left(\frac{189.94°\mathrm{K}}{1°\mathrm{K}} \right) = 22.79 \text{ dB K}$$

The resulting $(G/T)_{\mathrm{dB/K}}$ is then $(32 - 22.79) = 9.21$ dB/K.

3.5 *DISCERNIBLE LEVELS IN A NOISY SIGNAL*

In this section we consider the limitations on determining the value of an analog signal that has added to it a random Gaussian noise signal. Assuming that both the information signal and the noise have normal (Gaussian) distributions centered about zero voltage (see Figure 3-9) and that the impedance level R_o is 1 Ohm, we have that the rms voltage of either the information signal or the noise is equal to the standard deviation, σ, of its Gaussian distribution, and thus its power level is equal to σ^2. Furthermore, we will consider the total voltage range of either the signal or the noise to be six standard deviations so that $\Delta V = 6\sigma_S$ and $\Delta V_N = 6\sigma_N$. Now we will define the number of discernible levels n_{dis} to be equal to $\Delta V_{\mathrm{total}}/ \Delta V_N$, where since there is no correlation assumed between the information signal

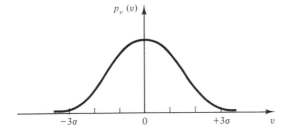

Figure 3-9 The probability-distribution function for a "Gaussian" random signal or noise voltage.

and the noise, we have $\Delta V_{\text{total}}^2 = \Delta V_S^2 + \Delta V_N^2$. This then gives

$$n_{\text{dis}} = \frac{6\sqrt{(\sigma_S^2 + \sigma_N^2)}}{6\sigma_N} = \sqrt{\frac{S + N}{N}} = \sqrt{\frac{S}{N} + 1} \qquad (3\text{-}49)$$

3.6 DIGITIZING NOISE

Whenever we digitize an analog signal (with theoretically an infinite number of possible levels) into a digital signal with a finite number of levels at the transmitting end of a communication system and then at the receiving end convert this digital signal back into an analog signal, we introduce a type of noise called *digitizing noise*. For the purpose of calculating this noise in a typical situation we are going to assume that the analog signal to be digitized is a Gaussian-distributed signal such as is shown in Figure 3-10a and that we are going to digitize it over a range of $\pm 4\sigma$ (it is, of course, possible to use other digitizing ranges, and we will later present the results obtained using another range). The noise signal that is generated results from assigning just one voltage value to any analog signal value that falls within the range ΔV_N centered about the assigned digital voltage value. This produces a random noise voltage whose probability-density function is approximately constant (when the total digitizing range is $>> \Delta V_N$) from $-\Delta V_N/2$ to $+\Delta V_N/2$ as is shown in Figure 3-10b. The height of this function will be equal to $(1/\Delta V_N)$ and if ℓ_d is the number of digitizing levels in the $\pm 4\sigma_S$ range, then $\Delta V_N = 8\sigma_S/\ell_d$. This then gives

$$S = \sigma_S^2 \qquad (3\text{-}50)$$

$$N = \sigma_N^2 = \tfrac{1}{12}\left(\frac{8\sigma_S}{\ell_d}\right)^2 \qquad (3\text{-}51)$$

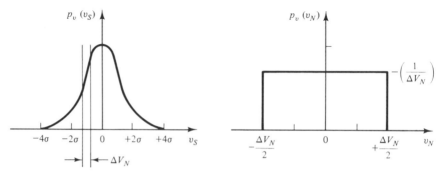

(a) Signal Voltage Distribution   (b) Noise Voltage Distribution

Figure 3-10   The voltage probability-density functions for a signal to be digitized and the digitizing noise.

where the standard deviation of the noise is given by

$$\sigma_N^2 = \int V_N^2 \rho_{V_N}(V_N) dV_N = \int_{-\Delta V_N/2}^{+\Delta V_N/2} V_N^2 \left(\frac{1}{\Delta V_N}\right) dV_N$$

$$= \tfrac{1}{12} \Delta V_N^2 \tag{3-52}$$

which gives $\sigma_N = \sqrt{\tfrac{1}{12}} \Delta V_N$. Now taking the ratio of the variances

$$\left(\frac{S}{N}\right)_d = \frac{\sigma_S^2}{\sigma_N^2} = \left(\frac{12}{64}\right) \ell_d^2 = 0.1875 \; \ell_d^2 \tag{3-53}$$

where if the digitizing range had been different from $8\sigma_S$, a different constant would have been obtained.   For instance, let $\Delta V_N = 6\sigma_S/\ell_d$

$$\Delta V_N = \frac{6\sigma_S}{\ell_d} \tag{3-54}$$

which leads to

$$\left(\frac{S}{N}\right)_d = \frac{\sigma_S^2}{\tfrac{36}{12}(\sigma_S^2/\ell_d^2)} = \tfrac{12}{36} \ell_d^2 = 0.333 \; \ell_d^2 \tag{3-55}$$

This gives a better signal-to-noise ratio for the same number of digitizing levels, but will result in the clipping of the $+$ and $-$ peaks of the signal (nonlinear distortion).   Up to now we have been considering *linear* digitization of the analog signal (usually used in instrumentation work).   In some cases (voice and music signals, for example) it may be desirable to use a nonlinear digitizing scheme wherein ΔV_N varies as a function of the signal level.   This will be discussed later in Chapters 7 and 8.

3.6.1 Digitizing a Noisy Signal

The discussion of digitizing has thus far assumed an analog signal with no noise on it.   We will now assume that the input analog signal has a signal-to-noise ratio of $(S/N)_i$ and let N_o' be the portion of the noise power on the reconstructed analog signal at the receiving end that is due to the input noise.   The output signal power S_o will then be given by $S_o = (S/N)_i(N_o')$. The total output noise power will be N_o' plus the added noise due to the digitizing/reconstructing process.   We will call this $N_{o,d}$, and it is equal to $S_o/(S/N)_d$.   Thus,

$$\frac{1}{(S/N)_o} = \frac{N_o}{S_o} = \frac{N_o' + N_{o,d}}{S_o} = \frac{N_o' + \left\{\dfrac{(S/N)_i N_o'}{(S/N)_d}\right\}}{(S/N)_i N_o'}$$

$$= \frac{1}{(S/N)_i} + \frac{1}{(S/N)_d} \tag{3-56}$$

It is now interesting to look at a digital transmission system in which a noisy analog signal is to be digitized into a pulse-code-modulation (PCM) signal that is then transmitted to the receiving end of the system where the analog signal is reconstructed through the use of a digital-to-analog converter (A/D converter) and is sent through a low-pass or band-pass filter with cutoff frequency BW_I (see Figure 3-11). The sampling rate at the transmission end must be at least twice the BW_I of the input analog signal. As a design criteria we are usually given the minimum required analog (S/N) ratio to be had at the receiving end (which must, of course, be less than that of the analog input signal) or the equivalent decibels of deterioration in the analog (S/N) ratio due to the digitization/reconstruction process.

At any rate, having $(S/N)_i$ and the required $(S/N)_o$ we can use Equation 3-56 to find the minimum required $(S/N)_d$. Then we can use Equation 3-53 or 3-55 to find the minimum needed number of digitizing levels ℓ_d. However, assuming that we are going to be working with a binary digital system, we would want to make ℓ_d to be equal to the next larger value of 2^m, where m is an integer (m equals the number of bits per digitizing sample). Then m times R_S will give us the *digital-data*-bit rate (on the digital channel) required to transmit the given analog signal with the specified S/N degradation allowed due to the digitizing process.

The digital-data-bit rate on the digital channel of a system such as is shown in Figure 3-11 is usually quite a bit larger than the *information-bit* rate of the analog input channel. This, coupled with the fact that the digital signal is usually only binary in nature, means that a much wider transmission bandwidth is needed on the digital channel than would be required were the data to be transmitted over an analog channel. This disadvantage is offset somewhat by the fact that S/N requirements on the digital channel are less stringent than on the equivalent analog channel. However, the real advantage to using the digital transmission scheme is that over long distances where many repeater stations are required the digital signal may be "reconstructed" at each repeater (effectively stripping off any noise that has been added during the transmission from the preceding repeater station), whereas in the analog case, the noise already present on the signal must be amplified along with the information signal and retransmitted toward the next repeater, along with any excess noise generated by the repeater amplifier itself.

Example:

> Let $(S/N)_i = 32$ dB, $BW_I = 5$ MHz, and let $(S/N)_o$ be required to be at least 30 dB. The minimum sampling rate will be 10 million samples per second, but let this be increased to $R_S = 12$ million samples/second to alleviate aliasing. The required digitizing S/N ratio will be
>
> $$\left(\frac{S}{N}\right)_d = \frac{1}{\left(\dfrac{1}{(S/N)_o} - \dfrac{1}{(S/N)_i}\right)} = 2710 = 0.1875\ \ell_d^2$$
>
> which gives $\ell_d > 120.22$ (make $\ell_d = 2^7 = 128$).
>
> Thus, with 7 data bits per sample and 12 million samples per second, we will need 84 million data bits per second to be transmitted over the digital channel. Using binary NRZ coding, this would be 84 Megabaud, which would require a channel bandwidth of 42 MHz (as compared to a BW_I of only 5 MHz). When the 84 million data bits

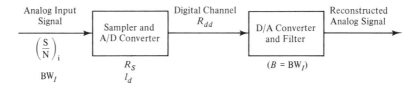

Figure 3-11   Transmitting an analog signal over a digital transmission path.

per second is compared with

$$R_I = 2(5 * 10^6) \log_2 \left(\sqrt{\left(\frac{S}{N}\right)_i} + 1 \right) = 53.1554 \text{ Mbps}$$

which is the information-data-bit rate for the analog signal, we see that the digital-data-bit rate is indeed greater than the analog signal's information-bit rate.

PROBLEMS

3-1. In the following cases, calculate how much noise power is delivered to R_L in each case if $B_N = 100$ MHz.   (*Note:* only the power from R to R_L, *not* the *net* noise power.)

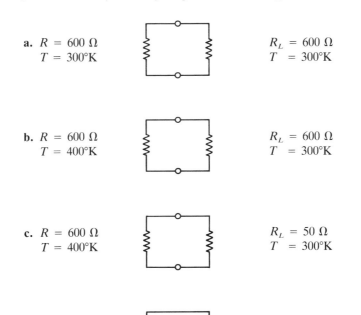

a. $R = 600 \, \Omega$
$\quad T = 300°K$

$R_L = 600 \, \Omega$
$\quad T = 300°K$

b. $R = 600 \, \Omega$
$\quad T = 400°K$

$R_L = 600 \, \Omega$
$\quad T = 300°K$

c. $R = 600 \, \Omega$
$\quad T = 400°K$

$R_L = 50 \, \Omega$
$\quad T = 300°K$

d. $R = 600 \, \Omega$
$\quad T = 400°K$

Attenuator
Loss = 5 dB
$T_a = 300°K$

$R_L = 600 \, \Omega$
$\quad T = 300°K$

3-2. a. How many Watts of noise power will a 50-Ohm wirewound resistor at 400°K deliver to the 50-Ohm input of a communications receiver in a noise bandwidth of 10 KHz? Express the answer in dBm.

    b. If the receiver input had been 300 Ohms instead of 50 Ohms, how many Watts of noise power (in dBm) would the same resistor deliver to the receiver input?

3-3. Consider a transmission line n feet long at T_a degrees Kelvin that has an attenuation of x dB per foot. Prove that this will behave the same with respect to its noise contribution as would a single attenuator with a loss of $(n \times x)$ dB at the same ambient temperature. (*Hint*: Prove it for n cascaded attenuators having x dB of attenuation each.)

3-4. Two signals are combined and fed into a two-input audio "mixing" amplifier where their voltages are *added* together with a gain of 30 dB for each channel. If the two signals that are applied each have an S/N ratio of 20 dB and if the information signals in each are the same (same time function, same voltage level, etc.), but there is no correlation between the noise signals on the two channels, what will be the S/N ratio at the output of the mixing amplifier? Assume that the noise power is the same in both channels and that the mixing amplifier itself introduces no excess noise (i.e., a 0 dB noise factor).

3-5. An audio amplifier terminated in a matched load has an output noise power of 10^{-4} Watt when its input is attached to a matched resistor that is immersed in a cup of icewater and an output noise power of $1.1 * 10^{-4}$ Watts when the input is connected to a similar resistor immersed in a cup of boiling water. What is the standard noise figure (in decibels) of this amplifier?

3.6. A helium-cooled microwave preamplifier has a standard noise figure of 1.3 dB and a gain of 35 dB. It is used in a satellite communication receiving system where the input noise temperature is 45°K (i.e, the noise temperature at the receiving antenna terminals). The system bandwidth is 10 MHz and the received signal level at the input of the amplifier is -90 dBm.

    a. What is the excess noise power contributed by the amplifier itself as measured at the amplifier's output terminals?

    b. What is the S/N ratio at the amplifier input and at its output?

    c. What is the excess noise power referred to the amplifier input terminals?

3-7. Find T_{N_1} N_o, S_o, and the output S/N ratio (in decibels) for the system shown below:

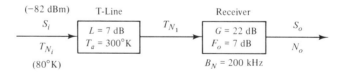

3-8.

    a. Find the output signal power in dBm and Watts.

    b. Find the output noise power in dBm and Watts.

    c. If A1 and A2 are considered as a single amplifier, what is the gain $(G)_{dB}$ and the standard noise figure $(F_o)_{dB}$ of this amplifier?

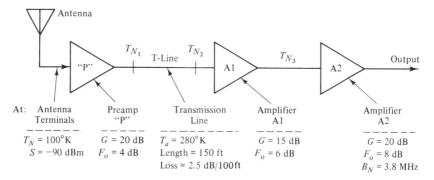

At: Antenna Terminals	Preamp "P"	Transmission Line	Amplifier A1	Amplifier A2
$T_N = 100°K$	$G = 20$ dB	$T_a = 280°K$	$G = 15$ dB	$G = 20$ dB
$S = -90$ dBm	$F_o = 4$ dB	Length = 150 ft	$F_o = 6$ dB	$F_o = 8$ dB
		Loss = 2.5 dB/100ft		$B_N = 3.8$ MHz

3-9. The following system is being used to receive a signal from a satellite. When the antenna is pointed at the satellite, the noise temperature at the antenna-output terminals is 100°K and the signal at these terminals is −70 dBm. The RF amplifier has a gain of 30 dB and a noise figure of 2 dB. What is S_1/N_1 in decibels? S_1 and N_1 are the signal power and the noise power in the bandwidth of interest at the RF amplifier output terminals (point 1). Similarly, find S_2/N_2 at the receiver IF output terminals. How does S_2/N_2 change if the position of the RF amplifier and the lossy T-line are interchanged in the system?

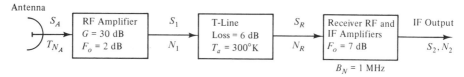

3-10. An analog communications signal has a signal-to-noise ratio of 17 dB and frequency components from 500 Hz to 3200 Hz.

   a. What is the maximum information rate of this signal?

   b. If this noisy signal is to be digitized so that the digitizing noise that is added is at least 10 dB less than the noise on the analog signal, what must be the *sampling rate R_S* and the *number of bits per sample* (use the next-larger integer number of bits)?

   c. What will be the *digital-data-bit rate?*

3-11. A voice signal exists between 300 Hz and 3 KHz and is to be sampled and digitized, transmitted as a digital signal, and then reconstructed into an analog signal as shown:

   a. What is the *minimum* sampling rate, R_S, that can be used?

   b. If an 8-bit A/D converter is used, what would be the effective *S/N* on the reconstructed voice signal due to the digitizing process alone?

   c. If the input voice signal has a 35-dB *S/N* ratio, what is the R_I of this analog input signal?

   d. Assuming a 35-dB *S/N* on the input voice signal, what is the overall *S/N* on the reconstructed voice signal?

3-12. A continuous analog signal has frequency components from 4 MHz to 5 MHz and a 25-dB signal-to-noise ratio.

   a. What is the signal's maximum information rate (in *information* bits per second)?

   b. If this continuous signal is sampled to form a discrete analog signal, what is the minimum sampling rate that is needed (in samples per second)?

   c. If the sampled signal is digitized using 64 uniformly spaced levels, what is the *S/N* ratio (in decibels) on the reconstructed output signal due to the digitizing process, and how many bits are needed to represent each sample?

3-13. An analog video signal whose BW_I is 4.2 MHz has a signal-to-noise ratio of 30 dB. It is to be transmitted as a digital PCM signal to a receiving site and reconstructed back into analog form. The *S/N* on the reconstructed analog signal must be at least 27 dB. The sampling rate to be used will be set at 1.3 times the Nyquist minimum. Find:

   a. How many samples per second and how many bits per sample must be sent over the digital path.

   b. What is the *maximum information rate* of the original analog signal.

   c. What is the *digital-data-bit rate* of the digital signal.

4

MODULATION TECHNIQUES

4.1 THE BASIC CONCEPT OF MODULATION AND DEMODULATION

The electrical or electronic process that we call *modulation* usually consists of, in some way, the multiplying of one signal by another signal. Since any multiplication of one time function by another time function is inherently a nonlinear process, and since all signals can be described as time functions, the process of modulation is therefore a nonlinear process, even when it is carried out in so-called "linear" modulators. The process of signal *detection* or demodulation, as we shall presently see, also falls into the same category except that it may also involve multiplying a signal by itself. There are a great many variations in the methods by which signals may be multiplied, and this results in a number of different types of modulation processes that have been developed over the years. As we study some of the more common of these methods, keep in mind that all modulation and demodulation is *some* sort of a multiplication process.

4.2 AMPLITUDE MODULATION (AM), SINGLE-SIDEBAND (SSB), DOUBLE-SIDEBAND SUPPRESSED CARRIER (DSBSC), AND VESTIGIAL-SIDEBAND MODULATION

One of the earliest forms of modulation and one of the ones most commonly used today is amplitude modulation. Today this is normally thought of as being generated by multiplying a quantity consisting of the sum of the information signal and a constant term by a signal of much higher frequency called the *carrier* signal. For example, let

$$s_m(t) = \text{the information signal (of max freq } f_H)$$

$$s_c(t) = \text{the carrier signal} = A \cos \omega_c t$$

where $\omega_c \gg 2\pi f_H$. Then the modulated output signal is given by

$$s_o(t) = s_c(t) \left[1 + (m_{AM})_p \frac{s_m(t)}{|s_m(t)|_{max}} \right] \tag{4-1}$$

where $(m_{AM})_p$ is called the *peak* AM modulation index.

As an example, let $s_m(t)$ be a very simple audio tone signal $s_m(t) = V_m \cos \omega_m t$ and let $(m_{AM})_p = 1.0$. Then

$$\begin{aligned} s_o(t) &= A \cos \omega_c t \, [1 + \cos \omega_m t] \\ &= A \cos \omega_c t + A(\cos \omega_c t)(\cos \omega_m t) \\ &= A \cos \omega_c t + \frac{A}{2} \cos (\omega_c - \omega_m)t \quad + \frac{A}{2} \cos (\omega_c + \omega_m)t \end{aligned} \tag{4-2}$$

$$\qquad\quad \Uparrow \qquad\qquad\qquad \Uparrow \qquad\qquad\qquad \Uparrow$$
$$\qquad\quad \text{carrier} \qquad\qquad \text{LSB} \qquad\qquad\quad \text{USB}$$

From Equation 4-2 we see that the multiplication of a sinusoidal information signal and a constant term by the carrier signal produces three sinusoidal components that we call the carrier, the lower sideband, and the upper sideband (LSB and USB, respectively). If we look at a plot of the amplitude of these signals versus frequency, we see that the *spectrum* of the signal $s_o(t)$ looks like that shown in Figure 4-1. When $(m_{AM})_p = 1.0$, we say that we have 100% peak modulation. If we let $(m_{AM})_p$ be less than 1.0, we have less than 100% modulation. Putting this into Equation 4-2 will show that the carrier term will remain fixed in amplitude at a level A but that the two sideband terms will be reduced to a level of $(m_{AM})_p(A/2)$. Under these conditions we say that the percentage of peak modulation is equal to $100 * (m_{AM})_p\%$.

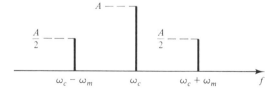

Figure 4-1 The spectrum of an AM signal with 100 percent modulation and a sinusoidal information signal.

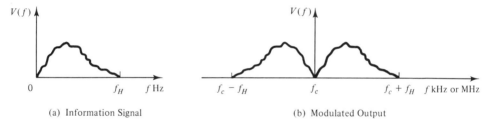

(a) Information Signal   (b) Modulated Output

Figure 4-2   The spectrums of an information signal and the modulated AM output.

Most of the information signals with which we work are not pure sinusoidal tone signals, but rather are complicated signals made up of many Fourier components. But since each Fourier component is a sinusoidal signal, each Fourier component present in the modulating information signal will produce a pair of sidebands in the manner just described. Thus, a periodic information signal with four Fourier components will produce a modulated output signal with four lower and four upper sidebands. For a nonperiodic information signal we cannot talk about the Fourier *series* components of the signal, but we can talk about the Fourier *integral* components (or the Fourier transform) of the signal. In such a case we have to talk about the voltage[1] *density* spectrum rather than about the voltage spectrum of the information signal and the modulated output signal. In this case we find that if we had an information signal whose voltage spectral density was given by Figure 4-2a, the voltage spectral density of the modulated output signal would appear as shown in Figure 4-2b (where Z = 1 Ohm). The carrier signal component appears as an impulse function on a spectral density plot. We see from a comparison of these figures that the upper sideband spectrum looks just like the information signal spectrum except offset in frequency by the amount $f_c = \omega_c/2\pi$ and that the lower sideband spectrum is just the mirror image of the upper sideband reflected about the carrier frequency f_c.

In addition to considering the appearance of the spectrum of the modulated output signal, it is informative to look at a typical modulation signal $s_m(t)$ and the modulated output that it produces as time-function plots. These are shown in Figures 4-3a and 4-3b. Notice here that we have about 90% maximum modulation, as indicated by the fact that the modulation envelope (the heavy line that encloses the shaded area representing the rapid oscillations of the carrier cycles) never goes to zero, but instead only down to within about $A/10$ of the zero axis on either side (at a point near the right-hand end of the signal segments shown).

In many cases we may wish to have $s_m(t)$ be composed of a finite number of sinusoids at different frequencies ω_{m1}, ω_{m2}, ω_{m3}, . . . , and so on, as given in Equation 4-3a (here we have used cosine functions for simplicity to represent the general sinusoidal functions composing $s_m(t)$).

---

[1] In the preceding chapter we discussed *power* spectral density. Voltage spectral density is just the square root of the product of the power spectral density times the impedance level being considered. It has the units of volt-$\sqrt{\sec}$.

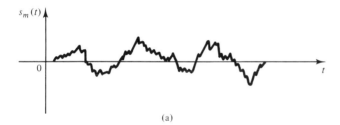

(a)

Figure 4-3a Information signal, $s_m(t)$.

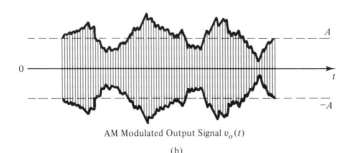

AM Modulated Output Signal $v_o(t)$

(b)

Figure 4-3b The modulation it produces.

$$s_m(t) = V_{m_1} \cos(\omega_{m_1}t) + V_{m_2} \cos(\omega_{m_2}t) + \cdots \qquad \text{(4-3a)}$$

$$(m_{AM})_p \frac{s_m(t)}{|s_m(t)|_{max}} = (m_{AM})_1 \cos(\omega_{m_1}t) + (m_{AM})_2 \cos(\omega_{m_2}t) + \cdots \qquad \text{(4-3b)}$$

From an examination of these relationships, we see that

$$\frac{V_{mk}}{|s_m(t)|_{max}} = \frac{(m_{AM})_k}{(m_{AM})_p} \qquad \text{(4-3c)}$$

When the ω_{mk}'s are *not* harmonically related[2] we then have

$$|s_m(t)|_{max} = \sum_k V_{mk} \qquad \text{(4-3d)}$$

and

$$(m_{AM})_p = \sum_k (m_{AM})_k \qquad \text{(4-3e)}$$

Let us now compute the average power contained in various parts (components) of the modulated output signal. For this we shall return to Equation 4-2 and Figure 4-1. Since all the components are sinusoidal, the average power contained in the carrier will be $A^2/2R_o$, where R_o is the impedance level where we are

---

[2] When the ω_{mk}'s *are* harmonically related, $|s_m(t)|_{max}$ must be determined by some other means (i.e., $(m_{AM})_p \neq \Sigma_k(m_{AM})_k$).

observing $s_o(t)$, when $s_o(t)$ is a voltage time function (we often let $R_o = 1$ ohm and work with the normalized case) and the average RF power in each sideband at 100% modulation will be $(A/2)^2/2R_o$. Thus, the carrier contains four times as much power as either sideband or twice as much power as in both sidebands. If the level of modulation is reduced so that instead of 100% modulation, we have only $M\%$ modulation, the power in the carrier will still be equal to $A^2/2R_o$, but the total sideband power will only be $(A \times M/2)^2/R_o$. Thus, at 50% modulation, the average power in each sideband will be reduced to 1/16 of the carrier power or the total average sideband power will be only 1/8 of the carrier power or 1/9 of the total average transmitted power. Thus, at low levels of modulation, most of the transmitter power is in the carrier (which actually conveys almost no information) and only a small fraction of the total average transmitted power is in the sidebands that carry the information to the receiver. The total average sideband power is sometimes referred to as the transmitter's "talk power." Thus, the standard AM signal is very inefficient when one considers the amount of RF power needed to send the important part of the signal. This situation becomes even worse when nonsinusoidal modulation is considered.

For example, if $s_m(t)$ is made up of two audio tones of equal magnitude, then the amplitude of each sideband at 100% modulation will be $A/4$, and the *total* average sideband power will be proportional to $4[(A/4)^2/2R_o] = A^2/8R_o$ or only 20% (rather than 33%) of the total transmitter power. A 50% level of modulation with two equal tones would give only 5.9% of the total transmitter power in all of the sidebands. Voice modulation produces even a much lower percentage of the total average transmitter output power appearing as useful sideband power. Thus, the AM transmitter is very inefficient for most modulating signals since a great deal of the input power to the transmitter is wasted in producing relatively useless carrier output power.

The power level of an AM transmitter is usually specified by giving its carrier power level without any modulation being applied. This may be given either in terms of the actual RF output power or in terms of the d.c. power input to the final RF amplifier stage. It is important to know which power measurement is being specified because the latter (d.c. input power) will usually be 25% to 50% greater than the RF output power, depending upon the efficiency of the final RF amplifier stage.

One of the reasons that AM has been so popular, despite its poor efficiency in the use of transmitter power, is the ease with which the AM signal may be demodulated. For example, all that is needed is to multiply the signal by some power (including fractional powers) of itself, and this can be easily done by applying the modulated signal to a nonlinear device such as a diode.[3] When this is done, a number of components are produced, but one of the components is the original information signal. For example, multiply $s_o(t)$ by itself raised to the first power,

---

[3] A more detailed discussion of the diode detector can be found in most electronic textbooks, although most of these look at the diode as a rectifier that removes the lower portion of the modulated signal (Figure 4-3b), leaving the upper portion to be put through a high-pass filter to remove the carrier-frequency variations to leave the envelope of the modulated signal.

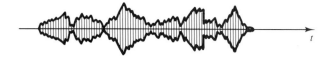

Figure 4-4 Typical DSBSC-modulated output signal (for the $s_m(t)$ of Figure 4-3a).

and from Equation 4-1 we get

$$s_o(t)^2 = [s_c(t)]^2 \times \left\{ 1 + 2(m_{\mathrm{AM}})_p \frac{s_m(t)}{|s_m(t)|_{\max}} + (m_{\mathrm{AM}})_p^2 \left[\frac{s_m(t)}{|s_m(t)|_{\max}} \right]^2 \right\} \qquad (4\text{-}4)$$

where, since $s_c(t)$ is a cosine function, we get from squaring it a quantity that is the sum of a constant and a double-frequency cosine term. The constant term is the one of interest, since when it multiplies the second factor in the square brackets in Equation 4-4, there is produced a signal that is proportional to $s_m(t)$, which is our information signal. The other terms which are produced are removed by filtering action and the output of the detector is just a signal proportional to the original information signal $s_m(t)$ plus some distortion products due to the last term in the square brackets in Equation 4-4.[4]

As mentioned previously, the carrier of the AM signal conveys very little information from the sender to the receiver. Also, since one sideband is the mirror image of the other, the information which is contained in one sideband is duplicated in the other. Thus, most of the information contained in the modulating signal could be conveyed by sending only one sideband of the AM modulated signal, and this is what is done in some modern communication systems, but it is done at the expense of more complexity at both the transmitter and the receiver since it is no longer possible to recover the original modulating information signal from the received modulated RF signal with a simple diode detector. In some cases it is found to be advantageous to compromise between the two extremes (transmitting the full signal [AM] or transmitting only one sideband [SSB]) and transmit both sidebands but with the carrier greatly reduced in power (double-sideband suppressed carrier [DSBSC] modulation). We will now discuss the DSBSC and SSB signals in more detail.

We could create a DSBSC signal merely by filtering out the carrier portion of an AM signal. This would have the effect of bringing together the $+A$ level and the $-A$ levels of Figure 4-3b to produce a signal which looks like that of Figure 4-4. Note that the bottom envelope of Figure 4-3b will overlap the top envelope so that the envelope of the signal shown in Figure 4-4 is formed by the area *enclosed* by the two signals $s_m(t)$ and $-s_m(t)$. A pure sinusoid for $s_m(t)$ will produce a DSBSC signal which looks like Figure 4-5a. It should be noted that the nulls in the waveform of Figure 4-5a are separated by a distance of $T_m/2$ where $T_m = 1/f_m$ and f_m is the frequency of the modulating sinusoid. Thus, the periodicity of the envelope in Figure 4-5a is $2f_m$.

[4] In the analysis of the AM diode detector as a filtered half-wave rectifier, these distortion products are not present.

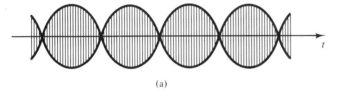

(a)

Figure 4-5a DSBSC-modulated ouptut when $s_m(t)$ is sinusoidal and the carrier is completely suppressed.

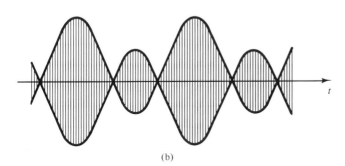

(b)

Figure 4-5b Same as Figure 4-5a except that the carrier is only partially suppressed.

Sometimes it is desirable to suppress only partially the carrier, transmitting a small amount of carrier signal which the receiver can use to synchronize an injection oscillator which is at the same frequency and in phase with the received carrier signal. The signal from the injection oscillator can then be used for the detection of the DSBSC signal sidebands. When this is done, the envelope of the modulated waveform appears as in Figure 4-5b. Here the periodicity of the envelope is f_m and the spacing between successive nulls is unequal.

For single sideband, both the carrier and one sideband are eliminated. If the eliminated sideband is the lower sideband, then we can think of the SSB signal as being created simply by adding the carrier frequency to the frequency of each Fourier component in the information signal. This means that if the information signal is a pure sinusoid the SSB modulated output signal will be a pure sinusoid also. This will mean that the SSB signal will appear as shown in Figure 4-6 (a high-frequency sinusoid of constant amplitude).

Next, let us consider some of the power relationships in AM, DSBSC, and SSB signals. We shall have to define several different power measurements:

Average power—the power of the modulated output averaged over a long period of time (at least over one cycle of a periodic modulating signal).

Maximum instantaneous power—the maximum peak power of the RF waveform.

Figure 4-6 SSB-modulated output signal when $s_m(t)$ is sinusoidal.

Peak envelope power—the maximum short-term average power where the average is taken only over approximately one to several RF cycles; it is always one-half of the maximum instantaneous power.

We will denote these three powers by P_{av}, P_{max}, and PEP, respectively.

As an example of calculating these powers, consider a 100% modulated AM waveform where $s_m(t)$ is sinusoidal. The average power is the sum of the average power of the carrier plus the average power in each of the sidebands, or

$$P_{av} = P_c + P_{LSB} + P_{USB} = \frac{A^2}{2R_o} + 2\frac{(A/2)^2}{2R_o} = \frac{\frac{6}{4}(A^2)}{2R_o} = \frac{3A^2}{4R_o} \qquad (4\text{-}5)$$

The maximum voltage of the AM signal at 100% modulation is $2A$, so

$$P_{max} = \frac{(2A)^2}{R_o} = \frac{4A^2}{R_o} \qquad (4\text{-}6)$$

and, of course, PEP is one-half of P_{max}. So, for the AM signal with 100% sinusoidal modulation, we have that the maximum power is $\frac{16}{3}$ times the average power.

The average power in the modulated output is important because it roughly determines the input power requirements of the final transmitter stage; the maximum instantaneous power is important because it largely determines the peak voltage which is impressed across the RF components in the transmitter final amplifier stage and feed line.

At this point, let us tabulate the different powers in an AM, DSBSC, and SSB signal with 100% sinusoidal modulation.

AM:

$$P_{av} = P_c + P_{LSB} + P_{USB} = \frac{A^2}{2R_o} + 2\frac{(A/2)^2}{2R_o} = \frac{\frac{6}{4}(A^2)}{2R_o} = \frac{3A^2}{4R_o} \qquad (4\text{-}7)$$

$$P_{max} = \frac{(2A)^2}{R_o} = \frac{4A^2}{R_o} \qquad (4\text{-}8)$$

$$PEP = \frac{(2A)^2}{2R_o} = \frac{4A^2}{2R_o} = 2\frac{A^2}{R_o} \qquad (4\text{-}9)$$

DSBSC:

$$P_{av} = P_{LSB} + P_{USB} = 2\frac{(A/2)^2}{2R_o} = \frac{A^2}{4R_o} \qquad (4\text{-}10)$$

$$P_{max} = \frac{A^2}{R_o} \qquad (4\text{-}11)$$

$$PEP = \frac{A^2}{2R_o} \qquad (4\text{-}12)$$

SSB (sideband amplitude $= A/\sqrt{2}$):

$$P_{av} = \frac{1}{2R_o} \text{ (sideband amplitude)}^2 = \frac{A^2/2}{2R_o} = \frac{A^2}{4R_o} \qquad (4\text{-}13)$$

$$P_{max} = \frac{\text{(sideband amplitude)}^2}{R_o} = \frac{A^2}{2R_o} \qquad (4\text{-}14)$$

$$PEP = \frac{\text{(sideband amplitude)}^2}{2R_o} = \frac{A^2}{4R_o} \qquad (4\text{-}15)$$

In this case, to keep the average sideband power the same in all three cases, the amplitude of the one remaining sideband in the SSB case was adjusted to be $\sqrt{2}$ times that of one of the two sidebands present in each of the other two cases. In other words, the three systems are comparable in terms of the actual transmitted information power.

Now it is interesting to run through the same type of analysis for the case of $s_m(t)$ being composed of two audio tones of equal magnitude. This is a common test signal used for testing and adjusting radio transmitting equipment. Again, we use 100% modulation.

AM:

$$P_{av} = P_c + P_{LSB_1} + P_{LSB_2} + P_{USB_1} + P_{USB_2} \qquad (4\text{-}16)$$

$$= \frac{A^2}{2R_o} + 4\frac{(A/4)^2}{2R_o} = \frac{A^2}{2R_o} + \frac{4(A^2/16)}{2R_o} = \frac{5A^2}{8R_o} \qquad (4\text{-}17)$$

$$P_{max} = \frac{(2A)^2}{R_o} = \frac{4A^2}{R_o} \qquad (4\text{-}18)$$

$$PEP = \frac{(2A)^2}{2R_o} = \frac{2A^2}{R_o} \qquad (4\text{-}19')$$

DSBSC:

$$P_{av} = \frac{4(A^2/16)}{2R_o} = \frac{A^2}{8R_o} \qquad (4\text{-}19)$$

$$P_{max} = \frac{A^2}{R_o} \qquad (4\text{-}20)$$

$$PEP = \frac{A^2}{2R_o} \qquad (4\text{-}21)$$

SSB (sideband amplitude $= A/\sqrt{8}$):

$$P_{\text{av}} = \frac{2(A/\sqrt{8})^2}{2R_o} = \frac{2A^2/8}{2R_o} = \frac{A^2}{8R_o} \tag{4-22}$$

$$P_{\text{max}} = \frac{(2A/\sqrt{8})^2}{R_o} = \frac{(A/\sqrt{2})^2}{R_o} = \frac{A^2}{2R_o} \tag{4-23}$$

$$PEP = \tfrac{1}{2} P_{\text{max}} = \frac{A^2}{4R_o} \tag{4-24}$$

These results, along with a sketch of the two-tone modulating signal and the modulated RF output signals which it produces and their respective spectrums are shown in Figure 4-7. Notice here that when one goes to DSBSC from AM, the spacing of the fine structure within the major envelope is halved because the mirror image of the portion of $s_m(t)$ below the zero axis (i.e., $-s_m(t)$) appears both above and below the zero axis in the DSBSC output waveform. In the SSB signal the major envelope is retained, but the fine structure within the envelope disappears completely. This occurs because there are no lower sidebands (in this example) for the upper sidebands to beat against. The major envelope is produced by the beat frequency of the two upper sidebands.

Tables 4-1 and 4-2 are matrices that show how the various powers in AM, DSBSC, and SSB waveforms are related for 100% sinusoidal modulation and 100% two-tone modulation, respectively. Note that in going from AM to DSBSC or SSB, there is a large savings in the total average power which needs to be transmitted (total sideband power is the same in all cases), and that going from DSBSC to SSB results in a 50% decrease in P_{max} and PEP, which means less chance of an RF voltage breakdown in the final amplifier stage or in the transmission line or antenna. P_{av} of the AM signal is used as the reference ($P_{\text{av,AM}} = 1$ Watt, the number entered in the upper left circle).

As previously mentioned, the total average sideband power in an AM signal is what is important in conveying the information contained in the modulating signal. Since for a complex modulating waveform such as was shown in Equation 4-3 this total average sideband power is the sum of the powers of the individual

TABLE 4-1

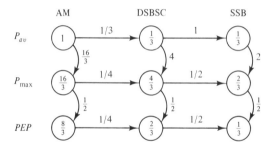

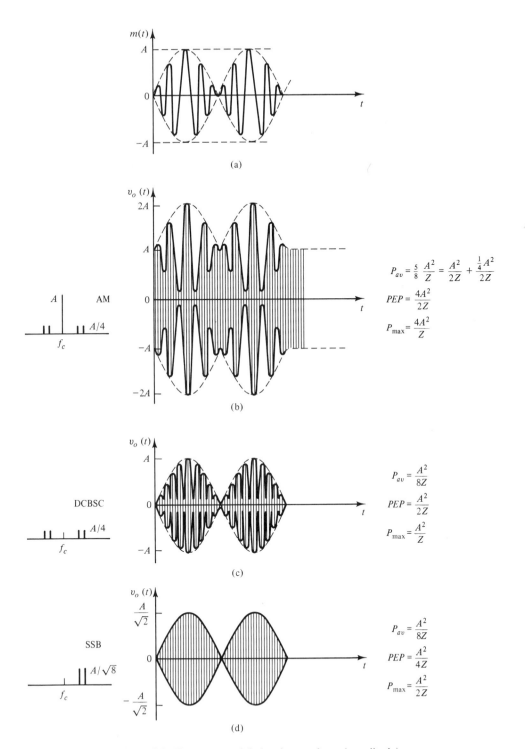

Figure 4-7 Two-tone modulation (tones of equal amplitude).

TABLE 4-2

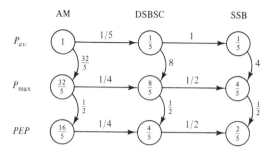

sideband components, it is convenient to define an *effective* AM modulation index

$$(m_{AM})_e = \sqrt{(m_{AM})_1^2 + (m_{AM})_2^2 + \cdots} \qquad (4\text{-}25)$$

which is related to the average total sideband power and the average carrier power by

$$P_{SB,total} = \left[\frac{(m_{AM})_e^2}{2} \right] P_c \qquad (4\text{-}26)$$

The total average AM transmitter output power is then given by

$$P_{AM,total} = P_c + P_{SB,total} = 1 + \left[\frac{(m_{AM})_e^2}{2} \right] P_c \qquad (4\text{-}27)$$

We shall find $(m_{AM})_e$ quite useful when discussing the *S/N* ratio of the demodulated AM signal.

When calculating P_{max} or the *PEP*, we want to use the *peak* modulation index. From Equation 4-1 and letting the amplitude of the carrier signal again be A, we can see that the peak amplitude of the modulated AM waveform is $[1 + (m_{AM})_p]A$ and that of a DSBSC signal of the same average total sideband power is $(m_{AM})_p A$. Either of these squared and divided by R_o gives P_{max} for the respective case (AM or DSBSC). The *PEP* is always just one-half of P_{max}, and P_{max} and the *PEP* for the equivalent SSB case are always just one-half of the DSBSC P_{max} and *PEP*.

4.3 ANGLE MODULATION (FM AND PM)

Angle modulation consists of changing the phase angle of the carrier wave rather than its amplitude. That this is still a form of signal multiplication can be seen from the relationship

$$\begin{aligned} s_o(t) &= V_{cm} \cos\left[\omega_c t + \Delta\theta(t)\right] \\ &= (\cos \omega_c t) \times \left[\cos \Delta\theta(t)\right] - (\sin \omega_c t) \times \left[\sin \Delta\theta(t)\right] \end{aligned} \qquad (4\text{-}28)$$

where in the second part of the equation V_{cm} has been set equal to unity for simplicity. The two basic types of analog angle modulation are FM and PM defined, respectively, as (with $V_{cm} = 1.0$) for FM:

$$s_o(t) = \cos [\omega_c t + \theta_o + K_f \int s_m(t) \, dt] \qquad (4\text{-}29)$$

and for PM:

$$S_o(t) = \cos [\omega_c t + \theta_o + K_p s_m(t)] \qquad (4\text{-}30)$$

As an example of the similarity between these two forms of modulation, let the modulating signal be a sinusoidal tone of the form $s_m(t) = V_m \cos \omega_m t$.   Then Equations 4-29 and 4-30 become for FM:

$$s_o(t) = \cos \left[\omega_c t + \theta_o + \left(K_f V_m \frac{1}{\omega_m} \right) \sin \omega_m t \right] \qquad (4\text{-}31)$$

and for PM:

$$s_o(t) = \cos [\omega_c t + \theta_o + (K_p V_m) \sin (\omega_m t + 90°)] \qquad (4\text{-}32)$$

Thus, we see that for the case of a single-frequency modulating signal the FM and PM output waveforms are about the same except for the constant ahead of the sinusoidal term at ω_m and the 90 degree phase shift which occurs in the PM function.

At this point it might be well to stop and define some of the notation we will be using as we discuss angle modulation.   We shall begin with phase modulation:

$$K_p = \frac{(\Delta\theta)_p}{|s_m(t)|_{\max}}$$

where $(\Delta\theta)_p$ is in radians and is the *peak phase deviation* (we will let $(\Delta\theta)_{\max}$ be the maximum *legally permitted phase deviation*).   This lets us define the *percentage of phase deviation* as

$$(\Delta\theta)_p = (\Delta\theta)_{\max} * \left[\frac{\% \text{ PM modulation}}{100} \right]$$

We will let the quantity defined by the square brackets in either Equation 4-29 or 4-30 be called $\Phi(t)$ and call it *the total phase* of the angle-modulated RF signal such that

$$s_o(t) = V_{cm} \cos \Phi(t) = V_{cm} \cos [(\omega_c t + \Delta\theta(t))] \qquad (4\text{-}33)$$

which then defines $\Delta\theta(t) = [\Phi(t) - \omega_c t]$ as the *instantaneous phase deviation* of the modulated RF signal.   The *instantaneous angular frequency* (in radians per second) of the signal is then defined as $d\Phi(t)/dt = \omega_i(t)$.   In terms of Hertz, this then becomes $f_i(t) = \omega_i(t)/2\pi$.   In turn the instantaneous frequency deviation then is $\Delta f(t) = [f_i(t) - f_c]$.

We will now define those notations which are specific to the FM (frequency modulation) case.   Let

$$K_f = \frac{2\pi(\Delta f)_p}{|s_m(t)|_{\max}}$$

where $(\Delta f)_p$ is the peak frequency deviation in Hertz that occurs when $s_m(t)$ is at its peak or maximum value. As in the case of phase deviation, we will let $(\Delta f)_p$ be some percentage of some legally permitted frequency deviation Δf and refer to this as the percentage of frequency modulation. At this point it might be well to note that in the case of AM modulation, the percentage of modulation has a definite physical interpretation, whereas in phase and frequency modulation, the percentage of modulation merely refers to some legally specified maximum phase or frequency deviation:

$$(\Delta f)_p = \Delta F * \left(\frac{\text{percentage of FM modulation}}{100} \right)$$

For any one modulating frequency $f_{m,k}$ we will define the individual FM modulation index $m_{f,k} = (\Delta f)_{p,k}/f_{m,k}$, while for the transmitting system as a whole we will define the transmitter's FM modulation index as $M_{FM} = \Delta F/B_{\text{audio}}$, where B_{audio} is the maximum modulating frequency of the system.

For a complex waveform where the modulating frequencies are nonharmonically related, we can define the peak frequency deviation as $(\Delta f)_p = \Sigma_k(\Delta f)_{p,k}$, and for any modulating waveform composed of sinusoidal components, the effective frequency deviation is defined by $(\Delta f)_e^2 = \Sigma_k(\Delta f)_{p,k}^2$ (Note: $(\Delta f)_e$ is used in somewhat the same way as $(m_{\text{AM}})_e$ in the AM case).

At this point we will now return to the discussion of the angle-modulated signal and begin by calculating its instantaneous frequency:

$$f_i(t) = \left(\frac{1}{2\pi}\right)\left(\frac{d}{dt}\right)[\omega_c t + \Delta\theta(t)] \tag{4-34}$$

For the FM case this becomes (from Equation 4-29)

$$f_{i,\text{FM}}(t) = \frac{1}{2\pi}[\omega_c + K_f V_m \cos \omega_m t] \tag{4-35}$$

and for PM (from Equation 4-30)

$$f_{i,\text{PM}}(t) = \frac{1}{2\pi}[\omega_c + K_p V_m \omega_m \cos (\omega_m t + 90°)] \tag{4-36}$$

These equations show that for FM, the maximum instantaneous frequency deviation, defined as $|f_i(t) - f_c|_p$, is determined only by the amplitude V_m of the modulating signal

$$|f_i(t) - f_c|_{p,\text{FM}} = \left|\frac{\omega_c + K_f V_m}{2\pi} - f_c\right|_{\text{max}} = \frac{K_f V_m}{2\pi} = (\Delta f)_p \tag{4-37}$$

while for PM it is also directly proportional to the frequency of the modulating signal f_m as well as its amplitude V_m

$$|f_i(t) - f_c|_{p,\text{PM}} = \left|\frac{\omega_c + K_p V_m \omega_m}{2\pi} - f_c\right|_{\text{max}} = K_p V_m f_m \tag{4-38}$$

It is interesting to note that the FM signal could be converted into a PM signal by preemphasizing (differentiating) $s_m(t)$ proportional to the modulating frequency

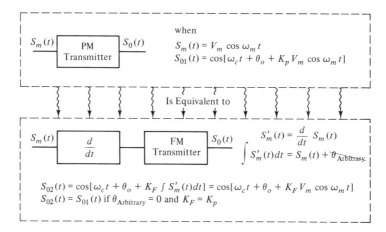

Figure 4-8   Producing a PM signal with an FM transmitter.

f_m and then applying $s'_m(t)$ to the FM modulator. This is illustrated by comparing the two drawings of Figure 4-8 where in the second case K_f can be made numerically equal to what K_p would be in the first case. It should be noted that the reverse is also true—a PM transmitter can be made to transmit "FM" by deemphasizing (integrating) $s_m(t)$ before applying it to the PM transmitter.

In conjunction with the idea of producing PM from an FM modulator, it is interesting to note here that the commercial "FM" broadcasting in the United States is actually a hybrid PM-FM system. It is produced by feeding the studio signal $s_m(t)$ into a preemphasis network with the characteristics shown in Figure 4-9 so that when $s'_m(t)$ is applied to the FM transmitter, FM modulation is produced

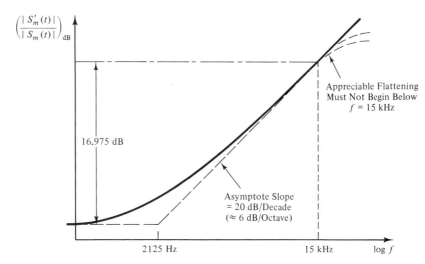

Figure 4-9   Commercial FM (USA) preemphasis curve.

for signals whose frequencies are somewhat below 2.125 KHz and PM is produced
for signals whose frequencies are somewhat above 2.125 KHz.   In the area around
2.125 KHz, the modulation is a combination of FM and PM.   The frequency of
2.125 KHz is the *corner frequency* of the preemphasis network.   It is equivalent
to a time constant $\tau_{RC} = 1/2\pi(2125) \approx 75$ microseconds.

Angle modulation (either FM or PM) produces a *set* of sidebands on either
side of the carrier for *each* sinusoidal modulating component.   This is in contrast
to the various types of amplitude modulation which produce only *one* component
one each side of the carrier for each modulating component.   For example, single-
tone FM will produce a spectrum given by[5]

$$
\begin{aligned}
s_o(t) = \; & J_o(m_f) \cos \omega_c t \\
& + J_1(m_f) [\cos (\omega_c + \omega_m)t - \cos (\omega_c - \omega_m)t] \\
& + J_2(m_f) [\cos (\omega_c + 2\omega_m)t + \cos (\omega_c - 2\omega_m)t] \quad\quad (4\text{-}39) \\
& + J_3(m_f) [\cos (\omega_c + 3\omega_m)t - \cos (\omega_c - 3\omega_m)t] \\
& + \cdots
\end{aligned}
$$

where the $J_i(m_f)$ are Bessel functions of the first kind and the ith order.   Tables
and plots of the Bessel functions are readily available in most comprehensive
mathematical reference books and tables of functions.

Let us now consider what happens to the spectrum of an FM signal when we
change the modulating frequency and the amplitude of the sinusoidal modulation.
To do this we will refer to Figure 4-10 and to the definitions for $(\Delta\theta)_p$ and m_f.   In
the top part of Figure 4-10 are plotted the spectrums of an FM signal where the
modulating frequency has been held constant but the amplitude of the modulation
was changed.   The result is that the spacing between the sidebands remains constant
and numerically equal to f_m but the number and the amplitude of the sidebands
changes.   Note also that the carrier amplitude drops as the level of modulation
increases.   As we go to large values of m_f, we will find that most of the power in
the signal is concentrated in an area just under the $f_c \pm (m_f \times f_m)$ points.   Since
$m_f \times f_m$ equals $(\Delta f)_p$, this means that most of the sideband power is concentrated
in a region just inside the $\pm(\Delta f)_p$ limits.   Note also that there are some sidebands
outside of the $\pm(\Delta f)_p$ limits (or in other words the bandwidth of an FM signal is
not limited to $f_c \pm (\Delta f)_p$.

The lower drawing shows what happens when the modulating frequency is
changed and the modulating signal amplitude is held constant.   In this case the
number of the sidebands increases as the spacing between the sidebands is reduced,
but the approximate distribution of the sideband power over the frequency spectrum
remains about the same.   The modulation index m_f increases in proportion to
$1/f_m$ if $(\Delta f)_p$ is held constant.

---

[5] To obtain Equation 4-39, let $\Delta\theta(t)$ in Equation 4-28 be $m_f \cos \omega_m t$ and apply the Bessel function
expansions given in Appendix E for $\cos (m_f \cos \omega_m t)$ and $\sin (m_f \cos \omega_m t)$ and then the trigonometric
identities for $\cos A \cos B$ and $\sin A \sin B$.

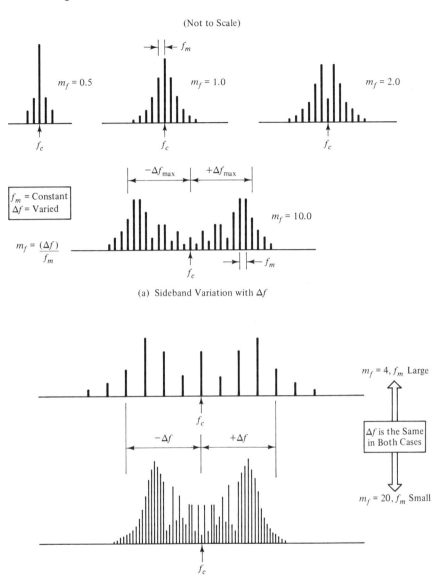

(Not to Scale)

(a) Sideband Variation with Δf

(b) Sideband Variation with f_m

Figure 4-10 Spectrums of FM signals with sinusoidal modulation.

So far we have considered an $s_m(t)$ composed only of one Fourier component at a frequency of f_m and have seen how a large number of sidebands are produced below and above the carrier frequency, separated in frequency from each other by f_m Hz. Now if we let $s_m(t)$ have many Fourier components, the spectrum of the FM or PM signal will contain components not only at frequencies separated from

f_c by integer multiples of each Fourier frequency present in $s_m(t)$, but also by integer multiples of all the possible sums and differences of the Fourier frequencies present.

Example:

> Let $s_m(t)$ have components at 30 Hz, 59 Hz, and 338 Hz. Then the PM or FM signal will contain components at f_c, $f_c \pm n \times 30$ Hz, $f_c \pm n \times 59$ Hz, $f_c \pm n \times 338$ Hz, $f_c \pm n \times (59 - 30)$ Hz, $f_c \pm n \times (338 - 30)$ Hz, $f_c \pm n \times (338 - 59)$ Hz, $f_c \pm n \times (59 + 30)$ Hz, $f_c \pm n \times (338 \times 30)$ Hz, and $f_c \pm n \times (59 + 338)$ Hz, where n is any integer.

Earlier we talked about commercial FM. The FCC sets standards for these stations, which are: $(\Delta f)_p$ is to be no more than ± 75 KHz and all appreciable components of the signal are to be maintained within ± 100 KHz of the carrier frequency. The maximum modulating frequency is about 15 KHz, so this means that at 100% (± 75 KHz) modulation, the *minimum* value of m_f will be given by

$$(m_f)_{\min} = \frac{(\Delta f)_p}{(f_m)_{\max}} = \frac{\Delta F}{B_{\text{audio}}} = \frac{75 \text{ KHz}}{15 \text{ KHz}} = 5 = M_F \qquad (4\text{-}40)$$

or in actual operation m_f is greater than or equal to 5 if f_m is less than or equal to 15 KHz and if 100% modulation is maintained.

Other applications of FM are in the public service and safety radio sector (such as police, fire, city vehicles, industrial, railroad, military, etc.). In these applications *narrow-band FM* is usually used where the ratio of ΔF to the maximum f_m (B_{audio}) is usually *on the order* of unity.

4.4 QUADRATURE MODULATION

Figure 4-11 shows a simplified block diagram of a quadrature modulation system. A system of this type is capable of transmitting two independent information (modulating) signals over the same path. At the transmitting end there are two balanced modulators, one of which produces a set of sidebands by modulating the carrier at f_c with the information signal $v_A(t)$. The other balanced modulator produces a set of sidebands by modulating the same carrier (but shifted in phase by 90°) with the second information signal $v_B(t)$. The two sets of sidebands are then added together and are transmitted along with the *original* carrier signal to the receiving site. At the receiving site, the carrier and the received composite sideband signal are fed into one phase detector, while the received composite sideband signal and the carrier shifted by 90° are fed into the second phase detector. The output of the first phase detector will be the information signal $v_A(t)$, and the output of the second phase detector will be the information signal $v_B(t)$. The system is very sensitive to phase-delay problems: delaying the sideband signal $v_o(t)$ or the carrier-reference signal $v_r(t)$ by different amounts will produce crosstalk between the two channels. In many practical systems the carrier itself is not transmitted, but rather the carrier at the transmitting end is used to produce a synchronization signal that is transmitted to the receiving end where it is used to phase synchronize an injection carrier oscillator.   Adjustments are usually provided at the receiving end to make slight changes in the phase of injection oscillator so that crosstalk can be minimized.

The system just described may actually be used for transmitting two completely independent information signals, in which case it is simply a two-channel DSBSC transmission system.   However, it also finds much use in cases where two *related* signals must be transmitted, such as in color TV.   In these cases it may be viewed either as a two-channel DSBSC AM system or, if the two information signals $v_A(t)$ and $v_B(t)$ represent the real and imaginary parts of some phasor function, the RF information signal $v_o(t)$ when analyzed is found to have a magnitude which is proportional to the magnitude of the phasor which produces $v_A(t)$ and $v_B(t)$, while the phase of $v_o(t)$ relative to $v_r(t)$ is found to be equal to the angle of the phasor producing the $v_A(t)$ and $v_B(t)$ signals.   In this case the quadrature signal $v_o(t)$ can be viewed as a hybrid-modulated signal, with both AM and PM on the same signal, where the amplitude modulation carries the information regarding the amplitude of the original complex modulating signal and the phase modulation carries the information regarding the phase angle of the original complex modulating signal.   In the case of color TV, the amplitude information conveys the degree of color saturation and the phase information conveys the hue (color). Another use which is proposed for quadrature modulation is in the transmission of discrete three-channel "stereo" FM broadcasts.

We now present a proof that the quadrature modulation process does indeed permit us to send two independent information signals $v_A(t)$ and $v_B(t)$ over the same carrier: referring to the block diagram in Figure 4-11, the outputs of the two balanced modulators are given by
for CH A:

$$v_A(t) \cos \omega_c t \tag{4-41a}$$

and for CH B:

$$v_B(t) \cos (\omega_c t + 90°) \tag{4-41b}$$

Assuming no delay in the equipment or on the signal path, the signals arriving at the receiver are

$$v_i(t) = v_A(t) \cos \omega_c t + v_B(t) \cos (\omega_c t + 90°) \tag{4-42a}$$

and

$$v_r(t) = \cos \omega_c t \tag{4-42b}$$

Then, the outputs of the phase detectors are

$$v'_A(t) = v_A(t) \cos^2 \omega_c t + v_B(t)(-\sin \omega_c t)(\cos \omega_c t)$$

$$= \underline{\tfrac{1}{2}v_A(t)} + \tfrac{1}{2}v_A(t)(\cos 2\omega_c t) - \tfrac{1}{2}v_B(t)(\sin 2\omega_c t) \tag{4-43a}$$

$$v'_B(t) = v_B(t)(\cos \omega_c t)(-\sin \omega_c t) + v_B(t) \sin^2 \omega_c t$$

$$= -\tfrac{1}{2}v_A(t)(\sin 2\omega_c t) + \underline{\tfrac{1}{2}v_B(t)} - \tfrac{1}{2}v_B(t)(\cos 2\omega_c t) \tag{4-43b}$$

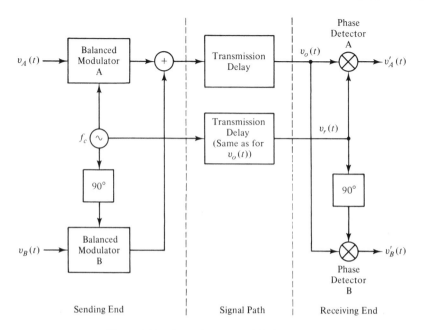

Figure 4-11 A quadrature modulation system.

The "audio" portions of each of the detector outputs are underlined in Equations 4-43a and 4-43b. All the other components occur at RF frequencies and can easily be filtered out with low-pass filters. Thus, we have left only one "audio" component in each output, and this is proportional to the respective modulating signal: $v_A(t)$ or $v_B(t)$.

Another application of the quadrature modulation system is the multiphase-shift keying system used to transmit multilevel (i.e., more than two levels) digital signals. This system is called quadrature-shift keying method (QSK) or quadrature-amplitude modulation (QAM) and is discussed in Section 4.6.

4.5 PULSE MODULATION SYSTEMS AND METHODS

When we talk about pulse modulation systems, we are not initially talking about the modulation of an RF system but instead of how analog or digital information is coded into a sequence of "d.c." pulses (a *pulse train*) by one of several techniques. This pulse train can then be used to modulate the RF carrier in some manner if so desired (see Section 4.6).

There are a number of different basic types of pulse modulation. Three of those are shown in Figure 4-12, and these will each be discussed first.

4.5.1 Pulse-Amplitude Modulation (PAM)

This type of pulse modulation involves a train of *uniform-width* pulses *uniformly* spaced along the time axis where the height of the individual pulses varies as a function of the information signal. The frequency at which those pulses occur (the

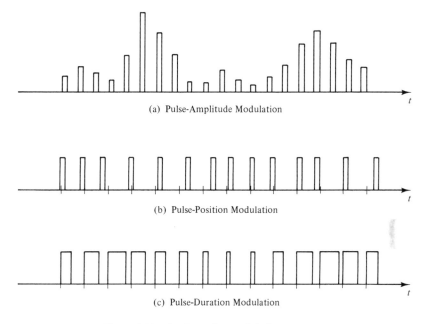

(a) Pulse-Amplitude Modulation

(b) Pulse-Position Modulation

(c) Pulse-Duration Modulation

Figure 4-12   Analog pulse-modulation systems.

pulse repetition frequency or PRR) must be at least twice the highest frequency contained in the baseband information signal.[6]   The equation for the type of waveform shown in Figure 4-12a (*unipolar* pulses where the pulse amplitude is always positive) is given by

$$v_o(t) = A \left\{ 1 + m_{pAM} \left[\frac{s_m(t)}{|s_m(t)|_{max}} \right] \right\} p(t) \qquad (4-44)$$

where $p(t)$ is the pulse function defined in Appendix E (which we again list here in slightly modified form):

$$p(t) = \frac{\tau}{T} \left[1 + 2 \sum_{n=1}^{\infty} \text{sinc}\left(\frac{n\pi\tau}{T}\right) \cos\left(\frac{2\pi nt}{T}\right) \right] \qquad (4-45)$$

where sinc $(x) = \sin (x)/x$.   It can be seen that Equation 4-44 bears a very close resemblance to that for standard AM modulation (see Equation 4-1), the only difference being that $p(t)$ replaces the carrier function cos $(2\pi f_c t)$.   Since this involves replacing a single sinusoid with the sum of a d.c. term and an infinite sequence of sinusoids, we would expect the spectrum of this type of PAM signal to resemble that of a number of AM signals having "carrier" frequencies given by

---

[6] One of the characteristics of pulse modulation methods in general is that in all cases the modulating signal must be sampled.   This is done at a uniform rate (even in the case of PPM) which is great enough to reproduce the information contained within $s_m(t)$.   This rate is given by the sampling theorem as 2 BW$_I$ (see Section 2.7).   Such modulation systems are referred to as "sampled data systems."

$n(f_{op})$, where n goes from zero to infinity. Each "carrier" will have an amplitude given by $(2A\tau/T)$ sinc $(n\pi\tau/T)$. The ratio of τ/T (i.e., the *duty cycle* of the pulse) will determine the relative amplitudes of these "carriers" with respect to each other. For short duty-cycle pulses, the power contained in the pulse train will be spread out over a wider range of frequencies than in the case where the duty cycle is relatively long (approximately equal to 50%, for instance). The working out of these is left as an exercise for the student.

If the "1" is left out of Equation 4-44, we then have *bipolar* PAM where the pulses can go either positive or negative as $s_m(t)$ goes positive or negative. This produces a result very much like continuous-time DSBSC modulation as can be proved by the working of some of the problems at the end of this chapter.

The type of PAM shown here is what is termed *natural* PAM in that the tops of the pulses will follow the $s_m(t)$ time function (i.e., the tops of the pulses are not flat). This produces a spectrum where the sidebands on each side of a given "carrier" frequency are of the same amplitude. Another type of PAM is called *flat-topped* PAM where the tops of the individual pulses are flat and are determined by the sampled value of $s_m(t)$ at or just prior to the point where the pulse begins (for analysis purposes, this may be taken to be the value that $s_m(t)$ has at the defined center of the pulse). In this case the values of each of a pair of sidebands are not equal but rather are determined by the values of the sinc function on either side of the "carrier" frequency.

4.5.2 Pulse-Position Modulation (PPM)

This type of discrete analog pulse modulation is shown in Figure 4-12b and consists of a train of pulses of uniform width and height whose positions are varied away from their nominal locations proportional to $s_m(t)$. For example, delaying a pulse until some time later than where it would normally occur might indicate a negative value for $s_m(t)$, while advancing it to an earlier position would then indicate a positive value of $s_m(t)$ at the sample time. The result is a train of nonuniformly spaced pulses such as is shown in Figure 4-12b. Demodulation of this PPM pulse train at the receiver requires some sort of synchronous timing or gating circuit to determine when each pulse arrives. In a sense, this is very similar to continuous–time-phase modulation. This can be seen by letting $p(t)$ be modified so that for PPM:

$$v_o(t) = A\,p_{\text{PPM}}(t) = Ap\left(t + (\Delta\tau_D)_p \frac{s_m(t)}{|s_m(t)|_{\max}}\right)$$

$$= A\tau/T\left[1 + 2\sum_{n=1}^{\infty} \text{sinc}(n\pi\tau/T)\cos\left(2\pi n\left(t + (\Delta\tau_D)_p \frac{s_m(t)}{|s_m(t)|_{\max}}\right)\right)\right] \tag{4-46}$$

where $(\Delta\tau_D)_p$ is the peak time advance that occurs at $|s_m(t)|_{\max}$. It can be seen that $\Delta\theta_n(t) = (\Delta\tau_D)_p[s_m(t)/|s_m(t)|_{\max}]$ is the phase deviation of the nth "carrier" frequency. Again, some of the problems at the end of the chapter lead to a deeper probing of this relationship.

4.5.3 *Pulse-Duration Modulation (PDM)*

This involves pulses of uniform height and uniform spacing of the pulses' leading edges, but where the width of each pulse is a function of the sampled value of the modulating signal $s_m(t)$. A train of PDM pulses is shown in Figure 4-12c. Detection of this type of modulation is simpler than the detection of pulse-position modulation since it requires no synchronous detector and can quite often be accomplished by a simpler integrator circuit. A variation on pulse-duration modulation is *pulse-width modulation*, where the *centers* of the pulses are uniformly spaced on the time axis rather than their leading edges.

These three pulse-modulation methods are all analog; in other words, the height of the pulses in PAM, the position of the pulses in PPM, and the width or duration of the pulses in PWM or PDM are all related in an analog manner to the sampled value of the modulating signal $s_m(t)$. In effect, we can classify these as *analog sampled-data* systems.

4.5.4 *Delta Modulation (DM)*

This is a type of differential pulse modulation where the pulse sent is used to indicate whether the analog signal being sampled has increased or decreased in value since the last sample was taken. Figure 4-13 shows a block diagram of a system of this type. The output is a series of two-level pulses that at this point are considered to be either positive or negative voltages of the same magnitude V_o. These are produced by sampling the quasi-constant output of the comparator that compares the information signal $s_m(t)$ with a reference signal that is produced by integrating the pulse output. Thus, positive pulses are generated and are sent when the reference signal is smaller than $s_m(t)$ and are continued until it becomes equal to $s_m(t)$. Similarly, if $s_m(t)$ falls below the reference signal, negative pulses are sent until the reference signal again is equal to $s_m(t)$. Thus, the reference

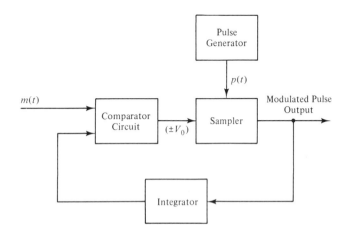

Figure 4-13   Simplified delta modulation system.

signal is made to "follow" $s_m(t)$. At the receiving end of the system, a similar integrator produces a signal that is the same as the reference signal and that therefore also closely resembles $s_m(t)$. It is interesting to note that for this system to work well, the sampling rate must be considerably greater than $2 f_H$.

4.5.5 Pulse-Code Modulation (PCM)

Thus far, the pulse-modulation systems we have been discussing have all been analog discrete-time systems and at the receiver will produce an analog output directly from the demodulation of the modulated pulse train. We will now consider a type of pulse-modulation system that works with input information which is in a digital rather than an analog form. This digital input signal may have been derived from the digitization of an existing analog data signal, or it may represent data already in digital form (such as character data or the data obtained from the memory of a digital computer or some other type of digital data device). To begin with, we shall discuss digital character data in which each character is represented by a binary sequence of some fixed number of bits (these coding systems are discussed in detail in Chapter 11, on data communications).

Figure 4-14 shows a five-bit[7] binary pulse-code system. The particular type of coding shown here for the binary pulse sequence is one-sided return to zero. That is, the two system levels are either positive or zero (on or off) and the signal returns to zero level in-between two adjacent pulses. In a nonreturn-to-zero system, the width of the pulses would be doubled so that in the case of two adjacent positive pulses there would be no returning of the signal level to zero. A pulse-code system such as is shown in Figure 4-14 requires a synchronous detector which knows which pulse is number one in a sequence of five, which is number two, and so on.

Figure (4-15) shows a variation of the pulse-code modulation system which uses doubled-sided nonreturn-to-zero–type coding. Here one level may be thought of as a positive voltage or current and the other level may be thought of as a negative voltage or current, and in-between pulses of the same polarity the signal is maintained at that polarity (i.e., there is no return to zero or to the other polarity). In addition, the system shown here does not require any external synchronization of the detector. This is handled through incorporating a start and a stop pulse in each code sequence. The system shown is the standard teleprinter code for a 5-bit 60-word-per-minute system. Normally, when a character is not being sent, the signal is at the negative (mark) level which holds the receiving teleprinter machine in the "ready-stopped" mode. The detector at the receiving machine is started by sending a "space" or positive-level pulse for 22 msec. This is followed by five 22-msec character-code pulses which actually send the character information to the receiving machine, telling it which character is to be struck. After the fifth code pulse, the sending machine again sends out a mark-level signal until the next

---

[7] Historically, the word "level" was used instead of "bit," but this is confusing because "level" is also used to mean signal voltage level (for example, the number of discernible *levels* in a noisy signal or the number of *levels* into which an analog signal is digitized and so on).

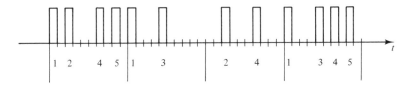

Figure 4-14   Five-bit pulse-code modulation.

character is to be sent.   There is a minimum of 31 msec of mark-level signal which must be sent before the next start pulse can be sent to initiate the sending of another character.   This minimum waiting period is used to be sure that the receiving machines' detectors have all had time to finish the job of interpreting the preceding character and are in fact ready to be started on the job of the decoding of the next character sent to them.   The mode of operation employed here is called the start-stop asynchronous mode of pulse-code modulation and comes in a number of different variations, the most common of which are the 60-, 75-, and 100-word-per-minute 5-bit code for "communications" (news services, telegrams, weather data) teleprinters and a number of 8-bit codes at several different speeds for computer input-output functions (8 code pulses plus a start and a stop pulse).   The most popular of these codes are the ASCII (*A*merican *S*tandard *C*ode for *I*nformation *I*nterchange).   Other binary codes which are used are several versions of coding decimal digits one-at-a-time into binary (binary-coded decimal or BCD), the computer industry standard EBCDIC (*E*xtended *B*inary-*C*oded *D*ecimal *I*nterchange *C*ode), and others.   It might be noted that the 5-data-bits teleprinter code offers a possibility of 32 different characters to be selected (this is almost doubled on a practical machine through using one of the characters to cause the machine to shift into a numerical mode and another character to cause it to shift back into the alphabetic mode; thus, we can have $2 \times (32 - 2)$ or 60 different characters or functions on the machine exclusive of the two shift functions).   With the trend toward digital electronics for almost everything, there will undoubtedly be many new special-purpose binary codes developed in the near future, some of which may be of importance in the communications area.

Although there are other (multilevel) *baseband* digital coding schemes, the binary nonreturn-to-zero (NRZ) is the most popular.   One of the reasons for this is that with a fairly low noise transmission channel (*S/N* ratio > about 20 dB), good decoding of the received NRZ signal may be obtained if only the fundamental component of the rectangular waveform is received.   For a series of alternating

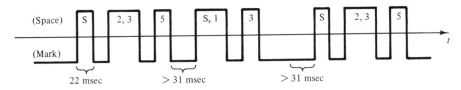

Figure 4-15   Standard 60-word-per-minute teleprinter code (5 data bits plus START and STOP pulses).

1's and 0's, this fundamental frequency (in Hertz) would be equal to one-half the Baud rate of the digital signal. In the case where high data-bit rates are needed at baseband, the usual technique is to employ several parallel paths (wires) so that a number of data bits may be sent simultaneously. This is usually more satisfactory than trying to transmit a single multilevel baseband digital signal.

At one time the baseband transmission of digital signals was conducted over long distances with the use of electromechanical or electronic repeaters which reshaped the pulse train before retransmitting it over the next link (this was the common technique on early telegraph and teleprinter circuits). However, in today's technology, the baseband transmission of digital signals is usually done over only very short distances, while for longer transmission distances, the baseband digital signal(s) is (are) used as the "modulating" signal to modulate an audio-frequency (AF) or radio-frequency (RF) carrier. This is discussed in some detail in Section 4.6 in the paragraphs dealing with FSK, PSK, and QSK modulation methods.

4.6 COMPOUND MODULATION

Several different types of pulse-code modulation have been discussed in the previous section. These pulse signals may be sent directly over a baseband hard-wire circuit as dc pulses, or they may be used to modulate the RF carrier of a radio transmitter by any of several different means. For example, in compound pulse-modulation systems (PPM, PWM, PDM, PCM, or delta modulation) with two levels of modulation, one level may be represented by the turning on of an RF signal, or it may be represented by the turning on of an audio tone. In the former case, the keyed RF signal could be fed into an antenna and transmitted, while in the second case, the keyed audio tone could be sent over a telephone circuit or used to modulate a radio transmitter or it could even be recorded on an audio tape recorder for future reference. In the case of a multiamplitude pulse signal (such as PAM), the amplitude of the RF signal or the audio tone is dependent upon the value of the information signal $s_m(t)$ at the time it is sampled. For the RF case, this produces a result very similar to AM modulation as shown in Equation 4-47, where the extra factor $p(t)$ is the unmodulated pulse train function.

$$s_o(t) = A(\cos \omega_c t)[p(t)][1 + K_a s_m(t)] \qquad (4\text{-}47)$$

The presence of $p(t)$ causes $s_o(t)$ to have groups of sideband components centered away from ω_c at each of the harmonics of the pulse-repetition frequency.

The techniques mentioned in the previous paragraph are just one form of *compound modulation* where first the information signal $s_m(t)$ was sampled and used to modulate by some means a train of pulses which were in turn used to control (modulate) an audio or RF signal. If, for example, we used the pulse train representing $s_m(t)$ to control an audio oscillator and then used the resulting audio signal to modulate an RF transmitter, we would have gone through three stages of modulation in preparing the transmitted signal. This means that at the receiving end, it would be necessary to have three processes of demodulation in order to recover the original information signal, $s_m(t)$.

Another example of compound modulation commonly used with two-ampli-tude-level pulse modulation is frequency-shift keying or FSK, where the modulated binary pulse train is used to shift the frequency of the transmitted wave from one RF frequency to another. In this case, one RF frequency is transmitted when the pulse level is positive, and the other RF frequency is sent when the pulse level is zero or negative. Also, it is possible to send a pulse signal over a telephone line or to record it on an audio tape recorder by sending one of two audio tones depending upon at which binary level the pulse signal is currently. This is called audio-frequency–shift keying or AFSK. The AFSK signal may also be used in some manner to modulate an RF transmitter to produce, for example, an AFSK-AM or an AFSK-FM signal. It is interesting to note that if the AFSK audio signal is used to single-sideband modulate an RF signal, the result of these two modulation processes (AFSK plus SSB) produces an RF signal equivalent to what would have been produced by the single process of FSK of the RF oscillator itself.

The FSK (or AFSK) signal may be generated in two different ways as shown in Figures 4-16a and b. Here we have let the binary signal be a bipolar NRZ signal representing alternate 1's and 0's as shown in Figure 4-16c. If this digital signal is used to generate the FSK output signal by causing switching between the outputs of two independent oscillators, one being at the *mark* frequency and the other being at the *space* frequency, then the FSK signal will have the characteristics of two AM signals added together (i.e., the spectral components will occur at $f_{\text{mark}} \pm nf_o$ and at $f_{\text{space}} \pm nf_o$. On the other hand, if the digital signal is used to FM modulate an oscillator whose carrier frequency is midway between the mark and space frequencies in such a way as to cause the instantaneous frequency to be at the mark frequency for one binary level and at the space frequency for the other binary level, then the FSK output will have the characteristics of an FM signal where the modulating signal is a square wave. This can be shown both theoretically and experimentally.

Another way in which a two-level pulse signal may be used to modulate an RF transmitter is phase-shift keying (PSK). Figure 4-17 shows a simplified PSK system employing coherent detection of the received signal. The phase shift in the RF signal corresponding to a change in the level of the pulse signal is 180 degrees. Such a system requires a stable reference signal at the receiving (demod-ulator) end of the system to recover the original pulse signal. In many cases, it is possible to derive this reference from the received signal itself.

Another version of phase-shift keying is differential phase-shift keying. As an example of how this works, let the two levels of the pulse train be encoded at $+1$ and -1 volts, respectively. If the next pulse to be transmitted is at the $+1$-volt level, a phase shift will occur in the transmitted signal, but if the next pulse is at the negative 1-volt level, the RF phase remains constant (no phase shift occurs). If the PRR is $1/T_p$, then by simply multiplying the received signal by itself delayed by T_p, the pulse train can be re-created at the receiving end of the system without the necessity of having a reference signal present. Figure 4-18 gives the block

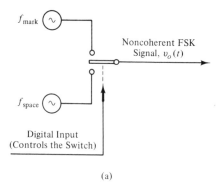

(a)

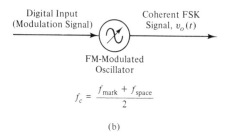

(b)

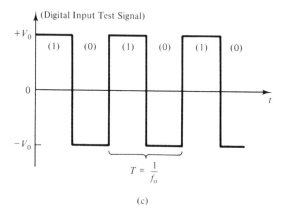

(c)

Figure 4-16 Two methods of FSK signal generation (a and b) and binal test signal.

diagram of a simple DPSK system where the flip-flop which controls the phase switch is triggered only by positive pulses from the sample gate.

Still another version of phase-shift keying is multiphase PSK. In this case we might, for example, be able to phase shift the carrier by $+90°$, $+180°$, or $+270°$ from its nominal phase $(0°)$. This would allow us to send 2 bits of information on each pulse and is called a quadrature (or four-phase) PSK system. Similarly, if we allow the transmitted signal phase to phase positions of $+45°$, $+90°$, $+135°$, $180°$, $-45°$, $-90°$, and $-135°$ from the nominal $0°$ position, we have an eight-

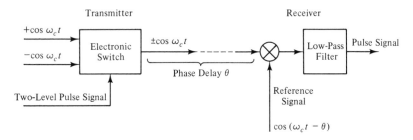

Figure 4-17   A simplified PSK system with coherent detection.

phase system which can convey 3 bits of information per pulse.   These multiphase PSK (or QSK) systems are used extensively for computer or data communications, especially in instances where the channel to be used has a good *S/N* ratio.

An extension of the multiphase technique involves both amplitude and phase-shift modulation of the carrier signal utilizing a technique very similar to analog quadrature modulation but with only a finite number of amplitude-phase combinations being permitted.   Several possibilities are shown in Figure 4-19 where the horizontal axis represents the voltage of the modulating signal v_A in Figure 4-11 and the vertical axis represents the voltage of v_B (i.e., the values the voltage of each of these two modulating signals fed into the quadrature modulation system) are set depending upon the bit combination to be sent on the "pulse."   The patterns in Figure 4-19 are referred to as the transmitted *constellation patterns*.   Due to the noise and distortion along the path over which the QAM signal thus generated propagates, the pattern as seen at the receiving end consists not of definite well-defined points but a scattering about the locations of the amplitude-phase points defined at the transmitting end as shown in Figure 4-20.

A *circle of acceptance* is defined about each of these points, and it is the job of the decoding unit at the receiving end to assign each of the received pulses to

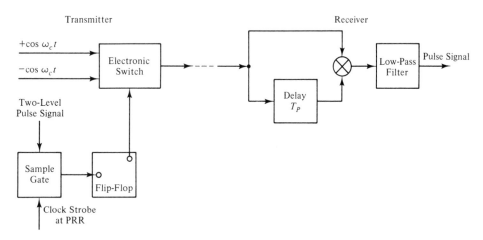

Figure 4-18   A simplified DPSK system.

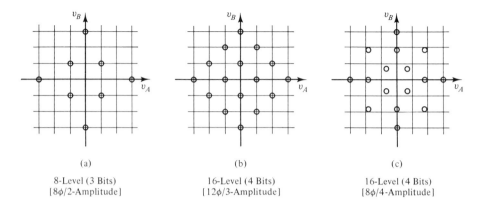

(a)

8-Level (3 Bits)
[8φ/2-Amplitude]

(b)

16-Level (4 Bits)
[12φ/3-Amplitude]

(c)

16-Level (4 Bits)
[8φ/4-Amplitude]

Figure 4-19   QAM phase/amplitude relationships (each "o" represents the phase/amplitude of the quadrature-modulated output (or the v_A/v_B input-voltage combination needed to produce it for one of the possible digital-data levels).

one of these circles.   If too many of the received pulses have amplitude-phase combinations which fall outside of any circle the received signal is judged unfit for use.   How many such unacceptable received pulses are permitted per minute is a function of the acceptable bit error rate established for the system.

4.7 SPREAD-SPECTRUM MODULATION TECHNIQUES

The term *spread-spectrum modulation* applies to a number of modulation methods which are utilized to transmit security messages or to attempt to defeat deliberate jamming attempts.   It may involve such techniques as frequency hopping the transmitted signal according to some preset sequence which the receiver has knowledge of, sending a rapid sequence of pseudorandom code for each bit of actual information (the intended receiver must know what the pseudorandom code is to understand the message), sending random bursts of information for random periods of time at random time intervals, and so on or a combination of such techniques.

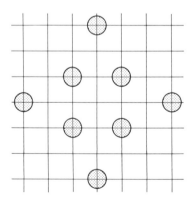

Figure 4-20   Received QAM constellation-pattern scatter plot and circles of acceptable received phase/amplitude combinations.

In general, a much longer period of time and/or a much larger bandwidth is needed than for conventional modulation techniques. The time-bandwidth product (often called the TW) of the spread-spectrum-modulated signal is usually much more than 10 times the time-bandwidth product of the actual message that is sent.

As an example of a spread-spectrum technique, consider the case where the frequency of a radio transmitter is frequency hopped among 25 different frequencies using some pseudorandom algorithm for determining the hopping pattern. The period of this algorithm (the length of time that elapses until it repeats the same pattern) may vary from a fraction of a second to several days in length. The receiving station can follow or "track" this frequency hopping if it also knows the algorithm or pseudorandom "code." It must also know, at least approximately, the point in time at which the period of the pseudorandom sequence begins. It can then hop the frequency of the receiver local oscillator according to the same sequence by which the transmitting station is hopping its transmitting frequency (delayed of course by some transmitter-to-receiver propagation time) and by doing so produce a constant carrier frequency out of the receiver's mixer into its IF amplifier (see Section 5.3.2 on superheterodyne receivers). If this frequency hopping is done at a faster rate than that at which the signal is modulated, the RF spectrum of the transmitter's output begins to look very much like random noise. Thus, the presence of other spread-spectrum transmitters operating in the same frequency range using other pseudorandom hopping sequence codes appears just as added noise power to a receiver attempting to receive from a particular spread-spectrum station. Thus, it is possible for a number of transmitters to operate simultaneously within the same range of frequency. The number that can so operate is determined by the degradation produced at each receiver by the amount of pseudonoise power spectral density added to the spectrum by each additional spread-spectrum transmitter. This type of spectrum sharing is a form of combined time-domain and frequency-domain multiplexing (each of these two basic multiplexing methods is described in the next section).

4.8 MULTIPLEXING OF INFORMATION

In many communication systems (such as on telephone trunk lines) there are many information signals available which need to be sent from point *A* to point *B*. One way of doing this is to build many communication channels, each capable of carrying one information signal, between points *A* and *B*. However, this is usually economically wasteful, and it is usually more advantageous to build one (or at least fewer) channel(s) which are capable of carrying more than one information signal simultaneously. This latter technique is known as *multiplexing*.

One common form of multiplexing is frequency-division multiplexing (FDM). An example of this is the standard CCITT (the French language abbreviation for the *International Consultive Committee for Telephone and Telegraph*, which is part of the International Telecommunications Union headquartered in Geneva, Switzerland) system for trunk telephone circuits where 12 telephone (audio) signals (each with a frequency range from 300 Hz to 3400 Hz) are combined into one low-frequency RF signal that occupies the region from 60.6 KHz to 107.7 KHz. This signal is generated as shown in Figure 4-21 by applying the 12 telephone signals to 12 single-sideband modulators where the 12 carrier frequencies are from 64 KHz to 108 KHz in steps of 4 KHz. The lower sideband is employed in each case, and the carrier and the upper sideband are suppressed at the output of each modulator. The 12 LSB-SSB signals are then added together to form a single FDM signal which can be sent over a single pair of wires in a cable. This 12-channel combination is the standard CCITT *group*. Instead of sending the *group* directly over a pair of wires, it can be combined with four other *groups* to create the CCITT *supergroup* which consists of five *groups* modulating five carriers at 420, 468, 516, 564, and 612 KHz. LSB-SSB modulation is again employed, thereby producing a band of signals which occupies the frequency range from 312 to 552 KHz. This *supergroup*, consisting of 60 telephone messages, may be put onto a cable, or it may be combined with four other *supergroups* by LSB-SSB modulating carriers at 1.364, 1.612, 1.860, 2.108, and 2.356 MHz to produce a *mastergroup* of 300 message channels extending from 812 KHz to 2.044 MHz. The *mastergroup* may be used directly, or it may be combined with two other *mastergroups* to produce a *super-mastergroup* of 900 channels. Again LSB-SSB is used, and the carrier frequencies of 10.56, 11.88, and 13.20 MHz produce a band of telephone signals extending from 8.156 MHz to 12.388 MHz. The creation of *supergroups*, *mastergroups*, and *supermastergroups* are all examples of compound modulation. CCITT also specifies other possibilities for combining *groups* and *supergroups*, but we will not go into these here.

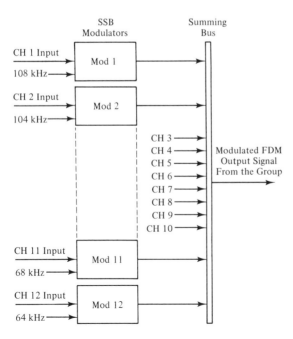

Figure 4-21   Creation of the CCITT-FDM *group* signal.

Another multiplexing technique commonly used is that of time-division multiplexing (TDM). This technique is based upon the sampling of information signals at a rate of at least $2BW_I$ (see Section 2.7). The samples thus taken can be encoded onto short pulses (using PAM, for example) occurring at a PRR equal to the sampling rate. However, if the duration of each pulse is much less than the time between pulses, the "dead" time can be occupied by other pulses derived from other sampled signals. As an example of this, consider Figure 4-22a (page 86), which shows an analog *sequential multiplexor* having eight inputs, one of which is connected to the output at any time. Whenever a clock pulse is applied to the multiplexor, it switches to the next higher input (or back to input #1 if it is on #8). If analog signals are applied to each of the eight inputs, then each input will be sampled at $R_S = f_{clock}/8$. Obviously, the clock frequency must be high enough so that R_S is twice the highest BW_I of any of the input channels. Very often, the case exists where the analog data channels have considerably different information bandwidths. In this case the sequential multiplexor may be "strapped" as is shown in Figure 4-22b. In this example, data channel A is sampled twice during each multiplexor cycle (of eight clock pulses), while channel B is sampled four times per cycle and channels C and D are sampled once per cycle as before. Thus, R_S for C and D will still be the same as before, but R_S for channel A will be equal to $2f_{clock}/8$ and R_S for channel B will be equal to $4f_{clock}/8$. The output for this strapped multiplexor might appear as in Figure 4-22c. This can be analyzed as four separate PAM signals as is indicated in Figure 4-22d. In this example τ/T for channels C and D will be 1/8, for channel A it will be 1/4, and for channel B it will be 1/2.

Other multiplexing techniques are also possible, such as the interleaving of the sidebands of two or more RF signals when each of the individual signals has a sideband structure which has large open spaces between the sidebands or groups of sideband components. This is the technique that is used to add the color signal to a standard black-and-white TV signal. Another technique is to employ the quadrature modulation technique discussed in Section 4.5 using $v_A(t)$ and $v_B(t)$ as two separate information signals. Many other multiplexing techniques are possible, and quite often a communication system will employ a combination of two or more different techniques to encode a number of information signals into one multiplexed modulated signal.

4.9 SPECIAL MODULATION AND CODING TECHNIQUES

One of the design objectives associated with communication systems is that of being able to send the maximum amount of useful information in a given bandwidth in a given length of time. This objective results from economic consideration and, in the case of RF signals which are radiated, from the limited amount of RF spectrum space available for all the different types of radio users. As an example of what can be accomplished to meet this objective, consider a long-distance telephone trunk. Since amplification of the signals is necessary, it is necessary to have a channel for each direction of the conversation. But, on the average, each channel is utilized less than 50% of the time since one party is usually listening while the

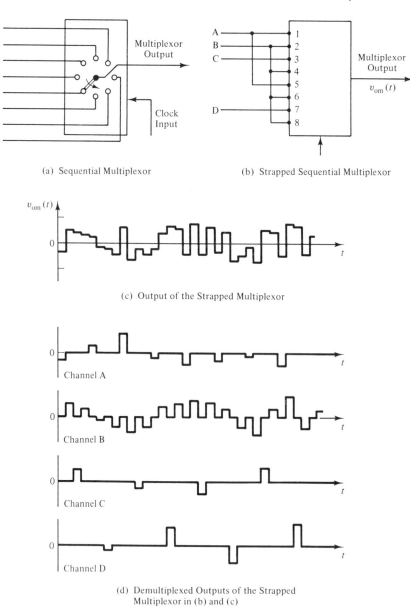

(a) Sequential Multiplexor

(b) Strapped Sequential Multiplexor

(c) Output of the Strapped Multiplexor

Channel A

Channel B

Channel C

Channel D

(d) Demultiplexed Outputs of the Strapped
Multiplexor in (b) and (c)

Figure 4-22   Sequential TDM multiplexing.

other party is talking.   Thus, if we had 1000 telephone conversations to convey
from point A to point B, we could actually get by with 500 or fewer channels if
we had some means of assigning an available channel to a person only when that
person was actually speaking.   This is something that is actually possible to do

using modern digital computer techniques. The actual implementation of the technique is far from being simple, since we not only have to find an unused channel and assign it to a user as soon as he or she begins speaking, but we also have to know to which trunk channel the listener's telephone line must be connected.

The subject of redundant information was discussed in Chapter 1, especially as it applies to TV pictures. The standard NTSC (National Television Systems Committee) color TV signal is based upon being able to transmit a very detailed picture which is changing very rapidly and where the color fill-in can take on many hues and levels of saturation (although the color detail has been sacrificed in the interest of conserving bandwidth). However, most of the time the picture being sent does not contain much fine detail, is not moving or changing rapidly, and contains a limited amount of color variation. Thus, most of the capabilities of the NTSC signal are wasted sending redundant information to the receiving set. Researchers have found that it is possible, by the use of digitizing techniques, to examine the TV picture and put the necessary information to re-create the scene into a relatively few bits of digital data which may then be sent to a special TV receiver which reconstructs the picture. Also, when changes do occur in the picture, it is only necessary to send information about the changes, and the special receiver makes the necessary changes in the picture which it has already constructed. The net result is a much slower flow of information from the transmitter to the receiver, and this of course means the capability to operate in a much smaller-bandwidth channel than is required by the NTSC signal.

Another device mentioned in Chapter 1 was the voice vocoder. This is a device which examines a human voice signal and determines its basic characteristics as a message is spoken (the basic vowel sound being voiced, the duration and rise and fall of the volume, unvoiced sounds such as hisses and tongue clicks, etc.) and codes this information into a digital message which may be sent to the receiving end of the system where a complimentary device decodes the digital signal and reconstructs an understandable voice signal. The channel bandwidth needed to send the message via vocoder techniques is much less than that required to send the actual voice signal. Vocoders, as well as other special audio and video techniques, are discussed in more detail in Chapter 8.

4.10. NOISE CONSIDERATIONS IN MODULATION SYSTEMS

The choice of a modulation system to be used is often affected by the considerations of the effects of noise added to the modulated RF signal as it passes through the system upon the signal-to-noise ratio of the postdetection reconstructed information signal. These effects are not only dependent upon the type of modulation employed, but may also depend upon the manner in which the signal is detected at the receiver. In this section we will discuss the results of analyzing these noise effects under different conditions. A more thorough discussion of the theory and mathematics behind these results may be found in any good communication theory textbook.

4.10.1 Amplitude-Modulation Systems

Probably the easiest system to analyze intuitively is the SSB system. The detection portion of such a system is shown in Figure 5-50. However, all that is really accomplished in such a detector is the heterodyning (converting) of the SSB signal down in frequency so that SSB spectrum is now based at $f_c' = 0$ Hz instead of at some RF frequency. Otherwise, there is no change in the signal, and the audio passband and the RF passband of the signal are the same. Thus, since the signal and all its noise components are simply slid down the spectrum by an amount f_c, there will be no difference in the S/N ratios of the postdetection (audio) and the predetection (RF) signals. However, this is assuming that the RF amplifier passband preceding the SSB detector is just wide enough to pass the one sideband. If instead it is wide enough to pass both the active and the suppressed sidebands, then the noise power on the postdetection signal will be doubled and the audio S/N ratio will be 3 dB lower than the S/N ratio in the half of the predetection bandwidth which actually contains the SSB signal. If we let f_{max} be the maximum modulating frequency as well as the cutoff frequency of the low-pass filter which follows the product detector (see Figure 5-50), then the only predetection noise which contributes to postdetection noise which passes through the low-pass filter is that which lies between $f_c - f_{max}$ and $f_c + f_{max}$ prior to the detector (where f_c is the frequency of the missing carrier at the detector input).

For a DSBSC signal the predetection bandwidth normally includes both the upper- and lower-sideband spectral components of the signal, and thus it is normally twice the bandwidth of the postdetection passband. As just mentioned, this gives twice the noise power as compared to the normal SSB arrangement. However, in DSBSC (or standard AM for that matter), the *voltages* of corresponding upper- and lower-sideband components add constructively in the detection process, and thus the signal power in the postdetection signal will be twice that of the SSB case if the total power in *both* the sidebands in the DSBSC case is equal to the average power in the one SSB sideband. Since the noise power is also only twice that of the SSB case, there is no overall difference in the S/N performance of SSB versus DSBSC. (The DSBSC postdetection S/N ratio is twice that of its predetection (RF) S/N ratio, while for SSB the two are the same, but it must be remembered that the predetection noise power in the DSBSC case is twice that of the SSB predetection noise power, or, in other words, for the same total average sideband power, the DSBSC predetection S/N ratio is one-half that of the SSB case.) The only advantages of SSB over DSBSC is that SSB does the job in half the spectral space required for either DSBSC or AM. Also, for communications-quality voice links, SSB does not require that a carrier-phase-synchronization signal be sent as is required for DSBSC (see Section 4.2).

Example:

Consider two systems, one SSB and the other DSBSC, where the total transmitter output power is 4 watts, the impedance level is 1 ohm, and the detector voltage gain is K_D:

	SSB	DSBSC
Signal:	$V_{rms} = 2$ volts (one sideband)	$V_{rms} = \sqrt{2}$ volts (two sidebands)
Noise:	Total RF noise power	Total RF noise power
	$= kT_N B_{audio} = N_1$	$= kT_N (2B_{audio}) = 2N_1$
Detector output:	$V_{out} = K_D \times 2$ volts rms	$V_{out} = K \times 2 \times \sqrt{2}$ volts rms
Detector signal power:	$S_{out} = K_D^2 \times 4$ Watts	$S_{out} = K_D^2 \times 8$ Watts
Detector noise power:	$N_{out} = N_1 \times K_D^2$ Watts	$N_{out} = 2N_1 \times K_D^2$ Watts
$\left(\dfrac{S}{N}\right)$ postdetection:	$\left(\dfrac{S}{N}\right)_{post} = 4/N_1$	$\left(\dfrac{S}{N}\right)_{post} = \dfrac{8}{2 \times N_1} = \dfrac{4}{N_1}$
$\left(\dfrac{S}{N}\right)$ predetection:	$\left(\dfrac{S}{N}\right)_{pre} = 2^2/N_1 = 4/N_1$	$\left(\dfrac{S}{N}\right)_{pre} = \dfrac{2 \times (\sqrt{2})^2}{2 \times N_1} = \dfrac{2}{N_1}$

Thus, we see that the pre- and postdetection S/N ratios are the same for the SSB signal while for the DSBSC signal the postdetection S/N ratio is twice the predetection S/N ratio, but the predetection S/N ratio for DSBSC is only half that for SSB for the same transmitter power, so that the postdetection S/N ratios are the same for SSB and DSBSC.

For AM modulation, the analysis just completed is valid if a synchronous detector is employed and if the predetection S/N ratio is given in terms of the total average *sideband* power to the total noise power in the predetection RF bandwidth (which is equal to two times B_{audio}). The analysis also holds for diode envelope detectors if the *carrier*-to-noise ratio is reasonable (larger than about 3.6 dB).[8,9] In many cases we may be interested in how the AM postdetection S/N ratio compares with the predetection AM carrier-to-noise ratio or to the predetection AM-total-average-power-to-noise ratio. Using Equations 4-26 and 4-27, we get

$$\left(\frac{S}{N}\right)_{post} = 2\left(\frac{P_{SB}}{N}\right) = 2\left[\frac{\dfrac{(m_{AM})_e^2}{2} P_c}{N}\right] = (m_{AM})_e^2 \left(\frac{P_c}{N}\right) \qquad (4\text{-}48)$$

or

$$\left(\frac{S}{N}\right)_{post} = 2\left(\frac{P_{SB}}{N}\right) = 2\left\{\frac{\dfrac{(m_{AM})_e^2}{2}\left[\dfrac{P_{total}}{1 + (m_{AM})_e^2/2}\right]}{N}\right\} \qquad (4\text{-}49)$$

$$= \frac{(m_{AM})_e^2}{\left[1 + \dfrac{(m_{AM})_e^2}{2}\right]} \left(\frac{P_{total}}{N}\right)$$

where $N = kT_N(2B_{audio})$.

[8] Simon Haykin, *Communication Systems*, 2nd ed., (New York: John Wiley, 1983), p. 330.

[9] David W. Gregg, *Analog and Digital Communication* (New York: John Wiley, 1977), p. 225.

Earlier we had seen that for the same total average transmitted sideband power, there was no S/N advantage of DSBSC over SSB. The same holds true for the comparison with AM if we consider only the total average sideband power which is transmitted. However, when we make the comparison on the basis of the total average power which is transmitted, then SSB (or DSBSC) has a tremendous advantage over AM:

$$\frac{\left(\dfrac{S}{N}\right)_{post,SSB}}{\left(\dfrac{S}{N}\right)_{post,AM}} = \frac{\left(\dfrac{S_o}{kT_{N_o}B_{audio}}\right)}{\left(\dfrac{(m_{AM})_e^2}{2kT_{N_o}B_{audio}}\right)\left[\dfrac{P_{total}}{1 + (m_{AM})_e^2/2)}\right]} = \frac{2[1 + (m_{AM})_e^2/2]}{(m_{AM})_e^2} \qquad (4\text{-}50)$$

Thus, for the case where $(m_{AM})_e = 1.0$ (100% sinusoidal AM modulation, we have 3:1 SSB-over-AM advantage (the total average AM power transmitted in this case is three times that of the SSB transmitter).

As the level of peak modulation decreases from 100% and as the modulating signal is made up of many Fourier components instead of being sinusoidal, $(m_{AM})_e$ becomes quite small and thus the ratio given by Equation 4-50 becomes quite large (201 when $(m_{AM})_p = 0.1$, for example).

It should be noted at this point that when the AM carrier-to-noise ratio drops below about 3.6 dB and diode envelope detection is used, we enter into what is termed the "threshold operation" region of the diode detector and the postdetection S/N ratio for the AM signal is even less than what has been predicted by the discussion thus far (which gives SSB even more of an advantage). However, system performance is so poor at these predetection S/N levels that this latter situation is almost never encountered in practice.

4.10.2 Frequency-Modulation Systems

One of the main reasons for using angle modulation is that under the right conditions, it can provide a great improvement in the postdetection S/N obtainable as compared with that obtainable from an amplitude-modulation system. In the following we will present an intuitive proof of this without going into a rigorous statistical proof of all the points involved (the latter is obtainable from any number of excellent tasks on statistical communications theory).

To begin with, we will define the system with which we shall be working: it will consist of an FM signal (as defined earlier in this chapter) applied to a Foster-Seeley discriminator via a hard-limiting amplifier that removes all the amplitude variations on the signal applied to the input of the discriminator. The operation of the discriminator is described in detail in Appendix D, but for the purposes of this discussion, all that we need to know is that its output voltage is proportional to both the amplitude of the RF sinusoid applied to its input and to the difference in frequency between the sinusoid's frequency and the frequency to which the discriminator is tuned. Defining the discriminator output voltage as $v_{so}(t)$ we have

$$v_{so}(t) = KV_{cm} \, \Delta f_s(t)$$

where

$$\Delta f_s(t) = (f_i(t) - f_c)$$

for the signal and K is the proportionality constant associated with the discriminator. We will also define the output signal power S_o of the discriminator in terms of the discriminator output impedance $R_o = 1\ \Omega$ and the mean-square value of the output voltage v_{so}^2 while the input signal power to the discriminator will be called P_{ci} and will be equal to $V_{cm}^2/2$ (again assuming that the input impedance level R_i is 1 ohm). The discriminator audio output noise power spectral density will be called p_{no} and will be a function of f_{audio} while the quadrature component of the RF input noise power spectral density will be called $p_{ni_q} = kT_{n_i}$ (constant with frequency). We will now begin the analysis by considering the mean-square value of the signal frequency deviation. It can be shown that this is

$$\overline{\Delta f_s^2} = (\Delta f)_p^2 \left[\frac{\overline{s_m(t)^2}}{|s_m(t)|_{\max}^2} \right] = \frac{1}{2}(\Delta f)_e^2 \tag{4-51}$$

Then,

$$S_o = \overline{v_{so}^2} = \frac{(K^2 V_{cm}^2 (\Delta f)_e^2)}{2} \tag{4-52}$$

To determine the audio noise power is a bit more difficult. We begin by assuming that since the hard limiter preceding the discriminator eliminates all amplitude variations, the only effects of a small noise voltage (of random amplitude and phase) added onto the signal voltage prior to going into the limiter will be to cause a random variation in the instantaneous phase of FM signal coming out of the limiter. This variation is caused by the part of the random noise voltage which happens to be in quadrature with the instantaneous phase of the FM signal at the point in time being considered. This is represented by the phasor diagram shown in Figure 4-23. In this time-dependent phasor diagram, the phasor of length V_{cm}

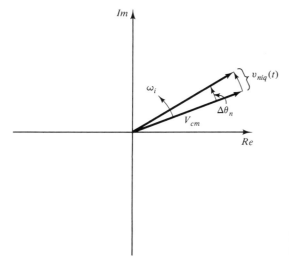

Figure 4-23 Time-dependent phasor diagram of the FM signal with noise.

represents the instantaneous phase of the FM signal (it is rotating counterclockwise at an angular velocity $(\omega_i(t)$, which is 2π times the instantaneous frequency in Hertz). The presence of the random noise voltage $v_{ni_q}(t)$ causes an "error" (or noise) in the instantaneous phase of the FM signal of

$$\Delta\theta_n(t) = \left(\frac{1}{V_{cm}}\right) v_{ni_q}(t) \sim \text{radians} \tag{4-53}$$

Combining this with the definition of the discriminator's audio output voltage, we get for the audio output noise voltage

$$v_{no}(t) = K \, V_{cm} \Delta f_n(t) = K \, V_{cm} \frac{\Delta\omega_n(t)}{2\pi}$$

$$= (K \, V_{cm}) \left(\frac{1}{2\pi}\right) \frac{d\Delta\theta_n(t)}{dt} = \left(\frac{K}{2\pi}\right) \frac{dv_{ni_q}(t)}{dt} \tag{4-54}$$

If $v_{ni_q}(t)$ were to be a sinusoid at some RF frequency separated from f_c by an amount Δf, then we could deal with this "noise" by means of phasors

$$|V_{no}| = \left|\left(\frac{K}{2\pi}\right)(j2\pi\Delta f)(V_{ni_q})\right| \tag{4-55}$$

or

$$|V_{no}|_2^2 = K^2 \Delta f^2 |V_{ni_q}|^2 \tag{4-56}$$

Now, without going into the statistical proofs, we will replace the noise powers (given by the rms phasor values squared) at one particular frequency with the noise power spectral densities

$$p_{no} = K^2 \Delta f^2 p_{ni_q} \tag{4-57}$$

and letting the magnitude of the frequency deviation of the noise component at RF be equal to the audio frequency it produces in the discriminator output, we get

$$p_{no}(f_{\text{audio}}) = K^2 (f_{\text{audio}})^2 [p_{ni_q}(f_c - f_{\text{audio}}) + p_{ni_q}(f_c + f_{\text{audio}})]$$

$$= 2K^2 (f_{\text{audio}})^2 (kT_{ni}) \tag{4-58}$$

If the discriminator is followed by an ideal low-pass filter of bandwidth B_{audio}, then the audio noise power will be

$$N_o = \int_0^{B_{\text{audio}}} p_{no}(f_{\text{audio}}) \, df_{\text{audio}}$$

$$= \int_0^{B_{\text{audio}}} 2K^2 (f_{\text{audio}})^2 \, (kT_{ni}) \, df_{\text{audio}}$$

$$= 2K^2 \, (kT_{ni}) \frac{B_{\text{audio}}^3}{3} \tag{4-59}$$

in which case the audio output signal-to-noise ratio becomes

$$\frac{S_o}{N_o} = \frac{3K^2\,(V_{cm}^2/2)\,(\Delta f)_e^2}{2K^2\,(kT_{Ni})\,B_{audio}^3} = \left(\frac{3}{2}\right)\frac{P_{ci}\,(\Delta f)_e^2}{(kT_{Ni}B_{audio})\,B_{audio}^2} \tag{4-60}$$

which for the case of 100% sinusoidal FM modulation where $(\Delta f)_e^2$ is equal to $(\Delta F)^2$ becomes

$$\frac{S_o}{N_o} = \tfrac{3}{2}\left(\frac{P_{ci}}{N_{i,AF}}\right)(M_{FM})^2 \tag{4-61}$$

where $N_{i,AF}$ is the predetection noise power within a Δf equal to the baseband-signal bandwidth B_{audio}.

If we now let the predetection RF bandwidth be given by Carson's rule as $B_{RF} = 2(M_{FM} + 1)B_{audio}$, then we can calculate the audio S/N ratio in terms of the RF predetection S/N ratio

$$\frac{S_o}{N_o} = \tfrac{3}{2}\left[\frac{2(M_{FM} + 1)\,M_{FM}^2\,P_{ci}}{N_{i,RF}}\right] \tag{4-62}$$

$$= 3(M_{FM} + 1)M_{FM}^2\left(\frac{S}{N}\right)_{RF,predetection}$$

which for $\Delta F = 75$ KHz, $B_{audio} = 15$ KHz, and 100% sinusoidal modulation is

$$\frac{S_o}{N_o} = \tfrac{3}{2}\left(\frac{P_{ci}}{N_{i,AF}}\right)(25) = 37.5\left(\frac{P_{ci}}{N_{i,AF}}\right) \tag{4-63}$$

$$= 450\left(\frac{S}{N}\right)_{RF,predetection}$$

As was discussed earlier in the chapter, standard broadcast FM uses audio preemphasis at the transmitter which is compensated for by the use of deemphasis at the FM receiver. The presence of this audio deemphasis filter following the discriminator output will cause a reduction in the audio power spectral density function at frequencies above $f_o = 2125$ Hz so that p_{no} will be given by

$$p_{no}(f_{audio}) = 2K^2 f_{audio}^2\,(kT_{Ni})\,\frac{1}{1 + (f/f_o)^2} \tag{4-64}$$

When this is integrated from 0 to B_{audio} we get

$$N_o = 2K^2(kT_{Ni})\left\{f_o^2 B_{audio} - f_o^3\left[\arctan\left(\frac{B_{audio}}{f_o}\right)\right]\right\} \tag{4-65}$$

When this is used to calculate the postdetection audio S/N ratio, we get (for 15 KHz audio bandwidth and $k = 1.381E - 23$)

$$\frac{S_o}{N_o} = (6.7033 * 10^{11})\left[\frac{P_{ci}\,(\Delta f)_e^2}{T_{Ni}}\right] \tag{4-66}$$

which for 100% sinsusoidal modulation gives

$$\frac{S_o}{N_o} = 31.243\left(\frac{P_{ci}}{N_{i,AF}}\right)M^2_{FM} = 781.1\left(\frac{P_{ci}}{N_{i,AF}}\right) \qquad (4\text{-}67)$$

or with the noise power given in terms of the predetection RF bandwidth as determined by Carson's rule

$$\frac{S_o}{N_o} = 62.49(M_{FM} + 1)M^2_{FM}\left(\frac{S}{N}\right)_{RF,predetection}$$

$$= 9373\left(\frac{S}{N}\right)_{RF,predetection} \qquad (4\text{-}68)$$

4.10.3 Noise in Pulse-Modulation Systems

In the PAM system where time-division multiplexing is used and the demodulation channel is only gated ON when a pulse is expected for that particular channel, the S/N ratio of the postdetection signal is theoretically equal to that of the predetection PAM signal.   In PDM or PPM modulation, the predetection bandwidth needed is dependent upon how sharply defined the pulse is, which also determines how precisely the output level may be obtained.   But the wider the predetection bandwidth, the greater the predetection noise.   From Haykin,[10] we can derive

$$\left(\frac{S}{N}\right)_{out} = 0.0514 \times \left(\frac{B_{pre}}{B_{post}}\right)^2 \times \left(\frac{S}{N}\right)_{pre} \qquad (4\text{-}69)$$

where $(S/N)_{pre}$ is inversely proportional to B_{pre}.   Usually (B_{pre}/B_{post}) is fairly large (on the order of 10 to 100).

PROBLEMS

4-1. Let an audio signal given by

$$v(t) = 9\cos(3141.6t) + 7\cos(4398.2t) + 5\cos(5654.9t) \text{ Volts}$$

be used to modulate an AM transmitter (carrier output power = 10 KW and carrier frequency = 900 KHz) to 80% peak modulation.
a. Give the *frequency* and the *power* of each sideband component.
b. Find the *effective* modulation index $(m_{AM})_e$.

4-2. Give the *procedures* for finding (calculating) P_{av}, P_{max}, and the *PEP* for AM, DSBSC, and SSB types of amplitude modulation.

4-3. Let a modulating signal consist of four nonharmonically related audio tones of equal amplitude.
a. For 100% AM modulation, find the P_{av} and the *PEP* for the AM signal and for

---

[10] Haykin, Communication Systems, pp. 392-397.

the DSBSC and SSB signals where the *total average sideband power* in these is the same as in the AM signal.

b. Repeat (a) for the case of 50% peak AM modulation.

4-4. A "standard AM" transmitter has a carrier output power of 10 KW. When it is modulated by two audio tones of equal amplitude, the total average RF output power increases to 11.5 KW. What is the peak AM modulation index $(m_{AM})_p$ and the peak percentage of AM modulation? What are the P_{max} and the *PEP* of the signal?

4-5. Calculate P_{av}, P_{max}, and *PEP* for the AM case, the DSBSC case, and the SSB case (making the total average *sideband* power equal in all three cases) for the following modulating signals and conditions in terms of $P_{c,AM}$:

Modulating signal $s_m(t)$	% Peak AM modulation
a. $1.0 \cos (\omega_{m_1} t)$	100%
b. $0.5 \cos (\omega_{m_1} t) + 0.5 \cos (\omega_{m_2} t)$	100%
c. $1.0 \cos (\omega_{m_1} t)$	50%
d. $0.5 \cos (\omega_{m_1} t) + 0.5 \cos (\omega_{m_2} t)$	50%
e. $0.333 \cos (\omega_{m_1} t) + 0.333 \cos (\omega_{m_2} t) + 0.333 \cos (\omega_{m_3} t)$	100%
f. $0.6 \cos (\omega_{m_1} t) + 0.3 \cos (\omega_{m_2} t) + 0.1 \cos (\omega_{m_3} t)$	100%
g. $0.6 \cos (\omega_{m_1} t) + 0.3 \cos (\omega_{m_2} t) + 0.1 \cos (\omega_{m_3} t)$	50%
h. $s_m(t)$ consisting of 10 nonharmonically related tones of 0.1 peak amplitude each	100%
i. the same $s_m(t)$ as in (h)	50%

4-6. Let $s_m(t) = 5.0 \cos (\omega_{m_1} t) + 3.0 \cos (\omega_{m_2} t)$ where the two frequencies are not harmonically related. This signal is then used to modulate an AM transmitter to 80% peak AM modulation. Let A be the amplitude of the RF carrier output of the transmitter with no modulation present, and let the RF output impedance level be 100 Ohms.

a. Give expressions for the *frequency* and the *average power level* of each component in the modulated RF output.

b. If this signal is transmitted and received by a receiver with a diode detector with a 30-dB carrier-to-noise ratio, what is the ratio between the postdetection audio *S/N* ratio and the predetection (RF) *S/N* ratio, assuming that the RF predetection bandwidth is twice the audio bandwidth. What is the audio *S/N* ratio? Let the predetection power used to calculate the predetection *S/N* ratio be the *total average RF signal power*, and let N be the predetection noise power in the RF bandwidth of the signal.

4-7. For a certain FM transmitter, $(\Delta f)_p$ is 75 KHz. What is $(\Delta \theta)_p$ when

a. $f_{audio} = 15$ KHz.

b. $f_{audio} = 20$ KHz.

Assume sinusoidal modulation in both cases.

4-8. If $s_m(t) = 1.0 \cos (25,120t)$ and the carrier frequency is 100 MHz and $(\Delta f)_p = 20$ KHz, for an FM signal

a. Find the value of $(m_{FM})_p$ and $(m_{FM})_e$.

b. Sketch the spectrum of the FM output signal giving the frequency of each component and its relative amplitude (let the amplitude of the unmodulated carrier be 1.0).

4-9. Let $s_m(t) = 7.0 \cos (1333t) + 2.5 \cos (3000t)$. This signal will be used to modulate an FM transmitter whose carrier frequency is 154 MHz and where $(\Delta f)_p = 10$ KHz. What frequencies are contained in the modulated RF output (give expressions for calculating the frequencies involved)? What is the approximate RF bandwidth of the RF signal? What is $(m_{FM})_e$?

4-10. What RF bandwidth is needed for an analog FM signal where $(\Delta f)_p$ is 25 KHz and the maximum audio frequency is 10 KHz? Compare the detected audio S/N ratio at the FM receiver with what would be obtained with a "standard AM" system having 100% peak modulation, the same total average transmitted RF power, the same noise temperatures, and so on.

4-11. Consider a *quadrature modulation system* where

$$v_A(t) = A(t) * \cos \theta(t) \quad \text{and} \quad v_B(t) = A(t) * \sin \theta(t)$$

Prove that the *amplitude* of the modulated RF signal is proportional to $A(t)$ and that the *phase* of the modulated RF signal relative to the unmodulated carrier phase is equal to $\theta(t)$.

4-12. Let $s_m(t) = 7.5 \cos (100t) + 6.3 \sin (219t)$. This signal is sampled at 80 times per second and a *natural* PAM pulse train is generated with 50% duty-cycle pulses. A 15-Volt d.c. bias is added to $s_m(t)$ before it is sampled.
a. What frequencies (in Hertz) are present in the PAM pulse train?
b. Write an expression for the time function of the PAM pulse train.

4-13. A *flat-topped* PAM system has an input *information* signal $s_m(t) = 1.0 \cos (5000t)$ and output pulses 150 microseconds long occurring every 500 microseconds. Sketch the frequency spectrum of the PAM output signal (specify the frequencies of at least 11 of the components and sketch the "envelope" function determining their amplitudes).

4-14. The information signal in a PAM/RF system is given by

$$s_m(t) = 4.0 \cos (30t) + 3.5 \cos (46t)$$

and the RF output is given by

$$v_o(t) = [10.0 + s_m(t)]p(t) \cos \omega_c t$$

where $p(t)$ is the pulse function defined in Appendix E, the pulse duty cycle is 50%, and the pulse repetition rate is just the minimum required to prevent aliasing error with the information signal given above. Determine the following:
a. The pulse repetition rate.
b. An expression for all the frequencies present in the RF output if the RF carrier frequency is 2 MHz.
c. If PCM is used instead of PAM and there are 256 evenly spaced digitizing levels, what would be the *digital-data-bit rate* of the digital data signal?

4-15. A *pulse-position modulation system* delays the pulses from their nominal positions by the amount $s_m(t) * 10^{-6}$ sec (where when $s_m(t)$ is negative, the pulses are *advanced* from their nominal positions). The modulating signal is $s_m(t) = 10.0 \cos (2000t)$ and the PPM pulses are one microsecond in width and occur at a repetition rate of

50,000 pulses per second. What component frequencies are present in the PPM pulse train? Give an expression or expressions which would enable one to calculate the amplitude of each such component.

4-16. Let $s_m(t) = 10.0 \cos (1000t)$ and let this signal be sampled and used to modulate a train of 1-millisecond pulses at a pulse-repetition rate of 500 pulses per second according to the relationship that the pulse amplitude is equal to $[15.0 + s_m(t)]$. This function forms the *upper envelope* of an RF signal whose frequency is 100 MHz. The lower envelope of this RF signal is formed by the negative of the PAM pulse-train function. Find an expression which gives the frequencies and amplitudes of the sideband components of this modulated RF signal.

4-17. A signal has frequency components from 300 Hz to 1.8 KHz. What is the *minimum* rate at which this signal should be sampled to generate a PCM signal?

4-18. A digital-data signal is an eight-level nonreturn-to-zero–type signal with 300 signal transitions (changes in level) per second.
   a. What is the *digital-data-bit rate* of this signal?
   b. What is its *Baud rate*?
   c. If this were a *binary* (instead of an eight-level) signal, what would be its *digital-data-bit rate* and *Baud rate*?
   d. What transmission channel bandwidth would be needed for the transmission of the binary signal in (c)?

4-19. An NRZ bipolar binary signal is used to FSK modulate a radio transmitter such that the transmitted frequencies are 20.0004 MHz when the binary signal is positive and 19.9996 MHz when it is negative. Assuming a sequence of alternating 1's and 0's, find the frequencies of the components in the FSK output if
   a. The FSK output is generated by switching between two free-running RF oscillators at 20.0004 and 19.9996 MHz.
   b. The FSK output is generated by the frequency modulation of a single 20.0000-MHz oscillator with $(\Delta f)_p = 400$ Hz.

4-20. Explain the difference between *continuous-phase* FSK and *noncontinuous-phase* FSK (i.e., draw block diagrams of the systems needed to generate each and sketch the frequency spectrums). *Why* do the two methods produce spectral components at different frequencies *in most cases*? Why might they produce components at the *same* frequencies in some cases (which cases)?

4-21. An analog telephone signal ($f_{max} = 3$ KHz) is to be converted into an 8-bit PCM signal.
   a. What is the *minimum* bit rate which would be needed?
   b. When the PCM signal is converted back into an analog signal at the receiving end, what will be the *S/N* ratio of this signal?
   c. If four-phase PSK is used to encode the PCM signal for transmission, what is the *Baud rate* of the transmitted data signal?

4-22. The 16-level (4-bit) QAM *constellation* shown here has 12 different phases and 3 different amplitudes. Each of the 16 different digital "levels" is designated by its 4-bit binary code.
   a. Assuming that the R axis shown on the drawing represents the carrier-oscillator phase and that the constellation is produced by quadrature modulation where $v_A(t)$ produces the modulation in phase with the carrier oscillator and $v_B(t)$ produces the quadrature-phase modulation, find the values for v_A and v_B needed to

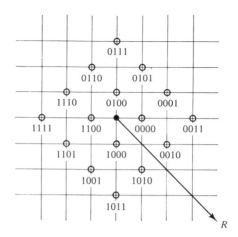

give each of the 16 different digital combinations if the $(0000)_2$ combination is represented by $v_A = +0.5$ volt and $v_B = +0.5$ volt.

b. Using voltage-comparator devices and the appropriate digital logic circuits, design a decoder that produces a parallel 4-bit binary digital signal from the two analog signals v_A and v_B.

4.23. A sequential multiplexor is *strapped* as shown to accept four analog input data signals. Each time the multiplexor receives a clock signal, it switches to next channel (or back to 1 if it is on 6). Let $A = 5 \cos (4000t)$, $B = 8 \cos (3800t)$, $C = 3 \cos (2200t)$, and $D = 6 \cos (1700t)$.

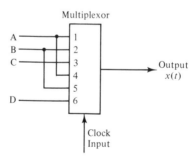

a. What minimum clock-pulse rate can be used (clock pulses per second)?

b. *If* the clock-pulse rate is set at 6000 pulses per second, what frequencies would be present in the multiplexor output?

c. Using 6000 clock pulses per second and letting input 1 be connected from $t = 0$ to $t = 1/6000$, input 2 connected from $t = 1/6000$ to $2/6000$, and so on, write an expression for the multiplexor output signal $x(t)$.

4-24. An FM system has $(\Delta f)_p = 10$ KHz and a maximum audio frequency of 4 KHz. Compare the postdetection audio S/N ratio of this system with that of an SSB system operating at the same average RF output power, the same predetection noise temperature, and so on.

4-25. In an FM system the predetection signal power is 1 mW, and the noise power spectral density 3×10^{-10} Watt/Hz. The peak deviation is 45 KHz, and the maximum audio frequency is 10 KHz. Find the postdetection audio S/N ratio.

4-26. The audio-modulating signal applied to AM and FM transmitters having the same total average RF output powers is given by

$$s_m(t) = A[\cos (6.28 \times 5 \times 10^3 t) + \cos (6.28 \times 11 \times 10^3 t)]$$

Let the maximum audio frequency which could be expected in either system be 15 KHz, the RF passband in the AM receiver be 30 KHz, and the RF passband in the FM receiver be 210 KHz. Find the relative improvement in the postdetection audio S/N ratios (FM versus AM) for the following two cases:

a. $(\Delta f)_{p, \text{FM}} = 75$ KHz, 100% AM modulation.
b. $(\Delta f)_{p, \text{FM}} = 37.5$ KHz, 50% AM modulation.

5

COMMUNICATION SYSTEMS CIRCUITS AND DEVICES

In this chapter we will be taking a closer look at some of the "building blocks" that make up the different types of communication systems. Some of this material will be a review of specific topics that should be covered in courses on electrical and electronic circuits and devices, so our coverage of it here will not be in depth, but merely to refresh the student's memory or to bring out specific details or characteristics with which we shall be concerned.

5.1 Electronic Amplifiers Used in Communication Systems

In this section we shall be taking a look at some of the characteristics of the various types of amplifier circuits used at the different points in typical communication systems. We shall begin by reviewing the various amplifier classes and their characteristics, then look at how we can control or reduce the adulteration of a signal as it is amplified, and finally consider some of the special-purpose amplifiers which can be made to perform some of the specific tasks required in communication systems.

5.1.1 Amplifier Classes

The most popular way to classify electronic amplifiers has been by dividing them into several categories dependent upon the quiescent point (also called the *Q point* or the *operating point*) of the active device(s) in the amplifier. This results in three major divisions: class A, class B, and class C, with the operation of many amplifiers falling part way between A and B into two additional transitional classes, AB_1 and AB_2. We shall discuss these very briefly. The class A amplifier has its active device operated at approximately the midrange of the linear region of its characteristic curves. This produces an amplifier which ideally has the most linear output-signal versus input-signal characteristics, especially when the level of the signal to be amplified is rather small. However, it is a very inefficient amplifier because it requires approximately the same amount of d.c. power input whether it is amplifying a small signal, a large signal, or no signal at all. Even when it is amplifying the largest signal which it can normally handle without excessive nonlinear distortion, its efficiency is usually less than about 30%. When it is amplifying a very small signal, its efficiency approaches zero. Because of its low efficiency, it is usually not used in applications where a large amount of output signal power is required.

The class B amplifier theoretically operates with its active device biased just at the *cutoff* point (where no device current flows when there is no input signal being amplified). Thus, this type of amplifier would only amplify the positive side of a bipolar (+ and −) signal if only one device were to be employed. This would have the effect of rectifying the signal to be amplified and thus would not be suitable in *most* applications.

To get around this problem, two class B–operated devices are employed in most class B amplifier applications, with one device amplifying the positive half-cycles and the other amplifying the negative half-cycles in what is known as a *push-pull* circuit arrangement. Ideally this then gives an amplifier which produces an output signal which is linearly related to the input signal and in which the d.c. input power to the amplifier is proportional to the output signal power (and also, of course, to the input signal power). The efficiency of a pure class B amplifier may be as high as 60 or 65% and is constant (i.e., it does not decrease as the level of the signal to be amplified decreases). The main problem with the pure class B mode of operation is that most electronic devices exhibit a fair amount of characteristic-curve nonlinearity as the cutoff point is approached. This produces what is known as *crossover distortion*. To get around this problem, a little bit of the efficiency of the class B amplifier is sacrificed, and the active device is biased just slightly away from the actual cutoff point. If the two devices employed have virtually the same cutoff region characteristics, then the crossover distortion is almost completely eliminated.

The third major type of amplifier class is the class C amplifier in which the active device is biased well into the cutoff region. Thus, when a small signal is applied, the device will not conduct current at all, and no output will be produced. Only when the peak amplitude of the signal is greater than the value of the beyond-cutoff bias will any output signal be produced at all and what is produced (for a sinusoidal input signal) is a series of peaked current pulses such as are shown in Figure 5-1. If these are then used to excite a high-Q–tuned resonant circuit, it is possible to obtain a nearly sinusoidal voltage waveform across the resonant circuit. This is nearly always the way in which a class C amplifier is operated. It is almost exclusively used as an RF amplifier with a constant (or nearly constant) RF input *drive* signal and a tuned resonant circuit on its output. Its output signal voltage is *not* proportional to its RF input signal, but it *is* proportional to its d.c. supply voltage. This makes it useful as a device to produce AM modulation since by varying the d.c. input voltage proportional to $s_m(t)$ plus a constant, the RF output voltage will be proportional to $s_m(t)$ plus a constant.

The other advantage of the class C amplifier is its higher efficiency (about 65 to 75% in some cases). Although it cannot be used to amplify signals which are already amplitude modulated, it can be used to amplify angle-modulated signals (which have a constant amplitude); FSK, PSK, or QSK signals; or binary ON/OFF signals (referred to as constant-wave or *CW* signals).

Another amplifier class is the *class D* amplifier. In this case the classification does not refer to where the active device's operating point is located, but rather

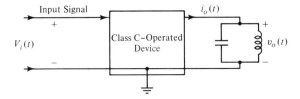

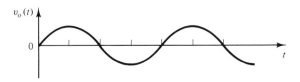

Figure 5-1   Waveforms of a class-C RF amplifier.

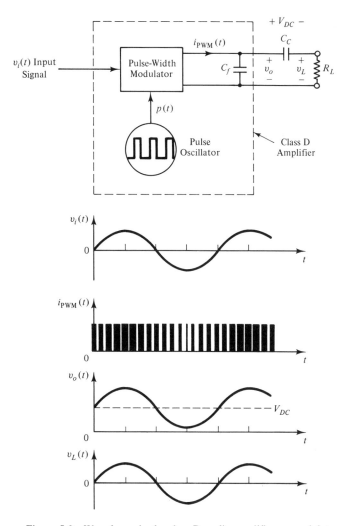

Figure 5-2   Waveforms in the class D audio amplifier or modulator.

to the mode of operation of the amplifier. Essentially, the amplifier consists of nothing more than a pulse-width-modulation generator whose pulse-repetition frequency is much higher than the highest frequency of the baseband signal to be amplified and whose output signal is fed into a capacitor (C_f in Figure 5-2) in parallel with the load for which the signal (usually an audio signal) is intended, where the R-C time constant $\tau = C_f R_L$ of the capacitor/load combination is long compared to the pulse period but short compared to the period of the highest frequency to be amplified. It can be shown under these conditions that the voltage across the capacitor/load combination will be proportional to $s_m(t)$ plus a constant (the constant can be eliminated via the use of a larger *coupling* capacitor, C_c, in series with the load as is shown in Figure 5-2). This type of amplifier is very efficient in its operation since the active device(s) do not operate in the normal (heat-dissipating) region of operation but instead operate alternately in the cutoff and saturation regions, with only very brief transitional passages through the normal region of operation when switching from cutoff to saturation. One of the main applications of the class D amplifier today is for the high-power audio-modulation amplifier in an AM radio transmitter. In this application, the *coupling* capacitor is not needed since the "constant" term is desired for input to the power-supply terminal of the class C RF amplifier to be modulated. The use of the class D amplifier as an AM modulation amplifier eliminates the need for the large, heavy, expensive audio-modulation transformer mentioned in Section 5.3.1. It should also be noted that the class D amplifier closely resembles what are known as *switcher-* or *pulser-*regulated power supplies.

5.1.2 Reducing Distortion and Noise in Amplifiers

In Chapter 2 we saw how important it is that all Fourier components of a complex signal be amplified or attenuated and delayed by the same amount if the output signal is to a faithful reproduction of the input signal. How we obtain this in practical devices is discussed in Section 5.1.2.1. Also, we saw that a signal could be "distorted" by the addition of noise to it, and in Chapter 3 we discussed some of the sources and effects of noise. In Section 5.1.2.2 we discuss some of the ways in which amplifiers can be built to reduce greatly the amount of excess noise which they generate internally. Finally, we mentioned that nonlinearities in the system could give rise to nonlinear distortion, producing the effects that are commonly known as harmonic and intermodulation distortion. Section 5.1.2.3 discusses some of the ways of reducing nonlinear distortion.

5.1.2.1 Amplitude-frequency and phase-frequency distortion considerations.   In a communication system one typically encounters two basic types of amplifiers: those designed to amplify signals in a narrow range of frequencies (small bandwidth) about some *center frequency* and those designed to amplify signals over a range of frequencies where the maximum frequency (f_H) is many times larger than the minimum frequency (f_L). The first type is usually referred to as a *passband* amplifier and quite often involves either tuned circuits consisting of inductance and capacitance or frequency-selective R-C feedback networks to select the range of frequencies to be amplified, while the latter type of amplifier is usually referred to as a *wideband* amplifier and involves techniques used to extend the range of frequencies over which amplification can take place.

5.1.2.1.1 Passband Amplifiers.   Very often these amplifiers are designed to operate at RF frequencies and involve L-C–tuned RF resonant circuits. In some cases the tuning of the resonant circuit is adjustable in the normal course of operation of the equipment; in other cases the tuning adjustments are made only at the time of manufacture or during maintenance operations. The tuned circuits involved in the RF amplifier, mixer, and local oscillator stages of a superheterodyne receiver are good examples of the situation where tuning is adjusted in the normal

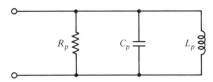

Figure 5-3 A parallel resonant circuit.

course of operation. Of importance here are two considerations: how large the bandpass ($f_H - f_L$) of the circuit is and how it and the center resonant frequency of the circuit vary with the means by which the circuit is tuned. For a parallel tuned circuit such as is shown in Figure 5-3 with the circuit losses and loading represented by a parallel resistance R_p, the circuit characteristics are given as follows:

$$f_c = \text{center resonant frequency} = \frac{1}{2\,\pi\,\sqrt{L_p\,C_p}} \tag{5-1}$$

$$B = \Delta f = \frac{f_c}{Q} = \frac{f_c}{R_p/\omega L_p} = \frac{1}{2\,\pi\,R_p\,C_p} \tag{5-2}$$

where B is defined as the 3-dB bandwidth or the difference in frequency between the lower and upper half-power points.

Sometimes the circuit of Figure 5-3 is a good model of the situation which actually exists for a single-tuned L-C-tuned circuit with high unloaded Q connected in parallel with other circuit elements which form the resistive component R_p (see Figure 5-4). However, in the case where mutual inductance coupling is employed, at least a part of the resistive portion of the circuitry may be modeled better by a resistor in series with the inductive branch such as is shown in Figure 5-5a. This may be converted to the equivalent parallel circuit of 5-5b by the relationships

$$R'_p = R_s \left(1 + \frac{\omega^2\,L_s^2}{R_s^2}\right) \tag{5-3a}$$

which gives, at resonance,

$$R'_p = R_s\,(1 + Q_s^2) \tag{5-3b}$$

and

$$L'_p = L_s \left[1 + \frac{R_s^2}{\omega^2 L_s^2}\right] \tag{5-4a}$$

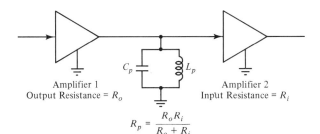

Amplifier 1
Output Resistance = R_o

Amplifier 2
Input Resistance = R_i

$$R_p = \frac{R_o R_i}{R_o + R_i}$$

Figure 5-4 The use of a single-tuned parallel resonant circuit between two stages of RF amplification.

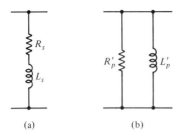

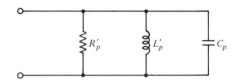

Figure 5-5   A series inductive circuit and its equivalent parallel circuit.

Figure 5-6   The equivalent parallel resonant circuit.

which gives, at resonance,

$$L_p' = L_s \left(1 + \frac{1}{Q_s^2}\right) \tag{5-4b}$$

For a high value of Q_s, we have

$$R_p \approx Q_s^2 R_s = \frac{\omega^2 L_s^2}{R_s} \tag{5-5a}$$

and

$$L_p \approx L_s \tag{5-5b}$$

When Q_s is not much greater than unity, we have an interesting situation in which, since R_p' and L_p' depends upon the frequency (see Equations 5-3a and 5-4a) the magnitude of the impedance of the equivalent parallel resonance circuit (Figure 5-6) reaches a maximum at a frequency greater than the frequency at which the phase angle of the impedance becomes equal to zero.[1]   Thus, in effect, we have two "resonance" frequencies, one defined with the magnitude of the impedance as the main consideration and the other with the impedance's phase angle as the main consideration.   It is interesting to note that *both* these resonance frequencies fall below the frequency defined by $(2 \pi \sqrt{L_s C_p})^{-1}$.   For Q values of less than 1.0, the zero-phase-angle resonance frequency becomes zero.   However, for Q values greater than about 2.0, there is very little difference between the impedance-magnitude resonance frequency and $(2 \pi \sqrt{L_s C_p})^{-1}$, and for Q values of about 10 or larger the zero-phase-angle resonance frequency also approaches this value.

As mentioned previously, the use of a resistor in series with the inductor quite often occurs in the case of inductive coupling.   Consider the situation where the resonant-circuit inductor consisting of L_1 and R_1 in Figure 5-7 is inductively coupled to an untuned secondary circuit consisting of L_2 and R_2 where R_2 comprises both the losses in L_2 and some resistive load which is being driven.   Applying normal phasor-circuit-analysis methods, we obtain for the equivalent circuit at the

[1] See Frederick E. Terman, *Radio Engineer's Handbook* (New York: McGraw-Hill, 1943), p. 147.

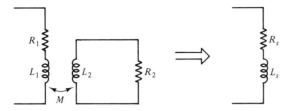

Figure 5-7   An inductively coupled circuit and its series-circuit equivalent.

right side of Figure 5-7

$$R_s = R_1 + \frac{\omega^2 M^2 R_2}{[R_2^2 + (\omega^2 L_2^2)]} \tag{5-6a}$$

$$L_s = L_1 - \frac{\omega^2 M^2 L_2}{[R_2^2 + (\omega^2 L_2^2)]} \tag{5-6b}$$

When the loaded Q is large, these become approximately

$$R_s = R_1 + \frac{\omega^2 M^2}{R_2} \tag{5-7a}$$

$$L_s = L_1 - \left[\frac{\omega M}{R_2}\right]^2 L_2 \tag{5-7b}$$

Another commonly encountered tuned-circuit configuration is the double-tuned circuit between two amplifier (or other) stages of a system.   Such a configuration is shown in Figure 5-8, where R_a is approximately the output resistance of the first amplifier stage and R_b is approximately the input resistance of the second amplifier stage.   For simplicity assume that $R_a = R_b$, $C_a = C_b$, $L_a = L_b$, that the L-C combination used gives the desired resonant frequency, and that the Q of the circuits is fairly large.   Then, using the standard definition for the coupling coefficient

$$k = \frac{M}{\sqrt{L_a L_b}} \tag{5-8}$$

we obtain curves for the voltage ratio $H = V_b/V_a$ that depend upon the value of k (and hence upon M) as shown in Figure 5-9 where k_c is called the critical-coupling

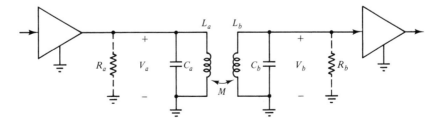

Figure 5-8   The use of doubly tuned resonant circuits between RF amplifiers.

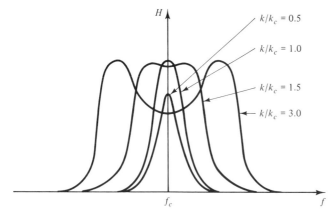

Figure 5-9 The response curves of doubly tuned resonant circuits as a function of the coupling coefficient.

value of k and is defined as $k_c = (Q_p \, Q_s)^{-1/2}$, where $Q_p = Q_s = (R_a/\omega \, L_a) = (R_b/\omega \, L_b)$.

As can be seen from Figure 5-9, going to slightly more than critical coupling produces a response with a broad top and steep side slopes. Wider bandwidths may be obtained with a cascade of two or more tuned-circuit/amplifier combinations. For example, cascading a circuit where $k/k_c > 1.0$ with one where $k/k_c = 1.0$ gives an overall response like that indicated by the solid line in Figure 5-10 when the ratio functions of the two curves are *multiplied* together and are normalized to 1.0 at f_c. Another technique for obtaining wide passbands is to employ *stagger tuning* in which one pair of resonant circuits is tuned to one frequency, the next to a slightly different frequency, and so on. Standard tables[2] exist for doing this for cascades consisting of two, three, four, and so on single-tuned or double-tuned circuits with different coupling coefficients and/or Q factors. Stagger tuning is also used to produce special shapes of bandpass-response curves. An example of this is the IF response curve which is required in a TV receiver which is receiving the standard NTSC vestigial-sideband composite video signal (see Chapter 7).

One of the considerations of using tuned circuits is the method by which the tuning adjustment may be accomplished. Historically, one of the first techniques was to construct the inductor so that a portion of the inductor's windings were movable and could be rotated with respect to the remainder of the windings, thus varying the coupling between the two sections. When the two sections were connected in series, the total inductance was then $(L_1 + L_2 + k\sqrt{L_1 \, L_2})$, where k could vary from -1 to $+1$. The second method which was used employed a coil with multiples taps or a sliding contact touching the coil windings. Many of the early radio sets were made using one of these two types of variable-inductance tuning. A little later the mechanically variable capacitor was introduced and became very popular. One of its big advantages was that it could be designed to have a nonlinear C versus shaft position function. This permitted, by using different C versus θ functions for the RF and LO tuning capacitors, the implementation

[2] Donald G. Fink, ed., *Electronic Engineers' Handbook* (New York: McGraw-Hill, 1975), Sections 7 and 13.

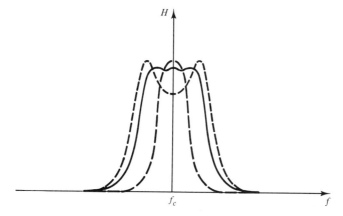

Figure 5-10   The total passband characteristic from combining two doubly tuned circuits in cascade with $k/k_c > 1$ and $k/k_c = 1$.

of almost perfect superheterodyne tracking.   In more recent years, since the development of the ferrite-ceramic materials which have high permeability and low losses at RF frequencies up into the VHF region, tuning has again been accomplished by varying the inductor component of the resonant circuit by providing a coil with a movable ferrite slug.   Slug-tuned resonant circuits have been very popular in IF transformer designs and have been used to some degree in the front-end circuits of superheterodyne receivers.   Again, the tuning function can be made a nonlinear function of the slug position by designing coils wherein the spacing of the turns of the winding are nonuniform along the length of the coil.

One of the more recent developments in tuning circuits has been the introduction of variable-voltage tuning, using a capacitor whose capacitance may be adjusted by means of the bias voltage applied to it (VARICAPS or diodes biased as discussed in Section 5.3.1.3).   This type of tuning has been very popular in systems where the tuning is controlled by some sort of computer or memory device (communication scanning receivers, etc.).

5.1.2.1.2 Broad-band RF Amplifiers.

The RF amplifiers discussed earlier in this chapter all employed some type of sharply tuned RF resonance circuit and hence had to be adjusted each time the frequency of operation was changed. However, the use of moderate-power-level ferrite-core transformers makes possible the design of an RF amplifier using either class A or B operation of the active device which will amplify over a very wide range of RF frequencies.   It is interesting to note that this technique cannot be applied to the class C RF amplifier because the latter depends upon the "flywheel" effect of the RF tuned circuit to form the sinusoidal output from what essentially amounts to pulse-type operation of the active device.

At the heart of the broadband RF amplifier is the ferrite RF transformer, and before discussing the amplifier circuitry, we will examine the construction of these devices.   One type of construction which is used is based upon a toroidal (doughnut-shaped) ferrite core.   This is wound simultaneously with two wires (for a 1:1 ratio) which are twisted together prior to winding them on the core and which form the primary and secondary windings of the transformer.   Figure 5-11 shows

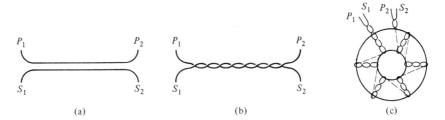

Figure 5-11   The construction of a bifilar-wound toroidal transformer with a 1:1 turns ratio.

the steps in the construction of a 1:1 transformer of this type and Figure 5-12 the construction of a 1:3 ratio transformer. The winding of a transformer in this manner (simultaneously winding the primary and secondary with twisted wires) is called *bifilar* winding. The use of a toroidal core and bifilar windings produces a transformer with low leakage (equivalent series) inductance and a good distribution of interwinding capacitance. The main disadvantage of the bifilar construction is that the turns ratio is limited to the ratio of small integer numbers (1:1, 1:2, 1:3, 1:4, 1:5, 2:3, 2:5, etc.).

Another type of ferrite-core construction is the *pot core* (sometimes called a *cup core*), which is shown in Figure 5-13. It consists of two cuplike ceramic shells, each with a ceramic center post. The windings are wound on a plastic or nylon bobbin which is then inserted into one of the core halves. The other half is then placed on the portion of the bobbin sticking out of the first half and the whole assembly bolted together with a machine screw which may also be used to hold the assembly to a printed circuit board or other mounting. The windings are completely enclosed and shielded by the core cup except for small openings where the winding leads are brought out. The pot core has two advantages over the toroid: almost complete shielding of the windings and ease of construction. Pot core transformers and inductors are used extensively at audio frequencies but may also be constructed for use in the RF region.

Figure 5-14 illustrates the use of ferrite-core transformers of the type just described in a broad-band push-pull class B amplifier. Two different types of transformers are illustrated here. A conventional split-secondary ferrite transformer (bifilar wound) is shown in the input circuit (T_{in}), and a similar, but larger, tapped-primary transformer could have been used on the output. However, the

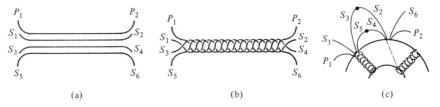

Figure 5-12   The construction of a bifilar-wound toroidal transformer with a 1:3 turns ratio.

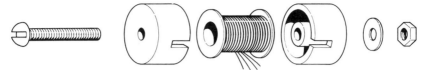

Figure 5-13   The assembly of a pot core transformer.

output circuit illustrates another method which can be used in place of the tapped-winding transformer.   This alternate method consists of using one 1:1 transformer (T1) to ensure that the collector voltage variation of both transistors is sinusoidal even though each transistor conducts only during one half-cycle of the RF signal. Transformer T2 is used to combine the power output of the two transistors, giving an output voltage equal to twice the a.c. voltage across one transistor's collector. The push-pull mode of operation causes at least partial cancellation of the even harmonics in the output, although the usual practice is to follow a broad-band amplifier with some sort of low-pass filter which will reject any harmonics appearing at the amplifier output.   In a broad-band amplifier designed to operate over a wide range of RF frequencies the base drive to the transistors is usually much greater at the low end of the range than at the high-frequency end, and to even this out over the band, some sort of compensation network is usually employed. Figure 5-14 illustrates two different schemes: a shunting network consisting of L_s and R_s across the input and frequency-dependent negative feedback networks consisting of L_f and R_f from the collector to the base of each transistor (the capacitor C_c is merely a d.c. blocking capacitor).

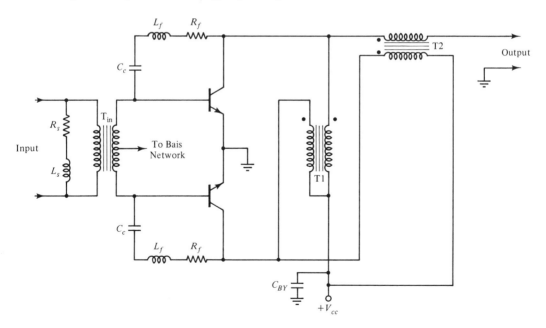

Figure 5-14   A broad-banded class B RF linear amplifier.

5.1.2.1.3 Wide-Band Baseband Amplifiers.   In the audio and video regions we are concerned with amplifiers which should have uniform gain and time-delay characteristics from very low (perhaps even to 0 Hz or d.c.) frequencies up to some maximum frequency f_H.[3] In a transformer or R-C–coupled amplifier, good low-frequency response implies either a large flux capability in the transformer's core or very large coupling capacitors. Either of these will add greatly to the cost, size, and weight of a piece of equipment. Consequently, the more recent trend has been to utilize direct (d.c.) coupling between amplifier stages, especially if part or all of the amplifier consists of integrated-circuit components. However, the use of d.c. coupling implies very careful circuit design to assure proper biasing of the active devices over a wide range of temperature and/or device-characteristic variations. In particular, care must be exercised in the design to prevent thermal runaway and to minimize the chances of device damage due to overloading or underloading (R_L too small or too large), the sudden application or removal of the load, to supply-voltage variations, or to too large a signal input.

At the high-frequency end of the amplifier range, the presence of shunt capacitance in the active device as well as the stray capacitance and inductance of the circuit will cause the amplifier gain to decrease and the time-delay characteristics of the amplifier to change. The frequency at which these effects begin to occur may be extended by the use of negative feedback in the amplifier (which however also lowers the midband gain obtainable from a given active device) or by the creation of a high-frequency resonant condition by adding inductances called *peaking coils* to the amplifier circuit which resonate with the device and/or circuit capacitances. Figure 5-15 shows the collector circuit of a video-amplifier stage employing peaking coils. L_1, the series peaking coil (in conjunction with R_1 which serves to lower the Q of L_1), isolates the shunt output capacitance of the transistor from the shunt capacitance, C_L, of the load being driven. L_2, the shunt peaking coil, should resonate with the shunt capacitances (primarily C_L) at some frequency near the desired extended f_H of the amplifier.

For details on the design of wide-band audio, video, or direct-coupled amplifiers, the reader should consult any good book on modern electronic circuit design.

5.1.2.2 Noise considerations in amplifiers.

When we consider the excess noise generated by amplifying devices such as transistors (either as discrete devices or as incorporated into integrated-circuits chips), vacuum tubes, and so on, we find that, in general, this noise source is not a great problem in the audio and low-RF regions, and what excess noise is produced is usually masked out by noise from other sources. When the excess noise is a factor at these frequencies, it can usually be handled through the careful selection and use of low-noise transistors and special low-noise vacuum tubes and the proper impedance matching of these devices. Most of the noise concerns in these regions comes from noise and extraneous signals (such as 60-Hz "hum," etc.) picked up prior to or within the amplifier. Some of

---

[3] See Section 2.7.

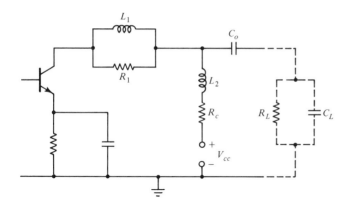

Figure 5-15 A video amplifier using series and shunt peaking coils.

this type of noise can be reduced or eliminated through the careful use of proper shielding and grounding techniques in the design of the amplifier.

In applications in the RF frequency regions below about 100 MHz, most of the noise on a received signal is due to noise picked up by the receiving antenna system itself. However, as we consider higher frequencies up into the microwave region, we find that usually the major contribution to the noise on a signal at the IF output[4] of a receiver is due to noise introduced by the first stage or two of the receiver itself. Thus, at these frequencies, it becomes desirable to go to special efforts to reduce the receiver noise contributions as much as possible. This can be done by techniques such as parametric amplifiers and cooled amplifiers (such as the maser, etc.). For example, a typical microwave receiver utilizing as its first stage nothing but a crystal mixer to mix the incoming signal with the first local oscillator signal will have a standard noise figure of about 7 dB. This means that the mixer stage will contribute about 1160° K to the noise output of the receiver (referred to the input). This is very large, considering that the noise temperature of the signal coming out of the terminals of the microwave antenna may only be on the order of about 150° to 200° K (from both noise on the signal and from losses in the antenna).

Now, if we place a maser preamplifier which is cooled with liquid helium and which has a gain of 20 dB and an effective excess noise temperature of only about 25°K between the antenna and the mixer, we find that the noise contribution of the mixer is reduced by a factor of 100 (or 20 dB). This means that now the mixer noise referred to the input of the maser preamplifier is only 11.6°K, and when this is added to the 25°K noise contribution of the maser itself, we have only 36.6°K total noise contribution from the receiver (which is much, much lower than the 1160°K we had without the maser preamplifier and is also now lower than the noise coming out of the antenna terminals). Thus, the use of a good low-noise preamplifier ahead of the mixer stage of a microwave receiver will do much to improve the noise performance of the system.

---

[4] See Section 5.3.2 for a definition of what is meant by IF (Intermediate Frequency).

Of course, the maser amplifier is an extreme example and is quite expensive to build and operate in the system. The maser at 25°K has a standard noise figure of only 0.359 dB, but for a relatively few dollars one can obtain a solid-state parametric microwave amplifier which has similar gain and a standard noise figure of 2 to 4 dB, which is still considerably less than the 7 to 8 dB of the crystal mixer. Thus, it is usually desirable to include some sort of amplifier ahead of the mixer stage of a microwave receiver to improve the noise performance of the receiving system. Further improvements in the noise performance of the receiving system can usually be accomplished by reducing the losses in the receiving antenna system. This will be taken up in Chapter 6.

5.1.2.2.1 The Parametric Amplifier. A schematic diagram of this type of low-noise amplifier is shown in Figure 5-16. It consists of a nonlinear element (such as a reverse-biased diode or a piece of yttrium-iron-garnet (YIG) material) attached across three series resonant circuits (or three sections of half-wave long transmission line or waveguide). These three sections are called the *pump* (resonant at $\omega_p = 1/\sqrt{L_pC_p}$), the *idler* (resonant at $\omega_i = 1/\sqrt{L_iC_i}$), and the *signal* (resonant at $\omega_s = 1/\sqrt{L_sC_s}$) sections of the parametric amplifier circuit, where ω_p is greater than either ω_s or ω_i and where $\omega_i = (\omega_p - \omega_s)$. When the pump portion of the circuit is excited from some signal source (oscillator) at the resonance frequency ω_p, it is found that the resistive portion of the impedance at ω_s looking into terminals A–A' is negative. Thus, if terminals A–A' were used to terminate a short length of transmission line, a signal of frequency ω_s propagating down the line and reaching this termination would be reflected back with an increased magnitude (i.e., the reflected signal would contain more power than the incident signal). As a result, amplification takes place. The only problem is that the reflected signal from the parametric amplifier must in some way be separated from the original incident (input) signal. Fortunately, this can be easily done as is discussed in the next paragraph.

However, before considering this problem, it should be pointed out that the parametric amplifier can also be operated as an amplifying frequency converter,

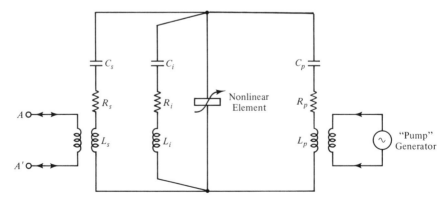

Figure 5-16 Parametric amplifier schematic.

because if a load impedance is coupled to the idler circuit, the signal applied to the $A-A'$ terminals will also appear at the idler output (converted in frequency to the idler frequency). However, since $\omega_i = (\omega_p - \omega_s)$, the sidebands on the input signal will be inverted in the idler-signal output.

The three types of low-noise RF amplifiers being considered here (the parametric amplifier, the tunnel-diode amplifier, and the MASER amplifier) are all single-ended devices. This means that they have a common input and output and that the output signal must in some way be separated from the input signal. Since all three types can be considered to be negative-resistance terminations on an RF transmission line, it is necessary to have some sort of device or devices on the single line from the amplifier which splits this line into two parts—the incoming line from the antenna or other input-signal source and the output line to the next part of the system (usually the receiver). One device which accomplishes this operation is the *circulator*. A four-port RF circulator is shown in Figure 5-17. The circulator is a device which passes a signal from one port to the next sequentially numbered port with little attenuation (less than 1 dB) but which has a high attenuation (20 to 80 dB) for passing a signal from a port to any port other than the next one ahead.

For example, almost all the signal entering at port 1 of the circulator in Figure 5-17 comes out at port 2, but very little signal from the antenna is transferred to any other port. The amplified reflected signal going back into port 2 comes out at port 3 and goes on to the receiver input. Any signal reflected from the input of the receiver goes back into port 3 and comes out at port 4 where it is absorbed by the matched load. Thus, the circulator makes an ideal device for coupling a single-ended amplifier into the transmission line between an antenna and a receiver. Other devices, such as directional couplers, uniline (one-direction) microwave devices, and so on can also be utilized.

5.1.2.2.2 The Tunnel-Diode Amplifier. This is another type of single-ended, negative-resistance, low-noise device often used as an RF amplifier. The tunnel diode (TD) has a characteristic such as shown in Figure 5-18. It is biased to operate in the portion of its characteristic where the current-voltage slope is negative (negative dynamic resistance) and the diode is used to terminate a section of

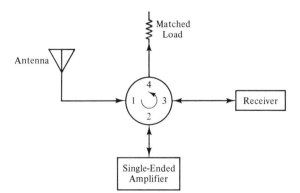

Figure 5-17 Connecting a single-ended amplifier using a circulator.

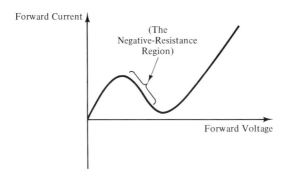

Figure 5-18 The tunnel-diode characteristic.

transmission line or waveguide. Because of its negative RF resistance, the magnitude of its reflection coefficient, like that of the parametric amplifier, is greater than unity (see Equation 5.9), and the reflected signal amplitude and power are greater than those of the incident signal.

$$|\Gamma_r| = \left| \frac{R_L - R_o}{R_L + R_o} \right| = \left| \frac{-R_{\text{neg}} - R_o}{-R_{\text{neg}} + R_o} \right| > 1 \qquad (5\text{-}9)$$

where

R_o = the characteristic impedance of the T line

R_{neg} = value of the TD's negative dynamic resistance

5.1.2.2.3 The MASER Amplifier.

The third type of low-noise, single-ended amplifier, the MASER amplifier, operates upon the principle of employing the energy absorbed and given off by an atom transitioning between different atomic energy states. Thus, its frequency of operation is strongly dependent upon which element and what energy-state transition is being used, although the frequency of the energy-state transitions can be "tuned" slightly by placing the active material in a d.c. magnetic field. The operation of the maser is very similar to the operation of the rubidium frequency standard discussed in Section 5.2.4 except that pumping is usually done at microwave rather than at optical frequencies. If the cavity in which the maser material is located has a high Q at the signal-frequency transition (see Figure 5-19), then the maser will oscillate at that frequency. However, if the cavity is properly loaded until the oscillations just cease, then any small amount of signal sent into the cavity will stimulate signal-frequency transitions and more signal-frequency power will come out of the cavity than went into it and a negative-resistance effect is observed. The maser cavity is usually designed to be resonant at both the pumping frequency and the signal frequency.

It should be noted that both the parametric amplifier and the maser amplifier require some source of RF pumping power at a frequency higher than that of the signal, while the tunnel-diode amplifier requires only a d.c. biasing circuit. For this reason the TD amplifier is commonly employed in many applications with the parametric amplifier being the next in popularity, and the maser amplifier, because of its operation at cryogenic temperatures, being used only in special-purpose

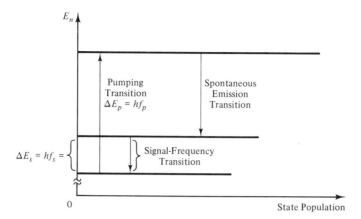

Figure 5-19   The maser energy-state diagram.

applications such as radio astronomy, space-probe communication systems, and some communication satellite receiving systems, although the tunnel diode or parametric amplifier is usually sufficient in the latter case.

5.1.2.3 Nonlinear distortion considerations.

One of the best ways to reduce nonlinear distortion in an electronic amplifier is by beginning with the correct design of the basic amplifier (i.e., selecting the proper device and operating it at an appropriate Q point for the amplitude of the signal to be amplified), having the correct driving-generator impedance and load impedance (which sometimes requires using impedance-matching transformers or networks), using appropriately sized bypass and coupling capacitors, and making sure that the amplifier is supplied from a well-regulated power supply (this is especially important for class B amplifiers where d.c. power input is dependent upon the level of the signal being amplified). Once all this has been done, other distortion-reducing techniques may be employed. One of these is the use of negative feedback to not only reduce the nonlinear distortion of the amplifier, but to also widen the amplifier's bandwidth. In the following we shall see how negative feedback can be used to reduce the nonlinear distortion products and noise that appear in an amplifier's output.

Figure 5-20 shows a basic amplifier in which the added distortion products (as well as added noise produced by or picked up within the amplifier itself) are represented by a generator time function v_{DN}, which we will assume is at a constant average output for a given signal waveform and level of desired output signal. The basic amplifier represented by the triangular amplifier symbol and having gain A_{vb} is considered to be distortionless in nature. We will add negative voltage feedback to the basic amplifier by means of a feedback network having a transfer function $\beta = v_f/v_o = -K_f$. The feedback voltage time function v_f out of this network is added to the input voltage v_i at the input summing node to produce v_1, the input to the basic amplifier. The output voltage of the amplifier with feedback is now given as

$$v_{of} = v_2 + v_{DN} = A_{vb}\, v_1 + v_{DN} = A_{vb}(v_i - K_f v_{of}) + v_{DN} \qquad (5\text{-}10)$$

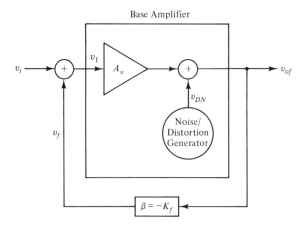

Figure 5-20 Feedback in amplifiers.

where the v_{of} at the left and the expression at the right can be combined to give

$$v_{of}(1 + A_{vb} K_f) = A_{vb}v_i + v_{DN} \qquad (5\text{-}11)$$

which then can be solved for v_{of}, giving

$$v_{of} = \left(\frac{A_{vb}}{1 + A_{vb}K_f}\right)v_i + \frac{v_{DN}}{(1 + A_{vb}K_f)} \qquad (5\text{-}12)$$

From Equation 5-12 we see that to maintain the same output level of desired signal by keeping the first term constant as the amount of negative feedback (as indicated by the magnitude of K_f) is increased, the input voltage v_i must be increased (the negative feedback reduces the overall gain of the amplifier system). However, as the negative feedback is increased, the second term in Equation 5-12, which is the distortion/noise term, gets smaller. Thus, by utilizing negative feedback, we have been able to reduce the nonlinear distortion and noise in the output signal (keeping the same amplitude of desired output), but this has been done at the expense of reducing the gain of the amplifying system. In practical system design, a trade-off must be made between the amount of nonlinear-distortion/noise reduction desired and the factors associated with providing additional gain to compensate for that lost due to including the negative feedback in the amplifying system.

5.1.3 Limiting Amplifiers, ALC, AVC, and AGC

This section discusses those types of circuits where the circuit gain, in some manner, is made dependent upon the level of the signal being processed by the circuit. For the purposes of this discussion, we will divide these circuits into four main categories:

1. *Hard limiters (or "clipper" circuits)*—These circuits essentially consist of a system wherein $v_{out}(t) = \{K\ v_{in}(t)\}$ as long as $|v_{in}(t)| \leq V_{max}$ while for $|v_{in}(t)| > V_{max}$, $v_{out}(t) = \{K\ V_{max}\}$ for $v_{in}(t) > V_{max}$ and $v_{out}(t) = \{-K\ V_{max}\}$ for $v_{in}(t) < -V_{max}$. In communication systems this type of limiter is most often

found in the RF portion of the system. Considerable generation of harmonics and intermodulation products is associated with the operation of a hard limiter.

2. *Companders*—This type of circuit is very similar to the hard-limiter in that it operates on an instantaneous basis. However, the relationship between $v_{out}(t)$ and $v_{in}(t)$ is nonlinear over the entire range of $v_{in}(t)$, although the relationship is usually symmetric with respect to the sign of $v_{in}(t)$. A compander is quite often used on an audio signal prior to digitizing it.

3. *Speech compressors, "soft" audio limiters and ALC (automatic level control) circuits*—These audio devices find use in broadcast studios, public address systems, and audio recording systems. Their primary function is to reduce the gain of the amplifying system whenever the *average* level of the input signal exceeds a certain level.

4. *AVC (automatic volume control) and AGC (automatic gain control) circuits*—An example of this type of circuitry is found in almost every radio and TV receiver. AVC and AGC are used to control the gain of the receiver RF and IF amplifier stages so as to prevent the overloading of these stages and to present a signal of the proper level to the detector stage of the receiver.

5.1.3.1 Hard limiters. In these circuits, the *instantaneous* output voltage is limited to a magnitude of $\{K\,V_{max}\}$ whenever the instantaneous magnitude of the input voltage exceeds V_{max} as shown in Figure 5-21. As long as the instantaneous value of the input never exceeds V_{max} in magnitude, the limiter behaves as a conventional amplifier with a voltage gain of K. It is when the positive and negative peaks of $v_{in}(t)$ begin to exceed V_{max} that the operation of the circuit becomes interesting.

For example, let $v_{in}(t)$ be comprised of a random noise signal plus an RF carrier which is angle modulated (FM or PM). Let the amplitude of the RF signal be adjustable from zero to some value many times V_{max}. Then, looking at the noise *power* output and the signal *power* output of the limiter as a function of the signal *power* input (proportional to the square of the RF voltage amplitude), we obtain a plot like that of Figure 5-22. Here it has been assumed that the noise input is maintained at a level just below the point where any appreciable clipping of the noise signal would occur (i.e., here we have set V_{max} equal to the 3σ limit of an assumed Gaussian noise signal). Also, Figure 5-22 is based upon the assumption that an RF resonant circuit or harmonic filter follows the output of the limiter.

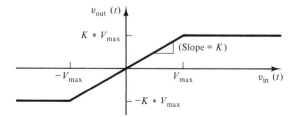

Figure 5-21 The instantaneous transfer characteristics of a hard limiter.

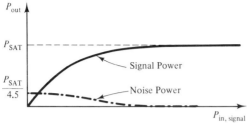

Figure 5-22 The suppression of a noise signal by a hard limiter.

From Figure 5-22 we see that the presence of the signal in the limiter reduces the noise output power and that above a certain threshold level the output *S/N* ratio is considerably greater than the input *S/N* ratio. This becomes quite important in an FM repeater system or an FM receiver. (*Note*: Even though the noise *power,* based upon *amplitude* variations caused by the noise, approaches zero for large signal inputs, the effects of the noise upon the zero crossings of the signal-plus-noise input prior to the limiting process still causes a certain amount of *phase noise* to appear on the limiter output signal—unwanted PM modulation.) For an FM broadcast receiver, one figure-of-merit that is used as a measure of the receiver performance is the *dB of quieting* (dB of reduction in the noise power) of the audio (postdetection) noise as a result of applying an unmodulated RF carrier signal of specified level to the receiver input.

Another interesting situation occurs when two RF signals within the RF passband of the limiter are being considered. As shown by Figure 5-23 when one signal is much stronger than the other at the input (by about 6 dB), practically all the limiter output power will go to the stronger signal. This phenomenon is known as the "capture" of the limiter amplifier and is important in the reduction of the effects of interfering signals in angle-modulation systems. Hard limiters find use at a number of points in communication systems, and some of these applications will be discussed further in the later chapters of this text.

5.1.3.2 Audio companders.
Like the hard limiter, the audio compander is an instantaneous type of limiting device in which $v_{out}(t)$ is some nonlinear function of $v_{in}(t)$. The relationship may be similar to that shown in Figure 5-24. Again, because of the nonlinear characteristics, there will be some degree of harmonic and intermodulation distortion introduced, but in many cases (telephone conversations for example), this may be tolerated to gain the advantages that companding

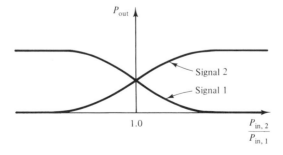

Figure 5-23 The signal/signal capture characteristics of a limiter amplifier.

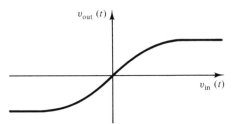

Figure 5-24 The instantaneous transfer characteristics of an audio compander.

can give.   A typical application of a compander is in a PCM system wherein it is desired to digitize a voice signal.   Since the response of the ear to audio levels is logarithmic rather than linear, it would be desirable to space the digital slicing levels on a logarithmic rather than a linear basis.   This could be done by the design of a logarithmic analog-to-digital (A/D) converter circuit, or it could be done by first preprocessing the audio signal with a compander whose output is proportional to the logarithm of the input and then using a standard linear A/D converter on the output of the compander.   This latter process is often the one that is done.

Figure 5-25 shows a PCM digitizing system of this type.   The 3-KHz LPF eliminates any of the compander's harmonic or intermodulation products which fall above 3 KHz.

5.1.3.3 Audio "soft" limiters.   These limiters are really gain-controlled amplifiers in which the amplifier gain is a function of the *average* signal level.   In many cases the limiting function may not take place until the average power level of the signal has exceeded some rather high level for some period of time (anywhere from a few milliseconds to a second or more).   Also, the gain of the amplifier may not be reduced abruptly, but may be brought down at some exponentially decreasing rate associated with a specific time constant.   This time constant, often referred to as the *attack time* of the limiting action, may be front-panel adjustable.   Similarly, after the large input signal has disappeared, there may be another time constant (the limiter's *release time*) associated with the rate at which the amplifier gain returns to the normal nonlimited level.   This, as well as the signal level at which

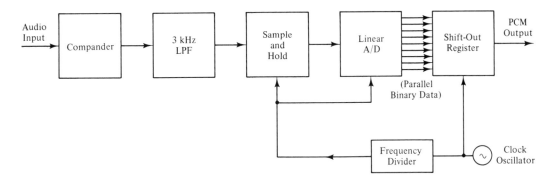

Figure 5-25 The use of an audio compander in a PCM telephone system.

the limiting action begins to take place (the *limiting* or *compression threshold*), may also be front-panel adjustable.

In some applications audio limiters or compressors are included primarily as protection devices to prevent overloading an audio power amplifier or overmodulating a broadcast AM or FM radio transmitter. In these applications, the limiting action takes place only under unusual conditions. In other applications such as communication-radio systems, the limiting action may be employed on a more regular basis (i.e., at a lower threshold) as a means of increasing the average signal level applied to the modulation system. This low-threshold limiting action is also quite often included (sometimes not even on an optional basis) in audio tape recorders. In this application the release time of the limiting action may be made quite long (a second or more) and the limiting threshold quite low, giving an effect in which the recording gain level, and consequently the background noise, may increase quite drastically a fraction of a second or so after a speaker stops talking or during a soft passage of music. This type of ALC action can become quite annoying on a recorded message and of course completely destroys the dynamic range of a piece of music.

5.1.3.4 AVC and AGC.

Automatic volume control and *automatic gain control* are two terms used to describe the same basic function (the control of the gain of a receiver's RF and/or IF amplifiers by the level of the received signal) although the first name is usually applied to the case of audio AM or FM receivers. AVC in vacuum-tube–type receivers consisted of obtaining a negative d.c. voltage proportional to the signal carrier level from the envelope detector circuit and applying it to the grids of the IF (as well as perhaps the RF and converter) stage tubes. By using "variable-mu" tubes in these stages (tubes whose amplification factor μ decreased with increasing grid bias voltage), the gain of these stages could be made to decrease as the magnitude of the negative AVC voltage increased. In transistor amplifiers, the h_{fe} of the transistor is usually dependent upon the amount of forward emitter-base bias current and hence upon the amount of forward-bias emitter-base voltage. By decreasing this voltage by making the base more positive (for PNP transistors) or more negative (for NPN transistors), the gain of the stage can be reduced. However, as this is done, the operating point of the transistor is moved into a very nonlinear region and nonlinear distortion tends to begin occurring, and this is one factor which must be taken into account when there is the possibility of receiving very strong signals (some receivers incorporate a manually operated attenuator between the receiver antenna terminals and the RF amplifier input). When receiving weak and moderate-level signals it is usually desired to have little or no AVC applied to the RF amplifier stage to have high gain at this point and a good overall receiver noise figure (see Equations 3-47 and 3-48). If AVC is applied to this stage at all, it is only on very strong signals. The applying of AVC only to the later stages in a receiver and little or none to the first RF amplifier is called *delayed AVC*. (Note: The term *delayed* here has nothing to do with *time delay*.)

If the AVC voltage were taken directly from the envelope detector diode in an AM receiver, the AVC would follow the AM modulation. Thus, it is necessary to pass the AVC signal through a low-pass filter whose cutoff frequency is less than the lowest frequency of modulation. However, if the filter cutoff frequency is made too low, the AVC action will not be responsive to sudden changes in the signal level. In some receivers the cutoff frequency (R-C time constant) of the low-pass filter is front-panel adjustable.

For other than audio AM or FM receivers, the term AGC is usually used, although this is sometimes also applied to audio receivers as well. In an audio AM or FM receiver, the AVC (or AGC) control voltage is proportional to the average received carrier level at the detector and can easily be obtained directly from a diode envelope detector. However, in receivers designed to receive other types of modulation, obtaining a desirable AGC control signal is not as easy. For example, in a TV receiver the average RF signal level depends upon the picture brightness, and we do not want the AGC control voltage to be dependent upon the picture brightness, so a diode detector followed by a low-pass filter would not be acceptable. However, the TV signal peak amplitude (the synchronizing-pulse level) is not picture dependent, and this is what we would like to have as the basis for the AGC control signal level. Thus, we would want to use a *peak* envelope detector (rather than an *averaging* envelope detector as in the case of the audio receiver) for obtaining the AGC in a TV set. Similarly, when receiving a DSBSC or SSB signal, there is no carrier on which to base the AGC level. What is often done here is to average (with some time constant) the *rectified* detected *audio* signal and base the AGC voltage upon this. Usually the averaging time constant is selectable to allow the operator to obtain the best AGC action for the characteristics of the information signal being received.

In most instances, adequate AGC or AVC can be obtained directly from a diode rectifier circuit, but in some cases, it may be desirable to incorporate a d.c. amplifier into the AGC control loop. Quite often an operator-adjustable AGC loop gain control is also included, and many times a signal-strength meter is made a part of the AGC or AVC system. This type of meter is useful not only for indicating the strength of the received signal, but also for tuning the receiver more precisely (although tuning meters for FM receivers are usually based upon the AFC (automatic frequency control) loop rather than upon the AGC or AVC loop).

5.1.4 Constant-Output-Voltage Amplifiers

Another effect of using negative feedback is effectively to reduce the output impedance of an amplifier and produce what is known as a *constant-voltage* amplifier. An explanation of how this occurs can be based upon the diagram in Figure 5-26 where the Thevenin output voltage is equal to $A_{vb}v_1$, where $v_1 = v_{\text{in}} + v_f = v_{\text{in}} - K_f v_{\text{out}}$ as in the case discussed in the previous section.

Now, writing the expression for the output time function v_{out} we get

$$v_{\text{out}} = v_{\text{th}} - R_0 i_{\text{L}} = A_{vb}v_1 - R_o(v_{\text{out}}/R_{\text{L}})$$
$$= A_{vb}(v_{\text{in}} - K_f v_{\text{out}}) - (R_o/R_{\text{L}})v_{\text{out}}$$

(5-13)

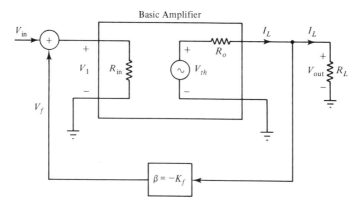

Figure 5-26 Use of feedback to reduce an amplifier's effective output resistance.

which when solved for v_{out} gives

$$v_{\text{out}} = \frac{A_{vb}}{[1 + K_f A_{vb} + (R_o/R_L)]}\, v_{\text{in}} \qquad (5\text{-}14)$$

Here we notice that when the middle term in the denominator is small, v_{out} is greatly dependent upon the ratio of R_o to R_L and becomes smaller as R_L becomes smaller.

To make the output voltage relatively independent of R_L, the second term $(K_f A_{vb})$ should be much larger than the maximum expected value of R_o/R_L. There are many cases where we might want to make the effective output impedance of an amplifier lower than it would normally be. One of these cases involves a public address system installed in a building such as a school or other place where it might be desired to feed the signal from the PA system output to one or two loudspeakers or to a large number of loudspeakers. The PA systems normally used in these applications are constant-voltage types employing this negative-feedback principle. They are designed so that with no load attached and a maximum-level input signal applied, the rms output voltage is either 25 or 70.7 volts (depending upon the model of the system used). Figure 5-27a shows a typical "70-volt" system installation. The loudspeaker loads are connected to the 70.7-volt rms line via matching transformers of the type shown in Figure 5-27b. The secondary of most such transformers have several taps to enable the transformer to be used with loudspeakers of different voice-coil impedance ratings, while the primary side usually has several taps which adjust the effective load impedance presented to the line and thus determine how much power each loudspeaker will draw from the line. This allows the sound level of each speaker to be adjusted to compensate for the size of the room and the background noise of the location where it is located. The constant-voltage amplifier will approximately maintain a constant 70.7-Volt rms (with the maximum specified input signal applied) until the load attached to the line is absorbing the amount of audio output power for which the amplifier is rated. This occurs at an effective parallel resistance value given by $5000/P_{\text{rated}}$ for a "70-Volt" system or $625/P_{\text{rated}}$ for a "25-volt" system. A good system, properly in-

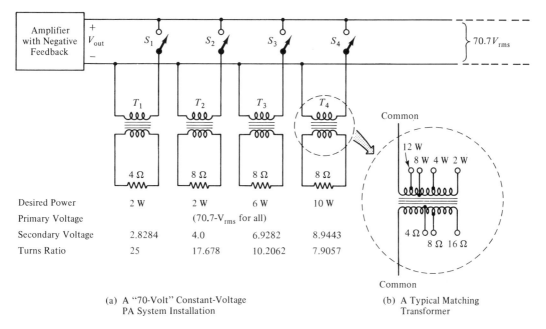

Desired Power	2 W	2 W	6 W	10 W
Primary Voltage		(70.7-V_{rms} for all)		
Secondary Voltage	2.8284	4.0	6.9282	8.9443
Turns Ratio	25	17.678	10.2062	7.9057

(a) A "70-Volt" Constant-Voltage
     PA System Installation

(b) A Typical Matching
     Transformer

Figure 5-27   Constant-voltage PA amplifier.

stalled, should maintain about the same audio output at any given speaker whether that is the only speaker being fed at the time or whether all the speakers on the system are being fed.

5.2 Radio-Frequency Signal Generation and Frequency Control

The generation of radio-frequency signals at precisely known and accurately controlled frequencies is an important part of communication systems design.   In general, RF oscillators fall into three categories: those that are controlled by tuned *L-C* resonant circuits (or their equivalents at UHF and microwave frequencies), those that are controlled by a quartz crystal or other precise device employing some type of physical resonance phenomenon (such as some type of atomic resonance), and those that do not depend upon either electrical or physical resonances (such as *R-C relaxation* oscillators).   Some of these devices may have their frequency of oscillation controlled by (or at least modifiable by) a controlling voltage or current so that they can be easily frequency modulated.   Such oscillators, when they are controlled by a voltage, are called *voltage-controlled oscillators* (or *VCO*s). We shall not spend a great deal of time discussing the various types of common oscillator circuits since that is properly a topic for a text on electronic circuits. However, we will mention a few of the common types and then introduce some of the not-so-common types of circuits.

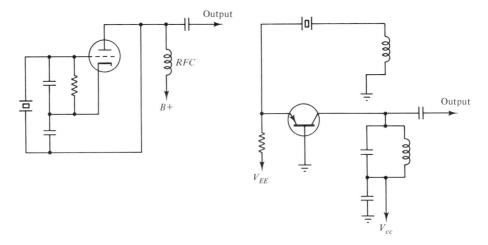

Figure 5-28 Typical crystal oscillator circuits.

5.2.1 Crystal-Controlled Oscillators

The crystal-controlled electronic oscillator has been very important to the communications industry for a long time. Figure 5-28 shows two typical crystal-oscillator schematic diagrams. The crystal itself is actually a small mechanically resonant device which includes an electromechanical transducer action called the *piezoelectric effect*. Looking into its input terminals, its electrical analog can be represented by a series-parallel electrical resonance circuit such as that shown in Figure 5-29, and as such has both parallel and series resonant frequencies. Which frequency is utilized depends upon whether the crystal is used in the oscillator circuit in place of a parallel tuned circuit or a series tuned circuit. It is important to know which resonant frequency is referred to by the frequency stamped on the crystal container can. Since the frequency of operation is also somewhat dependent upon the circuit capacitance of the particular oscillator circuit in which the crystal is to be used, it is important to know the nominal assumed circuit capacitance for

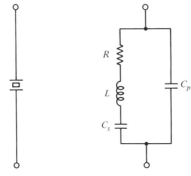

(a) Schematic Symbol   (b) Equivalent Circuit

Figure 5-29 The quartz crystal and its equivalent R-L-C circuit.

which the crystal was cut and tested. However, this secondary dependance of frequency upon circuit capacitance also allows the use of "trimmer" capacitors across the crystal for exact in-field adjustment of the oscillation frequency. It also allows the use of varactor diodes (see Section 5.3.1.3) for FM modulation of the crystal oscillator.

5.2.2 Frequency Synthesizers

Quite often it is desired to have communication equipment oscillators which have both good stability and frequency accuracy and which will operate at a large number of selected frequencies. For example, in an aircraft transmitter, it may be desired to have an oscillator which will cover the frequency range from 118.000 to 135.975 MHz in 720 steps of 0.025 MHz. To do this with individual crystals would require 720 crystals, and with crystals being several dollars apiece, this would be quite expensive, not to mention the size, weight, and complicated switching needed to utilize one of these crystals at a time in an oscillator. To get around this, the heterodyning or frequency-mixing principle is often used. A typical solution is illustrated in Figure 5-30. Here 26 crystals are used in three different oscillators to produce the 720 frequencies desired. The frequency of the second oscillator is subtracted from that of the first, and then the frequency of the third oscillator is either added to or subtracted from the frequency which results from the mixing of the first and second oscillators. Thus, by the use of this circuit, the number of crystals has been reduced from 720 to 26, or a factor of almost 30 times. This has been paid for by having to use three oscillators and two mixers in place of the one single oscillator. Usually one switch is used to control the selection of crystals in the first oscillator and a ganged switch is used to control the crystal selection in the second and third oscillators. It is also necessary to incorporate tuned circuits to reject the undesired mixer products (i.e., reject difference frequency if the sum frequency is desired, and vice versa).

 With the advent of modern digital circuitry, especially in integrated-circuit form, a different type of system for synthesizing a large number of accurately controlled frequencies has been developed: this is referred to as the digital fre-

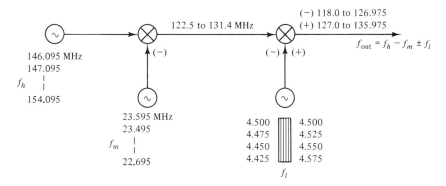

Figure 5-30 Multiple oscillator decade frequency synthesizer.

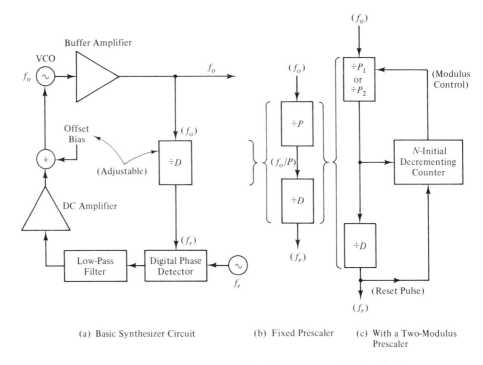

Figure 5-31   The phase-lock-loop (PLL) frequency synthesizer circuits.

quency synthesizer and uses one voltage-controlled oscillator whose frequency is controlled by phase-locking it to one of the harmonics of a stable crystal oscillator.

One version of this circuit that produces the range of frequencies needed for the aircraft transmitter is shown in Figure 5-31a.   In this case the output frequency of the VCO is fed into a programmable divider (digital integrated circuit), where the divide ratio is set equal to the quotient of the desired frequency divided by the frequency increment (25 KHz).   The result is a train of pulses out of the divider where the pulse repetition rate is 25 KHz when the VCO is at the correct frequency. This pulse train is then compared with the output of a stable 25 KHz crystal oscillator by feeding both signals to the inputs of a phase detector.   The output of the phase detector is amplified and is used to correct the d.c. voltage applied to the VCO so as to maintain it at the correct frequency (this technique is called a *phase locked loop*).   The frequency of the system is changed by changing the nominal d.c. bias voltage applied to the VCO and by changing the divide ratio of the programmable-divider circuit.   The actual VCO, rather than covering the range from 118 MHz to 136 MHz, might instead be designed to cover the range from 39 to 46 MHz and be followed by a frequency multiplier amplifier that multiplies this range of frequencies up to the 118- to 136-MHz range.

At this point it might be well to note that the "phase detector" shown in Figure 5-31a is usually a digital *frequency/phase comparator* which compares the output pulses of the divider with pulses out of a digital clock oscillator used as the reference frequency source. A number of different types of digital frequency/ phase comparators are available, most of them in the form of integrated-circuit chips. The type of phase/frequency comparator used, the bandpass of the low-pass filter in the feedback loop, whether a second-order feedback loop is also employed (i.e., creating a *type 2* servo system), and other design factors will determine how stable the VCO frequency will be (assuming a perfectly stable reference), how quickly the loop will "lock up" or bring the VCO to the correct frequency, how easily the loop may fall out of lock, how much *phase noise* is present on the VCO output, and so on. Most digital PLL (phase-locked-loop) frequency synthesizers designed to be switched among a number of different frequencies employ some type of circuit to disable the output of the synthesizer while it is in the *search mode* or locking up at the new frequency.

There are many possible variations in the design of a digital frequency synthesizer. For example, suppose that it was desired to have f_o in Figure 5-31a go from 500 to 600 MHz in steps of 100 KHz each. This could be done by having a 100-KHz reference oscillator and a variable divider going from 5000 to 6000, or we could keep the 25-KHz reference oscillator and precede the 5000-to-6000 divider with a fixed $\div 4$ *prescaler* divider so that in effect we would be dividing f_o by 20,000, 20,004, 20,008, and so on, to get the 25-KHz reference frequency. This is shown in Figure 5-31b. However, note here that we would not be able to get 25-KHz frequency spacing (even with the 25-KHz reference oscillator) if we used a *fixed* prescaler. This can be remedied though by using a *multimodulus* prescaler (a prescaler with two or more divide ratios). A two-modulus prescaler is often used with a one-count difference between its two divide ratios. This may be combined with the regular divider ($\div D$ in Figure 5-31c) and another decrementing counter (or divider) which counts down from N to 0. Let us analyze how this circuit works, assuming that we are starting at a point where all of the counters are set to their initial-count values.

The prescaler will count down from P_1 to 0, at which point it puts a pulse out to both the $\div D$ and the $\div N$ counters and resets back to P_1. This goes on until $N * P_1$ input pulses have been counted and a pulse is obtained from the $\div N$ counter. This pulse is applied to the prescaler and changes its modulus from P_1 to P_2 (the divide-by-N counter does not reset itself, so this pulse continues and keeps the prescaler in the $\div P_2$ state). From this point on, counter D will get one input pulse for every P_2 pulse applied to the input of the prescaler. It will finally output a pulse of its own after $(D - N)P_2$ *more* pulses have been counted (at which point one complete cycle has been completed and the $\div N$ counter is reset back to N, which removes the modulus-control pulse from the prescaler, putting it back into the $\div P_1$ mode). The number of input pulses (or cycles) which have occurred for one output pulse at the reference frequency will then be given by adding the two factors mentioned earlier:

$$\frac{f_o}{f_r} = P_1 N + P_2 (D - N) \tag{5-15}$$

If, as is true in many cases, P_1 and P_2 differ by only 1, then if we let $P_1 = M$ and

$P_2 = (M - 1)$, this becomes

$$\frac{f_o}{f_r} = MN + (M - 1)(D - N) = MN + MD - MN + N - D$$

$$= MD + N - D = N + D(M - 1)$$

(5-16)

Looking at the latter form of this expression, we see that the setting of the $\div N$ counter will determine the "units" and the setting of the $\div D$ counter will determine the second-order digit value in a numbering system whose base is equal to $(M - 1)$. For a decimally oriented frequency synthesizer, we would then want to have a prescaler whose divide ratios were 11 and 10 (or 9 and 8 for an octally oriented synthesizer, etc.). Additional prescalers of the same modulo pair could be cascaded to give a three- or four-digit decimally oriented divider, for instance.

Quite often in laboratory applications it is desired to have a synthesizer which can put out a wide range of frequencies in decimal increments. One method of approach is to use the phase-locked-loop method just described, and this is done quite frequently. Another approach is to use some method of direct frequency synthesis. Figure 5-32 suggests one possible method. Although the approach here is rather simple, some problems exist which make this type of circuit impractical in actual use.

Basically the circuit consists of a bank of frequency multipliers which provide nine reference frequencies at $m \times f_o$, where $m = 1$ to 9. These outputs are then

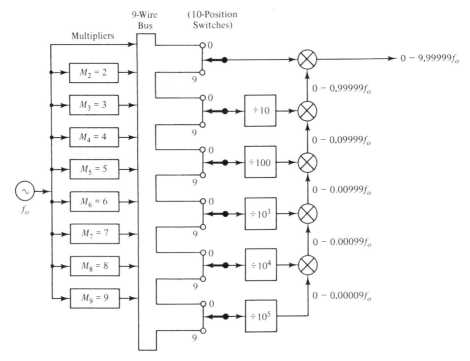

Figure 5-32 A possible method of direct frequency synthesis.

selected by switches followed by dividers and the divider outputs are then combined together in mixer circuits. The only problem is that the mixer circuits produce difference frequencies as well as sum frequencies, and these unwanted frequencies may be quite hard to filter out because they often fall within the band of desired frequencies from the output of the same mixer stage. A more practical method is shown in Figure 5-33.

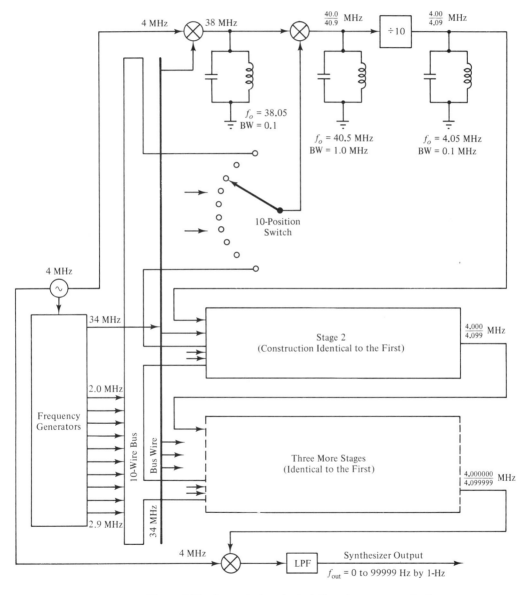

Figure 5-33 A practical method of direct frequency synthesis.

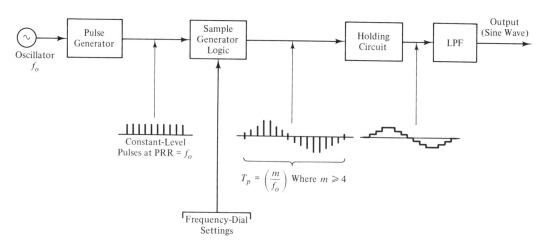

Figure 5-34   A type of direct digital frequency synthesis.

In this circuit the unwanted frequencies out of each mixer stage are far enough removed from the desired band of frequencies to permit very effective filtering and elimination of them.   The frequency generator which produces the 34 MHz and the 2.0- to 2.9-MHz reference signals may be of the harmonic-generator-divider type or may consist of a number of phase-locked generators which are locked to the 4-MHz master oscillator.

Another type of direct synthesis employs a digital logic circuit.   This type of system is illustrated in Figure 5-34.   Here the reference generator feeds a pulse generator which produces clock pulses which are then sent to the logic circuit. Depending upon the setting of the frequency dials, the logic circuit produces an output pulse train where the pulses are of varying height according to their pro-grammed values for the frequency selected.   By extending the pulse-height pro-gram over several cycles of the desired frequency, it is possible to design this type of circuit to operate at noninteger values of m, where m is the ratio of the reference frequency to the desired frequency (m is usually greater than or equal to 4).   The train of varying-height pulses goes to a sample-and-hold circuit which is followed by a low-pass filter, and the result is a sinusoidal output at the desired frequency. An alternate technique to the one shown in Figure 5-34 is to use parallel lines out of the logic generator into a D/A converter.   Then, instead of sending a pulse of some computed height over a single wire as in the original circuit, the proper combination of parallel (binary code) wires have pulses sent over them simulta-neously, thereby representing the pulse height as a binary number rather than as an analog amplitude level.   The holding circuit then becomes a D/A converter, which is again followed by the low-pass filter.

No matter which type of synthesizer circuit design is employed, the frequency stability of the synthesizer output cannot be any better than the stability of the reference oscillator driving the synthesizer.   In many applications it is desired to have very stable synthesized outputs.   In these cases a very stable master crystal oscillator in a temperature-controlled enclosure may be used. For even greater stability, the reference signal may be derived from some type of atomic-frequency-standard system. These systems will be discussed in Section 5.2.4.

5.2.3 Stabilizing UHF and Microwave Signals

Another problem which exists in the generation of RF signals for communication systems is the generation of stable signals in the high-UHF and in the microwave regions. The typical microwave oscillators such as the klystron or the magnetron are not especially noted for their inherent frequency stability. Heat, mechanical vibration, and supply-voltage variations all contribute to this frequency instability. However, the phase-locked-loop technique may again be used in this case to stabilize the frequency of oscillation of a UHF or microwave oscillator by comparing its output with the harmonics of a stable crystal oscillator operating at a lower frequency.

Figure 5-35 shows the block diagram of such an oscillator-stabilization system. In this case a sample of the microwave oscillator output (from a reflex klystron) is compared in a phase detector with a microwave signal at the desired frequency which is produced by feeding the output of a stable crystal oscillator through a string of harmonic generators. The d.c. signal from the phase detector represents the phase error of the klystron signal relative to the reference signal out of the harmonic generators. This phase-detector signal is then put through a low-pass filter and is used to control the klystron frequency through controlling the voltage supplied to the reflector of the klystron. Often, the reference oscillator is at 10 MHz, allowing the klystron to be locked onto any harmonic of 10 MHz, although any reference oscillator frequency may be used, with the klystron being locked onto the nth harmonic of that signal. This system is used extensively both for microwave transmitter frequency control and also to stabilize the first local oscillator of microwave superheterodyne receivers.

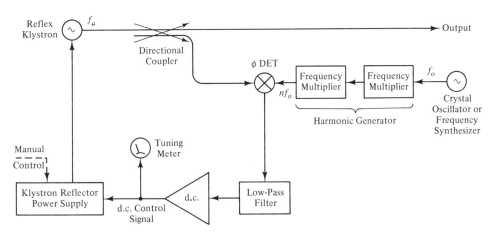

Figure 5-35 System for the stabilization of a microwave oscillator.

5.2.4 Atomic Frequency Standards

The need for very stable and accurate RF signals in many areas of instrumentation, communication, and navigational systems has led to the development of various types of atomic frequency standards.[5] These standards are used to control the master oscillators for a wide variety of applications in the areas just mentioned. Frequency standards are usually classified as either *primary* or *secondary*, depending upon whether the atomic frequency involved is or is not independent of various manufacturing and/or operating environment factors. A secondary standard may be very stable in frequency, but its frequency may be adjustable at the time of manufacture by controlling some physical parameter. Thus, it is necessary to calibrate it at the time of manufacture against some other (primary) frequency standard.

5.2.4.1 The Cesium-beam standard. The cesium-beam frequency standard is a primary frequency standard whose frequency of operation is relatively independent of external factors except magnetic fields. But with proper shielding, the effect of these fields can be reduced to a negligible degree. In the Cs 133 atom, there is a transition between two of the atomic energy states which result from the interaction between the nuclear spin and the electronic spin (these two states are part of the hyperfine structure of the lowest or "ground" energy state of the atom). The energy difference ΔE between these two states corresponds to radiation at a frequency of 9192.63177 MHz (using the relationship $\Delta E = hf$, where h is Planck's constant). The cesium-beam tube works in the manner illustrated in Figure 5-36.[6] A heated cesium oven produces a beam of cesium *atoms*, some of these atoms being in the higher hyperfine ground state (high atoms) while the remainder are in the lower hyperfine ground state (low atoms). The beam from the oven is passed through a region of nonuniform magnetic field, which causes the "high" and the "low" atoms to be deflected in different directions. As a result, two beams are formed, one composed of "high" atoms and one composed of "low" atoms. These two beams then pass into a microwave cavity which is in a weak uniform magnetic field and each beam interacts with the magnetic field and with the electromagnetic microwave radiation. The microwave radiation is at or near the atomic transition frequency of 9192.63177 MHz. The presence of the microwave field will cause some of the "low" atoms to absorb a quanta of microwave energy and transition to the "high" state while other atoms already in the "high" state are induced by stimulated-emission-of-radiation (MASER) action to give up a quanta of energy at 9192.63177 MHz and transition to the "low" state. Let us call these new states of atoms *which have undergone transitions* in the microwave cavity the "high (new)" and "low (new)" states, respectively.

There will now be four beams of atoms coming out of the far end of the microwave cavity (the "low" and "high" beams of atoms which *did not* undergo

---

[5]Steven F. Adam, *Microwave Theory and Applications* (Englewood Cliffs, N.J.: Prentice-Hall, 1969), Section 4.13.

[6] *Hewlett-Packard Journal*, March 1976.

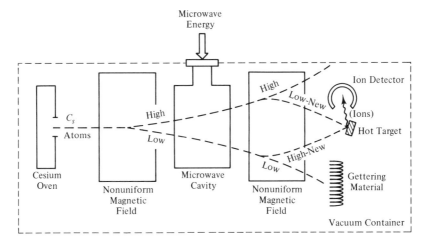

Figure 5-36   The cesium-beam tube components.

transitions in the cavity and the "low (new)" and "high (new)" beams composed of atoms which *did* undergo state transitions in the cavity).   These four beams then pass through another nonuniform magnetic field which bends the original "high" and "low" beams farther away from each other but which bends the "new" beams toward each other in such a way that they converge (focus) at a point outside of the magnetic field.   At this focus point is placed a hot target which ionizes the cesium *atoms* which strike it into cesium *ions*.   The *ions* are then collected by a photomultiplier type of ion detector, the output current of which then indicates the number of atoms per second arriving at the hot target in the two "new" beams and hence also indicates the number of atoms per second undergoing transitions in the microwave cavity.

The number of atoms per second undergoing transitions in the microwave cavity is dependent upon how near the frequency of the microwave signal fed into the cavity agrees with 9192.63177 MHz transition frequency.   If we plot the ion detector output versus the frequency of the applied signal, we get a response curve similar to that shown in Figure 5-37.

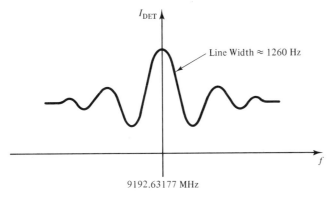

Figure 5-37   Cs-133 hyperfine transition line.

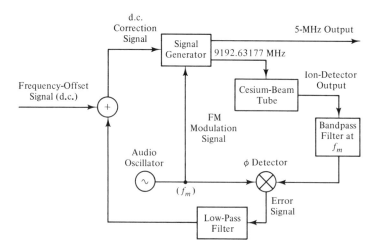

Figure 5-38 A Cs-beam frequency stabilization system.

To use the cesium-beam tube, the microwave cavity is supplied with a 9192.63177-MHz signal from a stable crystal-controlled signal generator whose frequency can be varied slightly by means of a d.c. correction signal applied to it as shown in Figure 5-38. Internally, the signal generator normally would consist of a 5-MHz crystal-controlled oscillator (whose output is also the main output of the whole system) followed by a digital frequency synthesizer which produces the 9192.63177-MHz signal. This microwave signal is FM modulated by an audio frequency of f_m, with the FM deviation being set at a couple hundred Hertz. If the center frequency of the microwave signal is exactly at 9192.63177 MHz, this FM modulation causes the ion-detector current to vary in such a way that the fundamental (i.e., f_m) Fourier series component is not present. However, if the center frequency is either below or above 9192.63177 MHz, there will be a Fourier series component at f_m present in the ion-detector current. The phase of this audio component will be in phase with the modulating signal if the microwave-signal center frequency is high and will be 180 degrees out of phase with the modulating signal if the microwave center frequency is low. Thus, by phase detecting the ion-

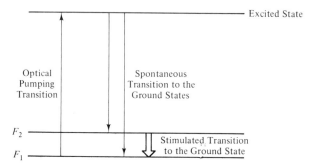

Figure 5-39 The transitions of rubidium-87.

detector current with the modulating signal as a reference, we can obtain a d.c. voltage whose polarity depends upon whether the microwave-signal center frequency is high or low.   This d.c. signal can be amplified and fed back to the signal generator frequency-control input, thus producing an automatic-frequency-control system locked onto the 9192.63177-MHz state-transition line of the cesium atom.

When a system such as that just described is used as the basis of an accurate clock system, it may be necessary temporarily to speed up or slow down the clock to synchronize it with some standard time.   This can be done by providing in the d.c. feedback loop a means for adding a d.c. frequency-offset voltage to the loop error voltage.   This is shown in the block diagram of the system.

In the operation of the cesium cell, the cesium material in the oven is constantly being vaporized to form the beam.   To preserve the vacuum in the tube, it is necessary to have a large amount of cesium-collecting (or "gettering") material in the vicinity of the hot target and ion detector.   Eventually, all the cesium in the oven will be vaporized and/or the gettering material will be unable to absorb any additional Cs atoms.   At this point the tube will cease to function.   Some of the more recent tubes will operate continually for up to five years.

The stability of such a cesium-beam frequency standard as just described can be made to be about ± 1 part in 10^{11} over a period of more than a day in length with short-term stability ranging from about 1 part in 10^9 on a millisecond basis to about 1 part in 10^{13} on a 30-minute basis.   Temperature variations, acceleration, shock, stray magnetic fields, and the like can degradate these accuracy figures to some extent depending upon the quality of the instrument construction and the severity of the environmental factors.

5.2.4.2 The Rubidium standard.   This *secondary* frequency standard uses a hyperfine transition in Rb-87 as shown in Figure 5-39.   Optical pumping with filtered light as shown in Figure 5-40 tends to depopulate the F_1 level, while the F_2-level population increases.   By stimulated emission of radiation, the F_2-to-F_1 transition is induced ($f \doteq 6835$ MHz) and increases the F_1 population.   The larger the F_1 population, the more optical pumping radiation is absorbed.   Thus, by using a stable light source shining through a cavity filled with rubidium vapor mixed with an inert gas and detecting the level of radiation not absorbed, we can detect the degree of population inversion occurring in the F_2 and F_1 levels.   The degree of population inversion is strongly dependent upon how near the frequency of the microwave field in the cavity is to the transition frequency.   Thus, a plot of the

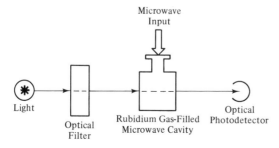

Microwave Input

Light

Optical Filter

Rubidium Gas-Filled Microwave Cavity

Optical Photodetector

Figure 5-40   The use of a rubidium cell for frequency stabilization.

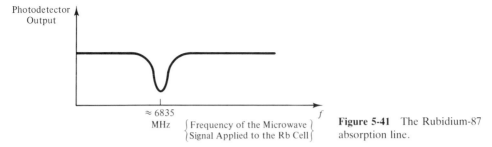

Figure 5-41 The Rubidium-87 absorption line.

photodetector output versus the frequency of the applied microwave signal might appear as shown in Figure 5-41.

The rubidium cell can be used in circuitry similar to that of the cesium cell. However, unlike the cesium cell, where the transition frequency is largely independent of external effects, the transition frequency in the rubidium cell is somewhat dependent upon the gas pressure in the cell. Thus, the rubidium cell is a secondary standard and must be calibrated against a primary standard at the time of its manufacture.

5.2.4.3 The Hydrogen maser oscillator.

In this device, shown in Figure 5-42, a beam of H_2 atoms is directed through a magnetic field which selects the high-energy-state atoms for injection into a quartz tube in a microwave cavity. The quartz tube is designed to give a long interaction time between the atoms and the microwave field. Random energy-state decay processes in the atom population start building up an electromagnetic field, and then stimulated emission of radiation begins to occur, and oscillation takes place at the hydrogen atom transition frequency to which the microwave cavity is approximately tuned. The exact accuracy of cavity tuning will determine the *amplitude* of the oscillation which occurs, but the atomic transition frequency determines the *frequency* of the oscillation ($\lambda \approx 21$ cm). Thus, the frequency of the radiation coupled out of the cavity depends only upon the hydrogen transition frequency. The hydrogen maser is therefore considered a primary standard if shielded from external magnetic fields.

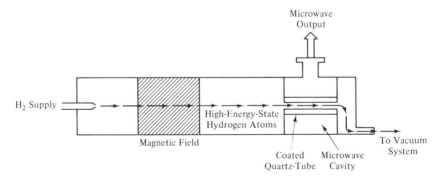

Figure 5-42 The hydrogen maser oscillator.

5.3 *TRANSMITTER AND RECEIVER DESIGN*

In this section we will be looking at some of the techniques and circuits which can be used to implement the operations discussed in the previous chapters (modulation, demodulation, amplification, frequency synthesis and carrier generation, mixing and frequency translation, filtering, etc.). Although a few circuits are discussed as examples, the objective is not to present the techniques of electronic circuit design. This is covered adequately in other books on electronics.[7] The main emphasis will again be on combining "black boxes" into systems or parts of systems.

5.3.1 *Modulation and Detection Techniques*

5.3.1.1 AM modulators and demodulators. One of the ways in which an AM signal may be formed is through changing the plate or collector voltage supplied to a class C RF amplifier so that the change in this voltage is proportional to the instantaneous value of the modulating signal $s_m(t)$. The class C RF amplifier, when supplied with a constant input signal, will produce an RF output whose voltage is very nearly proportional to the value of the d.c. voltage supplied to the plate or collector circuit (assuming common-cathode or common-emitter configuration). Such an amplifier and its basic waveforms are shown in Figure 5-43. The d.c. voltage supplied to the tube or transistor's plate or collector circuit is obtained from a well-filtered d.c. supply in series with the secondary of an audio transformer. The primary of the transformer is supplied with the modulating signal $s_m(t)$. Thus, the voltage supplied to the plate or collector circuit, rather than just being d.c., is now given by Equation 5-17:

$$V_{bb}(t) = V_{dc} + V_{audio} = V_{dc} + (N_s/N_p) \times s_m(t) \qquad (5\text{-}17)$$

Thus, the RF output is proportional to the quantity on the right-hand side of Equation 5-17.

If we divide this through by V_{dc}, then this quantity becomes just the quantity shown in the square brackets in Equation 4-1, where $(m_{AM})_p/|s_m(t)|_{max} = N_s[N_p \times V_{dc}]^{-1}$. The d.c. power supply will supply the power to produce the RF energy we find in the carrier, and the audio power coming through the transformer will be used to produce the RF power which we find in the total of all the sideband components.

It is interesting to note one thing about class C amplifiers at this point: the RF output signal is *not* proportional to the level of the RF input signal. Thus, a class C amplifier *cannot* be used to amplify a signal which is already amplitude modulated. The class C amplifier does have the advantage of high efficiency. This usually runs between 65 and 80% (RF output power versus the input power to the amplifier plate circuit). Thus, the RF carrier power will be about 65 to 80% of the d.c. power supplied to the plate circuit of the class C amplifier, and the sideband power will be about 65 to 80% of the audio power supplied out of

---

[7] See, for example, Mandl, *Principles of Electronic Communications*, and Fink, *Electronic Engineer's Handbook.*

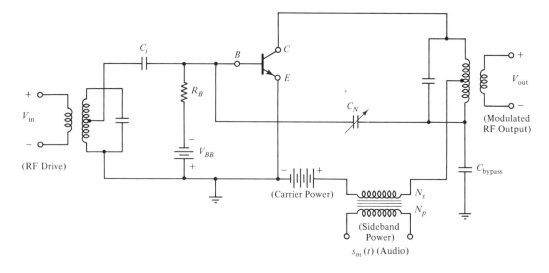

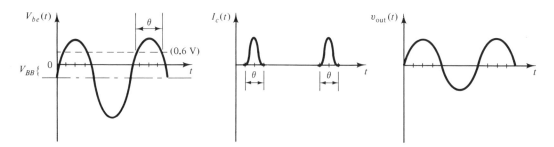

Figure 5-43   High-level class-C modulated RF amplifier and waveforms.

the secondary of the modulation transformer. To generate this audio power, the primary of the modulation transformer must be driven by an audio power amplifier. This is usually a class B or class AB_2 push-pull audio stage having an efficiency of about 50 to 60% (audio power into the modulation transformer to d.c. power into the audio amplifier final stage). Thus, to produce the modulated RF output, d.c. power must be supplied both to the modulator audio amplifier and also directly to the class C RF amplifier.

To give a feeling for what is happening here, consider the example in Figure 5-44. Note that both audio and d.c. efficiency figures are given for the modulation transformers. For the d.c. case, the only loss is the I^2R loss in the secondary winding, while for the audio signal the losses consist of I^2R losses in both the primary and secondary windings plus the hysteresis and eddy-current losses in the core.

Taking all these into account, we now have a complete AM transmitter with assumed stage efficiencies as indicated. If we assume 100% sinusoidal modulation

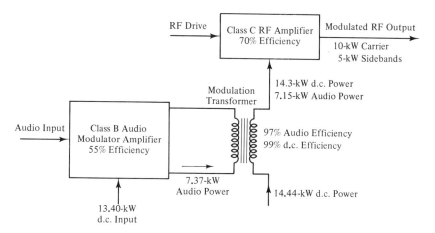

Figure 5-44   A high-level AM transmitter showing stage efficiencies and powers.

the various power levels of this 10 KW (carrier level) transmitter will be as shown in the drawing. The total power output (carrier plus two sidebands) will be 15 KW and the total d.c. power input is given by 14.44 KW from the d.c. supply connected to the class C amplifier and 13.40 KW from the d.c. power supply connected to the modulator, making a total of 27.88 KW input, or an overall efficiency of 53.8%, which is lower than the efficiency of *either* the class C stage or the modulator. The reason for this is that part of the RF power (that in the sidebands) suffers from the losses in both the RF stage *and* the audio stage. In other words, the 13.40 KW into the audio modulator amplifier only results in 5 KW of sideband power (an efficiency of only 37.3%).

Another way to generate a high-power AM signal is to generate the AM signal at a low level and then to amplify it by means of a class B RF power amplifier. Suppose that this was done as shown in Figure 5-45. Then the total d.c. input power needed to produce 15-KW total 100% sinusoidal-modulated output would be 27.25 KW, giving an overall efficiency just equal to that of the RF stage or 55%. This is comparable with the efficiency of the high-level modulation method already discussed at 100% modulation and would be the same for any level of modulation, whereas the efficiency of the high-level modulation method would improve with lower modulation levels or with multitone modulation because of a lower ratio of sideband to carrier power.

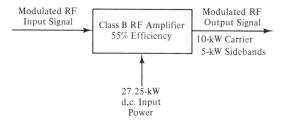

Figure 5-45   A low-level AM transmitter utilizing a class-B RF amplifier.

Both schemes are employed in practice, although the low-level scheme is sometimes preferred because it does not require the large, expensive audio modulation transformer needed in the high-level scheme. However, the low-level scheme requires RF amplifier tubes of higher plate dissipation rating because of the lower efficiency of the class B RF amplifier.

In recent years the use of high-level AM modulation with the class D modulator discussed in Section 5.1 has become popular. The class D audio amplifier is not only more efficient than is the class B type, but since the class D amplifier essentially eliminates the need for the modulation transformer, it is possible to reduce greatly the size, weight, and cost of the modulator section of the AM transmitter.

At this point it is perhaps well to consider the class B RF amplifier. A typical circuit is shown in Figure 5-46. The term "linear amplifier" is quite often applied to the class B RF linear, although looking at just the tube or transistor itself the circuit is electronically not linear. Conduction only occurs on the positive half-cycles (approximately) of the input signal, and thus the plate or collector current takes on the appearance of approximately a rectified half sine wave (which of course is not linearly related to the full sine wave signal of the input voltage). However, when this plate or collector current pulse is used to excite the resonant-tuned circuit attached to the plate, the "flywheel" effect of the tuned circuit will produce an approximately sinusoidal output voltage signal, and the *amplitude* of this output voltage signal *will* have a linear relationship to the *amplitude* of the voltage signal applied to the amplifier input circuit. The d.c. input current, and hence the d.c. input power, also depends linearly (usually with the addition of a constant no-signal "idle" current) upon the input voltage amplitude. Often the class B amplifier is used in the grounded-base or grounded-grid configuration (shown in Figure 5-47) to provide an amplifier which is easier to control insofar as the tendency to self-oscillate or to produce what are called "parasitic" oscillations.

The grounded-base or grounded-grid amplifier however requires much more grid-driving power than does the common-emitter or common-cathode type of stage shown in Figure 5-46. This is due to the driving source being in series with the emitter-collector or cathode-plate circuit via conduction through the transistor or

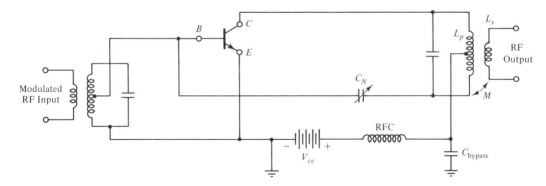

Figure 5-46   Transistor common-emitter class-B "Linear" RF amplifier.

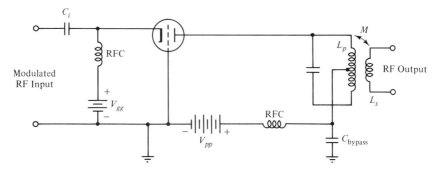

Figure 5-47 Vacuum-tube grounded-grid class-B "Linear" RF amplifier (biased almost to cut-off but with a small idle current).

tube.   Most of this extra driving power required from the preceding amplifier stage appears as a contribution to the RF power in the transmitter output.   Either type of circuit may be equipped with a pi-network output circuit in place of the conventional parallel L-C circuits shown in Figures 5-46 and 5-47. The pi-network circuit is shown drawn in two different configurations in Figures 5-48a and 5-48b.   From the latter drawing, it is easy to see that the two capacitors constitute a voltage divider across the resonant circuit which is formed by the inductor and the two capacitors in series.   This voltage divider makes it possible to match the fairly high output impedance of the transistor or vacuum tube to the usually low impedance of the transmitter output line (often 50 to 70 Ohms).   The tuning of the resonant circuit and the loading of the amplifier (adjusted by adjusting the impedance ratio between the amplifier output and the line) can both be accomplished by means of the two tuning capacitors C_1 and C_2.

In designing a pi-network, three things must be known to start with: the output impedance of the transistor or tube (call it R_t), the load or transmission-line impedance (R_L), and the desired value of loaded Q.   The latter can be determined from the center frequency of operation (f_o) and the desired 3-dB-down

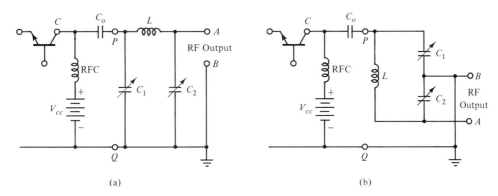

Figure 5-48 Pi-network output circuit for RF amplifiers.

bandwidth (Δf). Then the design proceeds as follows,[8] with the assumption that $R_t > R_L$ and $(Q^2 + 1) > (R_t/R_L)$:

$$Q = \left(\frac{f_o}{\Delta f}\right) \tag{5-18a}$$

$$X_{C_1} = \left(\frac{R_t}{Q}\right) \tag{5-18b}$$

$$X_{C_2} = R_L \left\{ \frac{R_t}{R_L \left[Q^2 + 1 - \left(\frac{R_t}{R_L}\right) \right]} \right\}^{1/2} \tag{5-18c}$$

$$X_L = \frac{QR_t + \left(\dfrac{R_t R_L}{X_{C_2}}\right)}{(Q^2 + 1)} \tag{5-18d}$$

If $(Q^2 + 1) >> (R_t/R_L)$, then it can be seen that

$$\frac{X_{C_1}}{X_{C_2}} = \frac{C_2}{C_1} \doteq \sqrt{\frac{R_t}{R_L}} \tag{5-19}$$

One of the main advantages of the pi-network is that no mechanical adjustments of the inductor components are needed (such as mechanically adjusting the mutual coupling between L_p and L_s in Figure 5-46 or 5-47).

The detection of AM signals is usually done with what is called an "envelope detector." This is essentially an RF rectifier followed by a low-pass filter. The rectifier rectifies the incoming modulated waveform, leaving a series of rectified half-sinusoids whose amplitudes are proportional to $[1 + (m_{AM})_p/|s_m(t)|_{max}s_m(t)]$. The low-pass filter then removes the sinusoidal impulses (smooths them out) to produce an output waveform proportional to $[1 + (m_{AM})_p/|s_m(t)|_{max}s_m(t)]$. The ability of the low-pass filter to do this well depends upon the separation of the carrier and the modulating frequencies and upon the design of the filter. Usually a simple R-C–type filter is all that is used. A series-blocking capacitor eliminates the d.c. component on the signal and produces an output signal which is proportional to just $s_m(t)$. The d.c. component, which is proportional to the strength of the carrier of the signal supplied to the detector, is often used for automatic-gain-control (AGC) or automatic-volume-control (AVC) circuits in the receiver. Figure 5-49 shows a typical AM diode envelope detector. If the incoming RF frequency is not much greater than the highest frequency of modulation, the choice of values for the R-C filter becomes very critical and even with a very careful selection of design values there still may be considerable amplitude-frequency and phase-frequency distortion of the postdetection signal.

---

[8] *Radio Amateur's Handbook.*

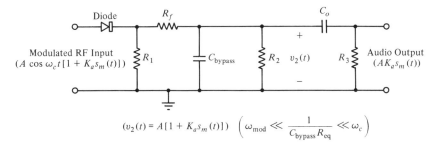

Figure 5-49   AM diode envelope detector circuit.

Another type of detector that can be used for the demodulation of either AM or DSBSC signals is the synchronous detector. A typical detector of this type is shown in Figure 5-50. Here the incoming signal is multiplied with a locally generated cosine signal which is at the same frequency and phase as the carrier of the received signal. In the case of the DSBSC signal, the multiplication of the received signal with the locally generated carrier produces an audio signal as well as a modulated RF signal at approximately twice the carrier frequency. This latter RF signal is filtered out by the low-pass filter, leaving the detected audio signal. If the incoming signal already has a carrier (standard AM signal), then a cosine-squared term is also produced which adds a d.c. term and a term at twice the carrier frequency to the output, but these are eliminated by the coupling capacitor and the low-pass filter, so again only the audio signal appears in the output.

The synchronous detector does have an advantage over the diode detector for the detection of standard AM signals in that when random noise is present along with the signal (and especially when the predetection ratio of carrier power to noise is small), there will be at least a 3-dB improvement in the audio output S/N ratio over what can be obtained with an envelope detector at low input S/N ratios (see Section 4.10.1). This effect comes about because with the synchronous detector, we do not have the higher-order carrier-noise intermodulation products that are produced with envelope detection. Also, the sideband-sideband intermodulation products present in envelope-detection output (see the comments regarding Equation 4-4 in Section 4.2) are eliminated from the audio output.

5.3.1.2 Single-sideband modulators and demodulators. SSB signals are generated by several different techniques. One of the most common is the use of a balanced modulator to produce a DSBSC signal followed by a filter of some type (usually a crystal lattice filter or a mechanical filter) which eliminates either the

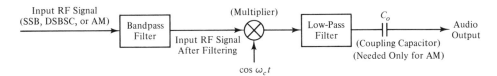

Figure 5-50   A coherent product detector for SSB, DSBSC, or AM.

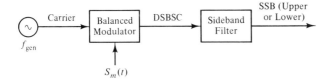

Figure 5-51 Filter method for generating a SSB signal.

upper or the lower sideband, leaving just the sideband of interest to be amplified and transmitted. Sometimes this desired sideband is translated (slid) to another frequency before being amplified and transmitted, but this will be discussed in the section on mixers and heterodyning techniques. A block diagram of this method of generating a SSB signal is shown in Figure 5-51. The basic circuit of the balanced modulator is shown in Figure 5-52. The balanced modulator is a device which produces a DSBSC signal; that is, it essentially is a circuit which multiplies an RF carrier signal by the modulating signal $s_m(t)$. (Remember, the class C RF stage, when modulated, multiplied the RF carrier not by $s_m(t)$ but by $1 + (m_{AM})_p / |s_m(t)|_{max} s_m(t)$, which can never go negative). A discussion of the operation of the balanced modulator is included in Appendix C. This appendix views the balanced modulator as a modulated switch or chopper.

The filtering needed to remove the unwanted sideband can be done by a crystal lattice filter such as is shown in Figure 5-53. Several stages of such filter sections may be cascaded to obtain the desired selectivity characteristic. Figure 5-54 shows a simplified drawing of an electromechanical filter. This type filter consists of a magnetostrictive rod upon which are mounted carefully machined

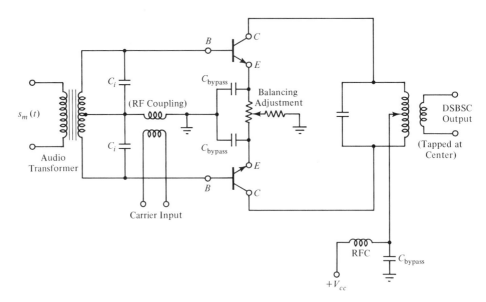

Figure 5-52 Basic circuit of a balanced modulator.

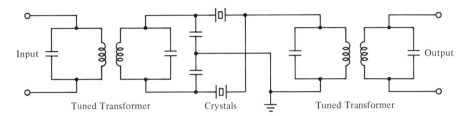

Figure 5-53   A simple crystal lattice filter.

disks which resonate at the desired frequencies. The rod is driven by a coil on one end and the portion of the mechanical wave which gets through the rod is picked up by the receiving coil at the other end of the rod. Mechanical filters are very compact, stable, and mechanically rugged. Another type of filter which may be used (especially with integrated-circuit techniques) is the active filter consisting of an operational amplifier with an appropriate R-C feedback network.

A second method of generating an SSB signal is by means of the phasing method. A system for doing this is shown in Figure 5-55. This method employs two balanced modulators whose outputs are added together. Because of the relative phasing of the carriers and the modulating signals fed into the two balanced modulators, the output signals are such that although both modulators produce a DSBSC signal, the phasing of the sidebands coming out of the first balanced modulator, when compared with the sideband phasing from the second modulator, will cause the sideband components on one side of the carrier to add and the components on the other side of the carrier to subtract or cancel, thus leaving only one sideband in the summed output.

The development of the equations for the time-function output signal of the system in Figure 5-55 is as follows and shows that when the signals from the two balanced modulators are added that the lower sideband is eliminated. If the carrier to the lower modulator had been shifted by -90 degrees instead of $+90$ degrees, then the upper sideband would have been eliminated instead of the lower one.

$$e_a = (\cos \omega_c t)[\cos (\omega_m t + \phi_\alpha)] \tag{5-20}$$

$$e_b = [\cos (\omega_c t + 90°)][\cos (\omega_m t + (\phi_\alpha - 90°)]) \tag{5-21}$$

$$e_a = \tfrac{1}{2} \cos [(\omega_c + \omega_m)t + \phi_\alpha] + \tfrac{1}{2} \cos [(\omega_c - \omega_m)t - \phi_\alpha] \tag{5-22}$$

$$e_b = \tfrac{1}{2} \cos [(\omega_c + \omega_m)t + \cancel{90°} + \phi_\alpha - \cancel{90°}] \tag{5-23}$$

$$+ \tfrac{1}{2} \cos [(\omega_c - \omega_m)t + \underbrace{\cancel{90°} - \phi_\alpha + 90°}_{(180° - \phi_\alpha)}]$$

$$e_a + e_b = \cos [(\omega_c + \omega_m)t + \phi_\alpha] + 0 \tag{5-24}$$

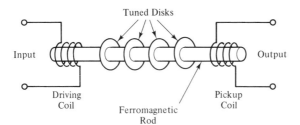

Figure 5-54   A mechanical RF filter.

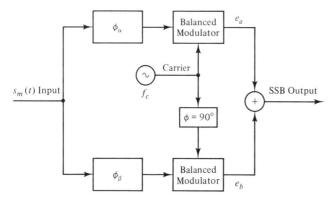

$[(\phi_\alpha - \phi_\beta) = 90°$ at all Modulating Frequencies Contained in $s_m(t)]$

Figure 5-55   The phasing method of SSB generation.

Single-sideband voice signals are usually detected with a product detector which multiplies the SSB signal with a locally generated "injection carrier" signal which is at or near the frequency that would normally be occupied by the missing carrier in the SSB signal itself.   For voice purposes (especially for communication-system quality requirements), it is only necessary that the frequency of this injection carrier be within a range of from several to perhaps about 20 Hz of the position of the missing carrier.   The only effect of this upon the demodulated voice signal is to change the tonal quality of the voice.   For a music signal, however, even a slight deviation from the original carrier frequency is quite noticeable because of the harmonic relationships which are present in music which are not present in a voice signal.

The circuit shown in Figure 5-50 may also be used to detect SSB signals. Essentially the product detector is a mixer that multiplies the SSB signal by the injection carrier to produce two output signals, one in the audio range which is the desired demodulated signal and one in the range of approximately twice the injection carrier frequency.   The low-pass filter rejects this latter signal, leaving only the desired audio signal as the output.

5.3.1.3 Angle modulators and demodulators.   The most common method of producing a direct-FM type of modulation is by means of some sort of reactance modulator.   A typical circuit of this type is shown in Figure 5-56.   Here an RF oscillator of some type whose frequency is controlled by a tuned circuit (or a quartz crystal resonator functioning as a tuned circuit) has connected across the tuned

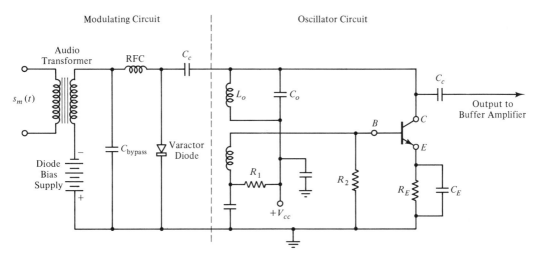

Figure 5-56   Varactor diode-type of FM modulator.

circuit a varactor diode.   The capacitance of the varactor diode can be altered by changing the current passing through it (if forward biased) or the voltage across it (if reversed biased).   This change in the diode capacitance, since it appears in parallel with the tuned circuit which determines the oscillator's frequency, will cause a change in the frequency of oscillation.   If the current through or the voltage across the diode is varied proportional to the modulating signal, then the frequency of oscillation will change proportional to the modulating signal and FM will be produced.

In Figure 5-56, the left-hand part of the circuit is the part that biases and couples the modulating signal to the diode, while the right-hand part of the circuit is the oscillator whose frequency is being varied.   The radio-frequency choke (RFC) is used to prevent RF energy from being coupled back into the biasing and modulating circuits, while the small coupling capacitor (C_c) is used to prevent audio and d.c. from being coupled into the oscillator circuit. C_c is normally on the order of only a few picofarads.   In many FM transmitters, the FM signal is generated at a lower frequency than the transmitting frequency and is multiplied up to the transmitting frequency by class C harmonic amplifiers.   In this case (shown in Figure 5-57), the deviation is also multiplied by the same factor as is the carrier frequency, but the percentage modulation $(\Delta f/f_c)$ remains constant.

Angle modulation can also be produced by any of the several different types of phasing methods.   One such method is illustrated in Figure 5-58.   In this system a balanced modulator is used to produce a DSBSC output to which is added the

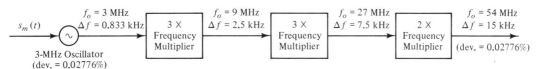

Figure 5-57   Frequency multiplication of an FM signal.

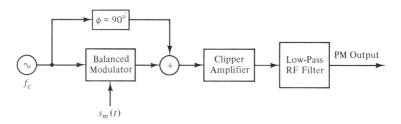

Figure 5-58   Phasing method of generating a PM signal.

carrier after the carrier has been given a 90-degree phase shift.   The amplitude of the shifted carrier is made much greater than the output of the balanced modulator. When the combined signal is passed through a clipper amplifier, the result is a rectangular waveform whose zero-crossing points are caused to shift proportional to the amplitude of the modulating signal $s_m(t)$.   If only the fundamental component of the square wave is examined, it is found to be a sinusoidal waveform which is phase modulated by $s_m(t)$.   The higher harmonics of the square wave are eliminated by the low-pass RF filter following the clipper amplifier.

The detection of FM signals is often done with a tuned-circuit type of detector called a discriminator.   The output of the discriminator is a signal which is proportional to the deviation of the instantaneous frequency of the FM signal from the carrier frequency to which the discriminator is tuned.   To reduce the response of the discriminator to any amplitude variations which may be present on the FM signal to be demodulated, the discriminator is usually preceded by a limiter amplifier stage which provides a constant-amplitude signal into the discriminator.   The operation of the discriminator is discussed in a number of textbooks on electronic circuits, but unfortunately the analysis given is not always correct or complete. Appendix D gives an analysis of the discriminator circuit as well as that of its companion, the ratio detector.

The main advantage of the ratio detector is that it is self-limiting and, therefore, does not need to be preceded by a limiter amplifier.   In the practical case, both the discriminator preceded by a limiter and the ratio detector will have some response to amplitude variations present on the FM signal to be demodulated, especially if these amplitude variations are very large compared to the normal signal level (as in the case of noise "spikes" resulting from auto ignition systems, lightning strokes, etc.), but the level of audio output produced by these amplitude variations is usually much less than the detector output signal produced by the FM variations on the signal being demodulated.   This is especially true if the deviation of the FM signal is large.   In other words, a wideband FM system (one with a large deviation, Δf_p) usually has a better signal-to-noise performance at the detector output than does a narrow-band (low-Δf_p) FM system.

PM signals can also be demodulated by a discriminator or ratio detector by following the detector with a deemphasis (integrator) circuit which attenuates the higher frequencies present in the modulation (using a discriminator or ratio detector alone causes the high-frequency components to be emphasized by a factor proportional to their respective frequencies).   PM signals may also be detected by phase-lock coherent detectors which compare the instantaneous phase of the incoming PM signal with the phase of a locally generated carrier signal which has been produced to have a phase which is the average of the phase of the signal to be demodulated, and which, therefore, should be approximately in phase with the original carrier signal which would be received if no modulation were present on the incoming signal.   Phase-locked detectors will be described in Section 5.3.3.1.

One of the problems of the discriminator or ratio detector is that the carrier frequency of the signal to be demodulated may vary away from the center frequency to which the detector is tuned.   This can be partially compensated for by providing an oscillator somewhere in the FM receiver whose frequency can be varied to cause a change in the frequency of signal received at the detector.   This oscillator then has its frequency controlled by the average (essentially d.c.) output voltage of the detector (which is proportional to the amount of offset between the carrier of the signal being demodulated and the frequency to which the detector circuit is tuned). This type of feedback control system is referred to automatic frequency control (AFC) and is quite widely used in all types of FM receivers (see Section 5.3.3.2).

5.3.1.4 Quadrature modulation methods.

Figure 4-11 shows an elementary modulation system in which two balanced modulators are used to generate the modulated signal and the reference (carrier) signal is transmitted along with the modulated signal (which really consists of two DSBSC signals).   This reference signal could be transmitted over a separate channel as shown in Figure 4-11 (page 72), or it could be transmitted along with the sidebands and separated out from them by a narrow bandpass filter providing that the sideband components do not come too close to the carrier frequency.   If this is not possible, the carrier phase reference can be transmitted by some other means (a pilot subcarrier or a synchronizing signal burst) as will be discussed in the sections on stereo FM and color TV in Chapter 7.

5.3.1.5 FSK and PSK methods.

Frequency-shift keying may be accomplished in either of two ways: by applying a binary plus-to-minus signal to a reactance modulator (or some other standard type of FM modulator) or by using the binary signal to cause switching between two continuously running oscillators at the two frequencies (MARK and SPACE).   An analysis of the frequencies at which the modulated-signal Fourier components occur will reveal that the signals produced by these two methods are not identical,[9] although they may be treated as such insofar as their demodulation is concerned.

In PSK transmission (two-phase or multiphase), some sort of reference signal is needed for the demodulation process.   If the modulating signal used to generate the PSK signal has little or no bias (i.e., approximately equal probability of being at any one level or phase), it is difficult to determine the reference phase from the signal itself unless a continuously updated reference oscillator is used.   The updating of the phase may be done in the manner described in the preceding section for the detection of quadrature modulation, or it may be done by sending a signal

---

[9] See Bennet and Rice, *Bell System Technical Journal*, May 15, 1963.

synchronizing group of pulses which when decoded allows a synchronizing burst to be applied to the receiver's reference oscillator.

FSK may be detected by a number of different methods: a standard limiter-discriminator or a ratio detector may be used (since FSK is really a form of FM), a pair of narrow-band *L-C* filters tuned to the MARK and SPACE frequencies, or special phase-locked-loop tone detectors (available as integrated-circuit chips[10]) which provide a d.c. output whenever the received tone frequency is close to the free-running frequency of the circuit's oscillator.

5.3.2 Superheterodyne Principles in Transmitters and Receivers

Most modern communications equipment makes use of the superheterodyne principle of operation. There are several reasons for doing this: to have a minimum number of tuned circuits that must be adjusted during the normal course of operation of the equipment, to permit various portions of the circuit to operate at nominal frequencies which simplify the design of these circuits, and to allow the production of standard parts which may be used in a wide variety of different models and types of equipment. Let us use the common home AM broadcast receiver as an example.

The block diagram of such a receiver is shown in Figure 5-59. The receiver must tune over a range of from 500 to 1700 KHz, but the only tuned circuits which must be adjusted to tune the different frequencies are the tuned circuit formed by the loopstick antenna and capacitor C_1 and the tuned circuit which determines the local oscillator frequency, consisting of L_2 and C_2. The local oscillator tunes over the range of 955 to 2125 KHz, which is 455 KHz above the frequency of the station which it is desired to be receiving.

Multiplying these two signals (the local oscillator signal and the signal from the desired station) together in the mixer amplifier produces two signals—one at 455 KHz and the other at a frequency equal to the sum of the signal and the local oscillator frequencies. The first signal is the desired signal which is amplified by the intermediate-frequency (IF) amplifier, while the second signal is the undesired signal which is rejected by the tuned circuits of the IF amplifier. In this example, the frequency of the signal which it is desired to receive and to convert to the IF frequency is given by subtracting the IF frequency from the local oscillator frequency. But at the same time the receiver is also capable of receiving the signal whose frequency is given by *adding* the IF frequency to the frequency of the local oscillator. This frequency is called the *image signal* frequency, and if it were not for the tuned circuit of the loopstick antenna, a signal present at the image frequency would be coverted to the IF frequency just as efficiently as would the desired signal. To prevent this, the tuned circuit which is ahead of the mixer must have sufficient selectivity to reject the undesired image signal. This is easiest to do if the desired signal frequency and the image frequency are widely separated in frequency. However, as we can see from the preceding, this separation amounts to twice the IF

---

[10] The type 567 integrated-circuit chip.

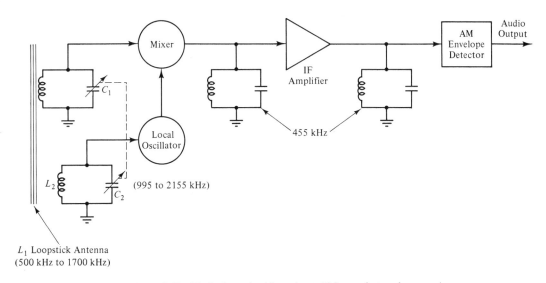

Figure 5-59   Typical standard-broadcast AM superheterodyne receiver.

frequency, so if the IF frequency is small, the amount of separation between the desired signal and the image signal frequencies will be small, and it will be difficult to reject the image signal from the receiver.

One of the problems of designing a superheterodyne receiver is that of "tracking" the tuning of the local oscillator with the tuning of the RF stages (loopstick antenna and any other tuned circuits at the signal frequency). Considering the broadcast-band superheterodyne, we see that the oscillator must be tuned over the range of 955 to 2125 KHz while simultaneously the loopstick antenna must tune over the range of 500 to 1700 KHz, always maintaining a difference of 455 KHz between the frequency to which the two circuits are tuned. If we try to do this with two linear variable capacitors operating from the same shaft, we find that in general we can obtain the exact desired frequency difference at only two points in the tuning range. Therefore, special nonlinear (with respect to degrees of shaft rotation) multisection variable capacitors are designed especially for use in super-heterodyne receivers tuning over a given range of frequencies. Use of these capacitors makes it possible (at least theoretically) to obtain exact tracking of the local oscillator and signal-frequency tuned circuits over the range of frequencies for which the capacitor is designed to operate.

The IF amplifier of our broadcast-band example has tuned circuits at 455-KHz center frequency. This is a frequency that makes it possible to easily establish an amplifier bandwidth of about 20 KHz without the use of extremely high-Q circuits and without using a large number of cascaded tuned circuits. If a higher IF frequency had been chosen, it would have been necessary to have used more and/or higher-Q–tuned circuits to achieve the same amplifier bandwidth. Going to a much lower IF frequency would have meant that the frequency of the desired RF signal and the frequency of the undesired image signal would have been too close together in frequency, thus making it difficult to separate the desired signal from its image. The problems of rejecting the image signal and of obtaining the desired IF selectivity (bandwidth) are two of the main factors in deciding upon the IF frequency for a receiver. The 455-KHz frequency has been a very popular frequency for most broadcast-band AM receivers, although some other frequencies have been used from time to time. Similarly, for broadcast FM receivers which tune from 88 to 108 MHz and need an IF bandwidth of about 200 KHz to pass the broadcast FM signal, the frequency of 10.7 MHz has been selected as a popular standard.

When we get away from broadcast-band receivers and start looking at special-purpose communication receivers, we may find that it is not so easy to select a single IF frequency which will satisfy both the need for a small IF bandwidth and good premixer image rejection. For instance, consider an SSB receiver which is to tune from 10 to 15 MHz in which we want an IF passband of only about 3 KHz. To get good selectivity here, we would need an IF frequency of 100 KHz or less, but this would mean that the desired-signal and the image-signal frequencies would be less than 200 KHz apart at 10 to 15 MHz (or only about several percentage points). This would make it difficult to design the input-tuned circuits prior to the mixer stage with enough selectivity to reject the image signal. Thus, for image-rejection purposes, it might be desirable to have an IF frequency of several Megahertz. This dilemma is resolved by building a double-conversion superheterodyne receiver which has two IF amplifiers (or more) at different frequencies.

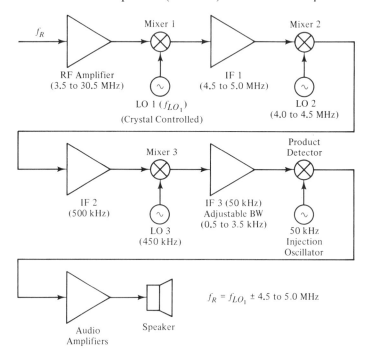

Figure 5-60 A triple-conversion single-sideband communications receiver.

Figure 5-60 shows a triple-conversion superheterodyne which employs *three* IF amplifiers, one of which is a tunable IF. Making the first IF amplifier tunable allows the use of a very stable crystal-controlled oscillator for the first local oscillator (it is often difficult to build a stable tunable oscillator at high frequencies). Tuning of the receiver over a 500-KHz range is accomplished by tuning the second local oscillator and the circuits in the first IF amplifier. By changing the crystals in the first local oscillator, the location of the 500-KHz range to be tuned can be changed (this particular receiver design was built to cover the range from 3.5 MHz to 30.5 MHz in 500-KHz steps). The image signal is 9.0 to 10.0 MHz removed from the desired signal. The 500-KHz signal out of the second mixer is amplified in the fixed-tuned second IF amplifier and is fed into the third mixer where it is mixed with the third local oscillator signal of 450 KHz to produce an IF signal of 50-KHz center frequency which is fed into a 50-KHz IF amplifier of adjustable bandwidth. The SSB signal out of the 50-KHz IF amplifier is detected through the use of a product detector consisting of a 50-KHz oscillator and another mixer. Thus, this receiver design gives good stability due to the crystal control of the first local oscillator, good image rejection because of the high frequency of the first IF, and good selectivity of the signal because of the low frequency of the third IF amplifier.

The superheterodyne principle was first applied to receiver circuits in communications systems, but in recent years, especially since the development of the single-sideband technique, the superheterodyne principle has also been applied to the transmitter end of the link. Figure 5-61 shows a typical SSB transmitter designed to operate over a large segment of the short-wave region. However, no matter what the frequency of operation, the generation of the SSB signal takes place at 4.0 MHz (for the missing carrier frequency), and the signal thus generated is heterodyned up or down to the desired transmitting frequency. Note that in the case where the 4-MHz signal is subtracted from the oscillator frequency, the sidebands are inverted. Thus if it is desired to transmit the *lower* sideband at 8 MHz, the *upper* sideband must be generated at the 4-MHz frequency. Again, two signals are present following the mixer, and it is necessary to include tuned circuits in the class B amplifier stages to eliminate the one that is not desired.

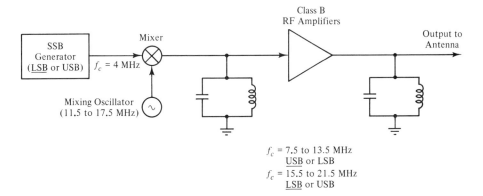

Figure 5-61 An SSB transmitter which uses the superheterodyne principle.

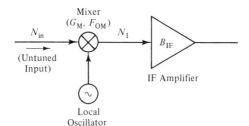

Figure 5-62 Superheterodyne without image rejection.

The superheterodyne principle is also encountered in many other places than radio transmitters and receivers. In Section 5.2.2 we looked at the mixing method of generating a large number of RF frequencies from a small number of crystals, and in Section 5.6 we shall look at the idea of phase-locked loops for receivers and detectors.

5.3.2.1 Noise contributions due to inadequate rejection of the super-heterodyne image signal. If a superheterodyne receiver does not have tuned circuits to reject the image signal, input noise in a passband B_{IF} at *both* the signal and image frequencies will contribute to the noise input to the IF amplifier. Referring to Figure 5-62, where $N_{in,L}$ is the noise in the passband B_{IF} at the frequency $(f_{LO} - B_{IF})$ and $N_{in,U}$ is the noise in the same passband at $(f_{LO} + B_{IF})$, we get

$$N_1 = G_M[(F_{oM} - 1)(kB_{IF}290°) + N_{in,L} + N_{in,U}] \qquad (5\text{-}25)$$

If an RF preamplifier is added with image-rejection tuning both prior to and following the preamplifier, we obtain (see Figure 5-63)

$$\begin{aligned} N_1 = G_M\{(F_{oM} - 1)\,(kB_{IF}290°) \\ + G_{RF}[(F_{OA} - 1)\,(kB_{IF}290°) + N_{in}]\} \end{aligned} \qquad (5\text{-}26)$$

where N_{in} is the input noise in the bandwidth B_{IF} at the desired signal frequency. However, if the tuned circuit between the preamplifier and the mixer is eliminated,

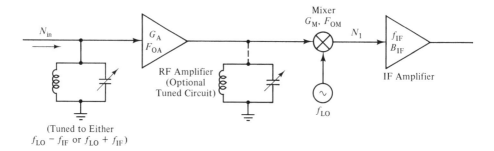

Figure 5-63 Superheterodyne with a tuned RF preamplifier.

the preamplifier's excess noise contribution is doubled, or

$$N_1 = G_M\{(F_{oM} - 1)\,(kB_{IF}\,290°)$$
$$+ G_{RF}[2(F_{OA} - 1)\,(kB_{IF}290°) + N_{in}]\} \tag{5-27}$$

In this analysis we have assumed that the presence of a tuned circuit at the signal frequency causes complete rejection of any *noise* at the image frequency, and in general this is approximately true, even if the complete rejection of *strong* image *signals* is not obtained. If a multistage RF preamplifier is used, it is desirable to have tuned circuits at least prior to the first preamplifier stage and between the last preamplifier stage and the mixer.

5.3.3 Special Receiver Designs

Quite often it is desirable to design a receiver that will track an incoming signal as that signal drifts or varies in frequency. This is especially true of situations where a large Doppler frequency shift may be present on the received signal (such as in the case of signals from satellites or space probes). Usually this tracking is done by controlling the frequency of one of the local oscillators in the superheterodyne receiver by means of a quasi-d.c. signal derived from a phase detector or FM discriminator which detects the signal out of the final IF amplifier stage.

5.3.3.1 Phase-locked receivers. These are receivers where the "d.c." frequency-controlling feedback signal is derived from a phase detector. Such a system is shown in Figure 5-64 as a microwave receiver, where the first local oscillator is a klystron which is phase locked to a fixed reference oscillator (a stabilized local oscillator or STALO) as described in Section 5.2.3. The first IF frequency is nominally 70 MHz, and there is a second local oscillator at 40 MHz, which is a

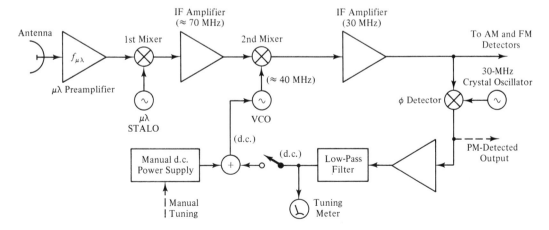

Figure 5-64   Design of a phase-locked-loop microwave receiver.

VCO that is controlled by the signal from the phase detector. Its frequency is automatically adjusted to maintain the center frequency of the received signal at *exactly* 30 MHz (the center frequency in the first IF is only nominally 70 MHz— it may be higher or lower due to Doppler shift, etc.). The output of the phase detector following the 30-MHz IF amplifier is fed through a low-pass filter (which prevents any deliberate FM or PM modulation on the signal from affecting the second local oscillator frequency) and is added to the nominal d.c. frequency-control signal supplied to the second local oscillator. The modulation on the 30-MHz IF signal may be detected by attaching conventional AM detectors or an FM discriminator to the 30-MHz IF output, or FM or PM on the signal may be detected from the phase detector output prior to its passage through the low-pass filter. The bandwidth of the low-pass filter must be less than the lowest frequency of modulation present on the signal, or else this part of the modulation will be eliminated from the 30-MHz IF signal due to the second local oscillator tracking the low-frequency FM or PM modulation.

5.3.3.2 Automatic frequency control (AFC).

Sometimes for simplicity where *exact* tracking of the *frequency* of the input signal is not needed, an automatic-frequency-control loop will be used in place of a phase-locked-loop in a receiver. The operating principles are very similar, except that in place of using an injection reference oscillator (such as the 30-MHz oscillator in Figure 5-64) and a phase detector, the "d.c." output from a discriminator tuned to the desired signal IF frequency is used. As the center frequency signal varies away from the discriminator's nominal center frequency, the d.c. output of the discriminator will increase and will *tend* to pull the signal back toward its center frequency by changing the frequency of the local oscillator. AFC loops are quite widely used in many different types of high-frequency receivers, even in the small pocket-sized FM broadcast radios.

At this point it should be noted that in a phase-locked-loop receiver, the *frequency* of the signal is tracked *exactly*, but there may be some *phase error* between the IF output signal and the reference signal phases. However, in the case of a receiver with AFC, there is a *frequency* error (i.e., the *frequency* of the IF output signal may differ from the desired nominal value).

5.3.4 The Use of Integrated Circuits in Communication Systems

Integrated circuits find application in three types of communication system circuits: analog circuits such as amplifiers, filters, oscillators, detectors, modulators, and timers; digital circuits such as logic gates, registers, flip-flops, memories, and micro-processors; and the interface area between analog and digital circuitry such as analog-to-digital (A/D) and digital-to-analog (D/A) converters, modems (FSK oscillators and detectors), digitally controlled analog switching matrices, and so on.

In the purely analog area most of the low-power functions formerly handled by transistors, small vacuum tubes, or diodes can be included in integrated circuits. The circuit chip incorporates the active devices and the resistors needed for the

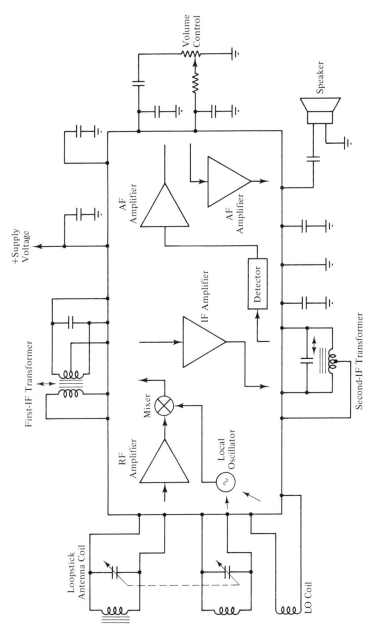

Figure 5-65 An AM radio on a chip.

circuit and some of the smaller capacitors.   However, the larger capacitors, inductors, resonant circuits, and so on needed by the circuit may be included as discrete components external to the chip.   For this reason, direct (d.c.) coupling is used as much as possible in the design of amplifiers and other analog circuits employing integrated circuits (I.C.s).   This entails more complicated biasing networks, but this is something that is easy to implement in designing the I.C. chip.

Figure 5-65 shows an AM radio with most of its active components and resistors on a single I.C. chip.   The chip has a number of points to which are attached the various discrete components (the RF and local oscillator resonant circuits, IF transformers, coupling and bypass capacitors, volume control, etc.).   Sometimes the L-C resonant circuit function is handled by resorting to an alternative circuit which does not require an L-C resonant circuit but rather uses some sort of R-C oscillator or a feedback amplifier with R-C feedback networks.   For example, most audio filtering functions are now done with operational amplifiers

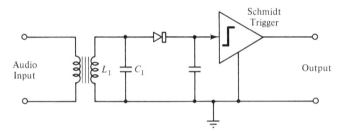

(a) L-C–Tuned Circuit Tone Detector

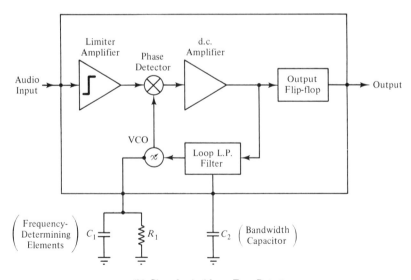

(b) Phase-Locked-Loop Tone Detector

Figure 5-66   Two methods of implementing a tone detector.

(I.C. chips) with R-C feedback. An audio-frequency tone detector can be implemented with a voltage-controlled oscillator, a phase detector, and a Schmidt trigger circuit as shown in Figure 5-66b. This I.C. tone detector (on a single chip, except for a couple of frequency-determining and bandwidth-determining capacitors) can replace the bulky L-C–tuned circuit shown in Figure 5-66a. The operation of the PLL tone detector is as follows: whenever the input signal's frequency is close enough to that of the VCO's free-running frequency, the PLL locks up and a d.c. output is produced (no d.c. output occurs if the input frequency is not within the bandwidth established around the oscillator frequency by the loop filter time constant).

The advent of modern digital technology implemented via MSI (medium-scale integration) or LSI (large-scale integration) integrated circuits has made available whole new concepts in communication system technology. A good example is the digital frequency synthesizer discussed in Section 5.2.2. Very often, because of the low cost of digital communication–link equipment and the inherent noise immunity of digital links, analog information is digitized and the digital signal is transmitted to the receiving end where it is converted back to an analog signal again. All the intermediate transmission, switching, and so on is handled digitally either via "hard-wired" logic circuits or through what is essentially a special-purpose digital computer (a central processing unit (CPU) plus communication peripherals and "programmed logic"). Also, the availability of digital communication links has encouraged the use of digital point-to-point communication rather than the use of voice links, especially as more and more business, technical, and governmental information is being handled in computer files rather than on ledger sheets, notebooks, and inventory cards. This trend will probably continue and accelerate in the future.

PROBLEMS

5-1. For each of the following categories, discuss what amplifier class or classes you might use and whether it is possible to use *an untuned single electronic device, a tuned single device, untuned push-pull devices,* and/or *tuned push-pull devices.*
  a. An audio or video amplifier.
  b. An RF amplifier for amplitude-modulated signals.
  c. An RF amplifier for angle-modulated signals.

5-2A. Identify the amplifier class in each of the following:
  a. To amplify an SSB signal.
  b. To produce an AM signal.
  c. To amplify a d.c. signal (what other requirement is necessary here)?

5-2B. Identify the type of limiter amplifier you would use
  a. In front of a Foster-Seeley discriminator in an FM receiver.
  b. To control the recording level in a portable tape recorder.
  c. As an audio compander (what *is* an "audio compander"?).
  In each case describe the output versus input voltage relationship and the response-time characteristics of the limiter.

5-3. What methods (name at least four) could you employ to reduce the various types of distortion that might occur in an audio amplifier?

5-4. Explain why inverse feedback will reduce both the nonlinear distortion and the internally generated excess noise in an audio amplifier. What else does the presence of the inverse feedback cause to happen to the amplifier's characteristics?

5-5. Why and how are series and shunt peaking coils used in a baseband amplifier where very wide bandwidth is required (such as in a TV video amplifier that has a bandwidth extending from about 30 Hz to about 4.5 MHz)?

5-6. In a cascade of several stages of tuned RF amplifiers, what two methods can be used to increase the total amplifier bandwidth?

5-7. Describe the operation of the tunnel-diode amplifier as an RF preamplifier. What other major component(s) (besides the tunnel diode itself) are needed to make it work as an amplifier?

5-8. If the nonlinear element in the *parametric* amplifier described in the text were to be a nonlinear capacitor, prove that the resistance seen looking into the signal port (terminal pair) is a negative quantity at resonance. Let $v_c(t) = K(q_c(t))^2$ where K is a constant.

5-9. Prepare a table showing the various types of limiters used in communication systems applications (give the limiter type or name, the amplitude-limiting characteristics, the response-time characteristics, the typical uses and/or applications, etc.).

5-10. An ideal transformer matches a 5-Ω load resistor to an amplifier whose output impedance is 1875 Ω. What does the turn ratio of the transformer need to be? If the high-impedance winding has 400 turns, how many turns are needed on the low-Z winding?

5-11. With regard to single-tuned RLC resonant circuits, discuss each of the following:
    a. Compare the current and voltage changes and the changes in the total impedance which occur in a series-resonance circuit as contrasted with those which occur in a parallel-resonance circuit as the frequency of the applied signal is varied through the resonance frequency.
    b. What is meant by the term *half-power frequencies*?
    c. What happens to the resonance frequency and the Q of a series-resonance circuit if we double the inductance value and halve the capacitance value? What about the same situation for a parallel-resonance circuit?
    d. What can be done to a resonance circuit to alter the shape and the passband width of its resonance curve?

5-12. If $f_o = 2$ MHz and $Q = 200$ for a series-resonance circuit, what is its -3-dB bandwidth?

5-13. In the circuit shown here, find the value of C necessary for resonance (i.e., to make $Z_{in} = R_{in} + j0.0$) and find the value of R_{in} at resonance. Circuit values:

$f_o = 10$ MHz
$k = 0.4$
$L_p = 5\mu H$
$L_s = 2\mu H$
$Q_{LP} = Q_{LS} = 100$

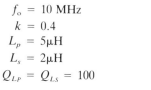
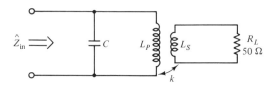

5-14.

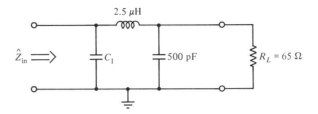

2.5 μH

$\hat{z}_{\text{in}} \Longrightarrow$ C_1 500 pF $R_L = 65 \, \Omega$

   a. What is this circuit *called*?   What is it used for?
   b. If the frequency of operation is 8 MHz, what should be the value of C_1?   Under these conditions, what would R_{in} then be?

5-15. Define what is meant by the following (in regard to two mutually coupled resonance circuits tuned to the same frequency).   What happens in each case?
   a. Critical coupling
   b. Loose coupling
   c. Tight coupling

5-16. With reference to the double-tuned resonance circuit,
   a. What determines the value of k at which *critical* coupling occurs (i.e., what determines k_c)?
   b. What determines the *actual* value of k which *does* occur?
   c. Sketch the amplitude-frequency response curves for the secondary voltage for $k/k_c = 1.0$, < 1.0, and > 1.0.

5-17. Sketch the equivalent circuit for a quartz frequency-determining crystal and analyze this circuit to get its reactance versus frequency characteristic curve.   Explain how you might utilize either the series-resonance or the parallel-resonance characteristic in an oscillator circuit.

5-18. Discuss how a phase-locked-loop frequency synthesizer works and tell how you would go about designing one for a specific application.   Discuss the use of the variable-modulo prescaler.

5-19. Design a PLL frequency synthesizer which produces an output covering the frequency range from 77.4 MHz to 97.2 MHz in steps of 200 KHz (you might want such a synthesizer for use in a digitally tuned broadcast-band FM tuner).   Use a single divider.
   Design a *direct-synthesis* frequency synthesizer to cover the same range of frequencies.

5-20. Discuss the difference between a *primary* frequency standard and a *secondary* frequency standard.   What determines into which class a particular frequency standard falls?

5-21. A *type 1* PLL frequency synthesizer maintains the correct frequency but does not have a zero phase error at the phase comparator, whereas a *type 2* PLL synthesizer will maintain zero phase error.   Sketch and describe a type 2 PLL control loop and explain why this is so.   (Hint: See a reference on basic servo control systems.)

5-22. A PLL frequency synthesizer utilizes a $\div 10/ \div 11$ prescaler.   Show how you would use this to obtain the following divide ratios:
   (a) 590   (b) 683   (c) 616   (d) 655

5-23. Draw the block diagram of the *phasing method* for generating a PM signal and explain how it operates by the use of a phasor diagram. Indicate what approximations you are using. Compare this with the *phasing method* for generating an SSB signal.

5-24. What is meant by *high-level* versus *low-level* AM modulation? Draw a block diagram of each and compare them as to the type of equipment required (amplifier classes, etc.) and the advantages and disadvantages of each.

5-25. What is a balanced modulator? Describe its input and output signals (frequency ranges, sinusoidal information signal, modulated signal, carrier signal, etc.). What are some of the things it is used for? Both the *filter* and the *phasing* methods of SSB signal generation employ balanced modulators. Discuss the similarities and the differences between these two methods and show that each does indeed produce an SSB signal.

5-26. In an AM transmitter using high-level modulation, the various parts of the transmitter have the following efficiencies:

the class C RF amplifier	70% efficient
the modulation transformer	97% audio-signal efficiency
	99% d.c. efficiency (secondary)
the class AB_1 audio modulator	45% efficient (exclusive of modulation transformer)

The *unmodulated* carrier RF output of the transmitter is to be 15 KW. What is the d.c. power input to each section of the transmitter (i.e., through the secondary of the modulation transformer to the class C stage and to the audio modulator amplifier) when it is being modulated by a 2-KHz tone at 80% peak modulation? What is the total d.c. input power and the total average RF output power? Repeat for the case of two audio tones of equal amplitude and 90% peak AM modulation.

5-27. Design the RF and local oscillator tuning circuits for a superheterodyne receiver for the broadcast band. The specifications are:

f range: 500 to 1700 KHz
C_{RF} range: 10 to 400 pF
C_1 and L_{RF} to be determined

f range: 955 to 2155 KHz
C_{LO} range: 10 pF to C_{max}
C_{max}, C_2 and L_{LO} to be determined (C_2 nominal value 25pF)

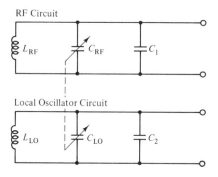

    a. Assume that C_{RF} and C_{LO} are on the same shaft and vary linearly from 10 pF to their respective maximum values. Determine the component values asked for.

    b. What is the *maximum* tracking error which occurs with your design (i.e., f_{RF} $- f_{LO} + 455$ KHz)?

    c. Assume that you could design your own ganged tuning capacitor instead of being limited to linearly varying capacitors. Design a circuit for perfect tracking. Give plots of C_{RF} and C_{LO} versus the angle of shaft rotation so that f_{RF} varies linearly with the rotation angle.

5-28. **a.** What is meant by the *image frequency* of a superheterodyne receiver? How do you calculate the image frequency?

    b. What frequency or frequencies might be involved when considering interference to other equipment outside of the receiver itself caused by the operation of the receiver (a single-conversion superheterodyne)? How do you calculate these frequencies?

5-29. An FM receiver is to be designed to cover the FM broadcast band (88 to 108 MHz). It uses an IF with a center frequency of 10.7 MHz, and the local oscillator frequency range is to be less than the range of received frequencies.

    a. Over what frequency range should the local oscillator tune?

    b. If the station being received is at 103.3 MHz,

        (1) What is the local oscillator frequency?

        (2) What is the *image* frequency?

        (3) Radiation from the receiver which might interfere with other equipment will occur at what frequencies?

5-30. Sketch the block diagram of a *double-conversion* superheterodyne receiver to receive the frequency range from 1.5 MHz to 30.0 MHz. Specify suitable IF frequencies and the local oscillators' frequency(ies) and/or frequency range(s).

6

ANTENNAS AND PROPAGATION

In this chapter we shall be considering the subject of transmitting, propagating, and receiving an information-bearing electromagnetic wave (or "signal") as just another "black box" in the communications system through which the information is passing (see Figure 6-1). As shown in Figure 6-1a, the modulated RF signal from the transmitter is applied to the terminals (or *port* if a waveguide-fed microwave antenna is being considered) of the transmitting antenna, T, which in turn "launches" an electromagnetic wave signal, which travels along the *propagation path* toward the receiving antenna, R. The receiving antenna collects a small portion of the launched signal, along with a certain amount of electromagnetic noise radiation and interfering signals, and "funnels" this conglomeration into the terminals of the receiving system, which is attached to the antenna's terminals or *port*. The net result of this operation can be viewed somewhat in the manner indicated by Figure 6-1b, where the rest of the communications system is viewed by the transmitting end of the system simply as a load resistor $R_{\text{term},T}$ attached to the output of the transmitter and driven with a current $I_{a,T}$.

From the receiving end of the communications system, the transmitting end, the antennas, the propagation path, the noise introduced, and any interfering signals picked up by the receiving antenna are viewed simply as an equivalent Thevenin

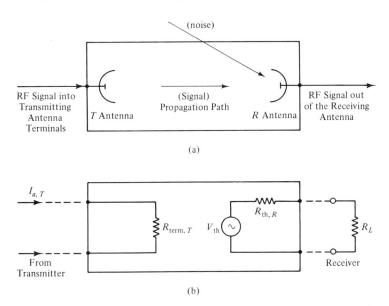

Figure 6-1   Representation of two antennas and the intervening propagation path as a "black box" in a communications system.

or Norton generator whose terminals replace the terminals of the receiving antenna. The equivalent Thevenin resistance value is referred to as the *terminal resistance* of the antenna, and the Thevenin voltage generator is a combination of several voltage generators representing the various voltages induced in the receiving antenna by impinging information and noise signals plus a random noise generator representing the thermal noise signal due to the resistive losses in the antenna itself. The particular Thevenin generator that represents the desired information signal will have an amplitude proportional to the $I_{a,T}$ into the transmitting antenna at the other end of the communications system link.

It might be well to note that thus far we have restricted the discussion to the use of the word *resistance* instead of the more general term *impedance*. What we have implied by doing this is that the reactive components of the various impedances involved have been set to zero by "tuning" the antennas, the receiver inputs, and so on to the exact operating frequency. We shall continue this assumption to simplify the notation, except in those cases where it is actually necessary to include the reactive effects as part of the then current discussion.

6.1 DIFFERENT WAYS OF MODELING AN ANTENNA

Turning our attention to one antenna, we find that a useful model for a low-frequency antenna (such as a wire or rod antenna) is that shown in Figure 6-2, where, once again, a simplification has been made by not showing the equivalent inductive and capacitive effects which are present when the antenna is not properly

tuned to the operating frequency. This equivalent circuit includes a voltage generator which represents the effective voltage induced in the antenna conductors by
impinging electromagnetic field (EM field) radiation representing the desired incoming signal, unwanted signals, and radiated EM noise which is incident upon
the antenna. Since most antennas are bilateral in nature (i.e., they can transmit
or receive equally well) some means must be provided to represent a "sink" for
power that is radiated outward from (or "leaves") the antenna. This is taken care
of in Figure 6-2 by the inclusion of the *radiation resistance*, R_{rad}, as part of the
antenna model.

The other resistance shown, R_{loss}, represents the ohmic and dielectric losses
in the conductors and the insulators (and in some cases the loss in a ferromagnetic
core) which make up the antenna structure. Since R_{loss} represents actual resistances
or resistive effects, it can be represented as a source of thermal noise power by
including a mean-square noise-voltage or noise-current generator (as described in
Chapter 3) into the model. This thermal noise source is often important when
the model of the antenna is used to describe a receiving antenna, but is usually
not considered when describing a transmitting antenna. Also, in most (but not
all) cases, the voltage generator representing induced signal voltages is usually of
no interest when talking about transmitting antennas.

Another type of antenna model is one that begins with the assumption of an
incident plane EM wave having a power density of ρ Watts per square meter. If,
as is shown in Figure 6-3, this wave is incident upon an antenna structure, the
antenna will cause a disturbance in the propagating wavefront, absorbing, scattering, or reradiating a certain portion of the power in the incident wave. The amount
of power in the incident wave which is thus affected is called the *captured* power,
and it is represented by the product of ρ times an area called the *capture area* of
the antenna, as defined in Equation 6-1 which breaks A_{capt} down

$$P_{capt} = \rho A_{capt} = \rho \left[A_{scat} + A_{eR} + A_{loss} \right] \qquad (6\text{-}1)$$

into three components: A_{scat} representing the power scattered or reradiated from
the antenna, A_{eR} representing the power delivered to the load attached to the
antenna's terminals, and A_{loss} representing the power lost (absorbed) in the ohmic-
type losses of the antenna itself. A_{scat} is called the antenna's *scattering aperture
area*, A_{eR} is called the *effective receiving aperture area* of the antenna, and A_{loss} is
called the antenna's *loss aperture area*.

It is important not to try to visualize A_{capt} as a physical cross-sectional area
in the plane wave. The product $\rho \times A_{capt}$ merely is an indication of how much

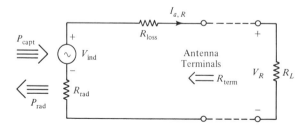

Figure 6-2 Circuit-model
representation of an antenna.

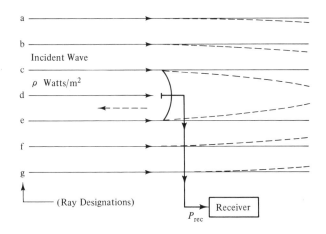

Figure 6-3 The "capture" of an electromagnetic wave by an antenna.

power in the wavefront is disturbed from its normal direction of propagation. The effects of placing an antenna structure in a plane wave actually extend out to an infinite distance away from the antenna structure, with the magnitude of the effects becoming less and less pronounced as we go farther away from the antenna. A_{capt} actually represents the areal integration of a "wavefront disturbance factor," which is less than unity over the infinite cross-sectional area perpendicular to the plane wave and which has its maximum value located at the location of the antenna.

To some extent (at least for a receiving antenna), we can make a correspondence between the Thevenin circuit model of Figure 6-2 and the aperture model of Equation 6-1, since from Figure 6-2 we have

$$
\begin{aligned}
P_{capt} &= V_{ind}\, I_{a,R} = P_{rad} + P_{eR} + P_{loss} \\
&= (I_{a,R})^2 [R_{rad} + R_L + R_{loss}] \\
&= \left(\frac{V_{ind}}{R_{capt}}\right)^2 [R_{rad} + R_L + R_{loss}]
\end{aligned}
\tag{6-2}
$$

In Section 6.3, we will be developing a relationship between the various *gains* of an antenna and some of the aperture areas we have just discussed. At that point, we will try to establish further interrelationships among the various antenna models.

6.2 BASIC ANTENNA RADIATION THEORY

Having briefly considered antennas and propagation as parts of a communications system, and having looked at a couple of different ways in which we might model an antenna, we now turn our attention to a more radiation-oriented view of antenna theory. We won't be discussing in any great detail the electromagnetic field theory of antennas, since this is properly a topic for an entire book and many excellent books have been written on the subject. Rather, we will look at the concepts of antennas and their radiation from the geometric and systems point of view.

6.2.1 *The Isotropic Radiator*

Consider an antenna which radiates equally well in all directions, and at a distance ℓ from the antenna, let there be a receiving surface of area A_{eR}. If the total RF power into the antenna is given by P_T, the power radiated from the antenna per unit area of a sphere of radius ℓ is given by

$$\rho = \frac{P_T}{A_{tot}} = \frac{P_T}{4\pi\ell^2} \text{ watts/m}^2 \tag{6-3}$$

and the power intercepted by the receiving area is

$$P_R = P_T\left(\frac{A_{eR}}{4\pi\ell^2}\right) \text{ watts} \tag{6-4}$$

where A_{eR} is the area of the receiving antenna *effective aperture* in square meters. This value may be somewhat different from the actual physical area A_R. Figure 6-4 illustrates the situation being described.

Normally a system such as that illustrated in Figure 6-4 will be bilateral (that is, the position of the transmitter and the receiver can be interchanged, and the receiver will still receive the same fraction of the transmitted power as in the original situation). If this is done, it can be proven by field theory that the effective aperture area of the isotropic antenna is equal to $(\lambda^2/4\pi)$. The effective aperture area of an antenna may be either larger or smaller than its physical cross section as viewed from the direction of the incoming wave. For the case of thin-wire antennas, the effective aperture A_{eR} is much larger than the physical cross section, while for a microwave dish antenna, it may be somewhat smaller because of surface inaccuracies.

6.2.2 *Effective Aperture*

The effective aperture area of a receiving antenna can be defined as

$$A_{eR} = \frac{P_R}{\rho} \tag{6-5}$$

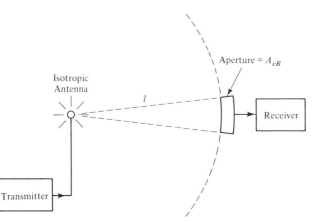

Figure 6-4   An isotropic radiator antenna.

where P_R is the power out of the antenna terminals into a matched load R_L and ρ is the radiation intensity in watts per square meter incident upon the antenna. The power into a load resistor is V_R^2/R_L, where V_R is the rms (root-mean-square) voltage at the antenna terminals, but if the system is impedance matched for maximum power transfer, R_L is made equal to $R_{\text{term}} = R_{\text{rad}} + R_{\text{loss}}$. The radiation intensity is given by $\rho = E_f^2/Z_o$, where E_f is the rms field intensity in volts per meter and Z_o is the impedance of free space (equal to the magnitude of the E-field vector divided by the magnitude of the H-field vector $= 120\,\pi = 377$ ohms). For the case where $R_{\text{loss}} = 0$ we then get

$$A_{eR} = \frac{V_R^2/R_{\text{rad}}}{E_f^2/Z_o} = \left(\frac{V_R}{E_f}\right)^2\left(\frac{Z_o}{R_{\text{rad}}}\right) \tag{6-6}$$

which can be applied to any lossless antenna to find its effective aperture area. For example, for a very short loaded dipole (one whose length $\mathscr{L}$ is $<< \lambda/2$), $R_{\text{rad,SD}} = 80\,(\pi\mathscr{L}/\lambda)^2$, and $V_{R,\text{SD}} = (\mathscr{L}E_f)/2$. Thus,

$$\begin{aligned} A_{eR,\text{SD}} &= \left(\frac{\mathscr{L}E_f}{2E_f}\right)^2\left[\frac{120\,\pi}{80(\pi\mathscr{L}/\lambda)^2}\right] \\ &= \left(\frac{\mathscr{L}^2}{4}\right)\left[\frac{3(1/\pi)\lambda^2}{2\mathscr{L}^2}\right] = \frac{3\lambda^2}{8\pi} = 0.119\,\lambda^2 \end{aligned} \tag{6-7}$$

and for the half-wave dipole where $R_{R,\text{HD}} = 73$ ohms and $V_{R,\text{HD}}$ is given by $(E_f\lambda)/\pi$, we get

$$A_{eR,\text{HD}} = \tfrac{1}{4}\left(\frac{E_f\lambda/\pi}{E_f}\right)^2\left(\frac{120\,\pi}{73}\right) = \frac{1}{4}\left(\frac{\lambda^2}{\pi^2}\right)\left(\frac{120\,\pi}{73}\right) = \frac{30\,\lambda^2}{73\,\pi} = 0.1308\,\lambda^2 \tag{6-8}$$

6.2.3 Effects of Ohmic Losses in Wire Receiving Antennas

If ohmic losses exist, they can often be represented as a loss resistance R_{loss} in series with the radiation resistance R_R and the load resistance R_L with an equivalent Thevenin voltage source $V_{\text{ind}} \propto E_f\mathscr{L}$. In this case the maximum power into the load comes when R_L is made equal to $R_{\text{term}} = R_{\text{rad}} + R_{\text{loss}}$. The effective aperture then becomes

$$\begin{aligned} A_{eR} &= \left(\frac{I_L^2 R_L}{\rho}\right) = \left(\frac{V_{\text{ind}}}{R_{\text{rad}} + R_{\text{loss}} + R_L}\right)^2 R_L\left(\frac{Z_o}{E_f^2}\right) \\ &= \left[\frac{E_f^2\mathscr{L}^2}{4(R_{\text{rad}} + R_{\text{loss}})^2}\right]\left[\frac{(R_{\text{rad}} + R_{\text{loss}})Z_o}{E_f^2}\right] \\ &= \frac{\mathscr{L}^2 Z_o}{4(R_{\text{rad}} + R_{\text{loss}})} = \frac{30\pi\mathscr{L}^2}{R_{\text{rad}} + R_{\text{loss}}} \\ &= \frac{120\pi\mathscr{L}^2 R_L}{(R_L + R_{\text{rad}} + R_{\text{loss}})^2} = \frac{30\pi\mathscr{L}^2 R_L}{(R_{\text{rad}} + R_{\text{loss}})^2} \end{aligned} \tag{6-9}$$

A_{eR} represents the power which is captured by the antenna and is delivered to the load, while some of the other captured power is lost in the ohmic losses in the antenna. This latter power is proportional to the *loss area* A_{loss} (or loss aperture as it is usually called). It is given by

$$A_{\text{loss}} = \frac{I_{a,R}^2 \, R_{\text{loss}}}{\rho} = \frac{30\pi \mathcal{L}^2 R_{\text{loss}}}{(R_{\text{rad}} + R_{\text{loss}})^2} \tag{6-10}$$

We can also consider a *scattering aperture* (proportional to the power which hits the antenna and is reflected back or scattered off in some other direction). The scattering aperture is given by

$$A_{\text{scat}} = \frac{30\pi \mathcal{L}^2 R_{\text{rad}}}{(R_{\text{rad}} + R_{\text{loss}})^2} \tag{6-11}$$

Summing all three apertures (without the requirement that $R_L = R_{\text{rad}} + R_{\text{loss}}$) gives what is called the *capture* aperture, which represents the amount of the power in the incident wave which is removed from the wave and which is either delivered to the load, dissipated as heat or scattered off in some other direction from which it was originally traveling. The capture aperture is given by

$$A_{\text{cap}} = A_{eR} + A_{\text{loss}} + A_{\text{scat}} = \frac{V_{\text{ind}}^2 (R_L + R_{\text{loss}} + R_{\text{rad}})}{\rho (R_L + R_{\text{loss}} + R_{\text{rad}})^2}$$

$$= \frac{120\pi \mathcal{L}^2}{R_L + R_{\text{loss}} + R_{\text{rad}}} \tag{6-12}$$

$$= \frac{60\pi \mathcal{L}^2}{R_{\text{loss}} + R_{\text{rad}}} \quad \text{IF } R_L = R_{\text{loss}} + R_{\text{rad}}$$

One problem with antennas that are physically small compared to the wavelength of the radiation for which they are designed is that R_{rad} is small. Thus, we find that R_{loss} is often almost equal to R_L (when R_L is made equal to R_{term}). Therefore A_{loss} is often comparable in size to A_{eR}, and the ohmic efficiency of the antenna is poor (only about 50%), while A_{eR} is small as is indicated by Equation 6-9, resulting in a small effective aperture area for a receiving antenna (or a poor gain for a transmitting antenna as will be explained in the following sections).[1]

6.3 *ANTENNA GAIN AND FRII'S TRANSMISSION EQUATION*

Suppose now that an antenna radiates only over a hemisphere as is illustrated in Figure 6-5. Then the power density in watts per square meter will be twice as great at the receiving area than if the transmission antenna were an isotropic

---

[1] Note that this discussion does not apply to "aperture" antennas such as horns, dish antennas, and the like which are essentially impedance-transforming devices which match the impedance of free space (≈ 377 ohms) to the characteristic impedance of a waveguide or a transmission line. However, we could still think of $A_{\text{capt}} = A_{eR} + A_{\text{scat}} + A_{\text{loss}}$ but with A_{scat} and $A_{\text{loss}} \ll A_{eR} \sim A_{\text{capt}}$.

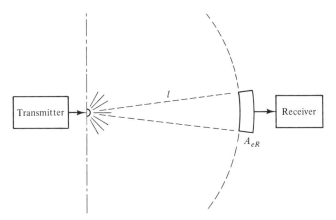

Figure 6-5 A hemispherical radiator antenna.

radiator, and we would say that the transmitting antenna has a gain of 2. Thus,

$$P_R = P_T G_T \left(\frac{A_{eR}}{4\pi \ell 2} \right) \text{ watts} \qquad (6\text{-}13)$$

where $G_T = 2$ in this example. In this case we have used a gain ratio G_T to characterize the transmitting antenna and an effective aperture area to characterize the receiving antenna. For bilateral symmetry considerations where we put a signal into the terminals of the receiving antenna and determine what signal we get out of the terminals of the transmitting (formerly) antenna, we would surmise that G_T and A_{eR} are somehow related to each other for a particular antenna. To find this relationship, we must go to EM field theory.

For example, the maximum gain of a short dipole, calculated from EM field theory, is 1.5, and that of a half-wavelength dipole is 1.64. Comparing the ratios between the lossless effective aperture and the gain for each of these, we find

$$\frac{A_{eR,\text{SD}}}{G_{\text{SD}}} = \frac{3\lambda^2/8\pi}{3/2} = \frac{\lambda^2}{4\pi} \qquad (6\text{-}14)$$

for the short dipole and

$$\frac{A_{eR,\text{HD}}}{G_{\text{HD}}} = \frac{30\lambda^2/73\pi}{1.64} = 0.25 \, \lambda^2/\pi \qquad (6\text{-}15)$$

for the half-wave dipole. And doing this for other common antennas where the parameters may be easily calculated from field theory, we find that for any antenna

$$A_{eR} = \left(\frac{\lambda^2}{4\pi} \right) G_{\text{max}} \qquad (6\text{-}16)$$

Thus, for the isotropic antenna where $G_{\text{max}} = 1.0$, we get

$$A_{eR,\text{iso}} = \lambda^2/4\pi \qquad (6\text{-}17)$$

Also, for the transmitting-end antenna in Frii's transmission equation in Equation

6-13, we can define an *effective transmitting aperture*

$$A_{eT} = \left(\frac{\lambda^2}{4\pi}\right) G_T \tag{6-18}$$

as well as a gain ratio for the receiving-end antenna

$$G_R = \left(\frac{4\pi}{\lambda^2}\right) A_{eR} \tag{6-19}$$

Thus, we can write the Frii's equation for P_R in several different forms:

$$P_R = P_T \left(G_T G_R \left(\frac{\lambda}{4\pi\ell}\right)^2\right) \text{ Watts} \tag{6-20a}$$

$$P_R = P_T \left(G_T \frac{A_{eR}}{4\pi\ell^2}\right) \text{ Watts} \tag{6-20b}$$

$$P_R = P_T \left(G_R \frac{A_{eT}}{4\pi\ell^2}\right) \text{ Watts} \tag{6-20c}$$

$$P_R = P_T \left(\frac{A_{eT} A_{eR}}{\lambda^2\ell^2}\right) \text{ Watts} \tag{6-20d}$$

These four forms are four different forms of Frii's transmission formula for the lossless case (i.e., where there is no absorption along the propagation path). If losses are involved, these will reduce the power out of the terminals of the receiving antenna. The propagation-path losses are usually accounted for by multiplying each of the above by a path transmission factor H_P, giving, for example,

$$P_R = P_T G_T G_R \left(\frac{\lambda}{4\pi\ell}\right)^2 H_P \tag{6-21}$$

We shall see later that for the most part G_R and G_T are actually functions of the spherical coordinate angles θ and ϕ defining the direction from the source to the small receiving area. Thus, the foregoing equations apply only for one particular orientation between the antennas of the transmitter and the receiver. Usually this is taken to be the orientation for which both $G_R(\theta, \phi)$ and $G_T(\theta, \phi)$ are at their maximum values. We will assume this unless stated otherwise.

If we look at the right-hand side of Equation 6-21, we see that it is composed of five factors, one being the transmitter power and the others being dimensionless gain factors. G_T and G_R and H_P have already been discussed as the gain factors of the transmitting antenna, the receiving antenna, and the propagation path, respectively. The quantity $(\lambda/4\pi\ell)^2$ is a dimensionless quantity depending upon the spacing between the transmitter and the receiver and upon the frequency of operation. It is often referred to as the free-space loss factor (although as expressed here it is actually a gain factor of less than unity). Since the Frii's equation can be written as a product of factors, the calculation can often be handled through the use of the decibel system.   This is illustrated below:

Power transmitted	(dBm units)
+ $(G_T)_{dB}$	(dB)
+ $(G_R)_{dB}$	(dB)
− free space loss	(dB)
− propagation losses	(dB)
= the power received	(dBm)

(6.22)

This method of solving a system problem is called a *power budget.*

Example:

Let a ground station transmitting a signal to a satellite in synchronous orbit 22,300 sm above the earth have a 26-in. antenna with a gain of 1450 (31.5dB) and let it be transmitting at 8.5 GHz.   The transmitter power into the antenna is 10 KW and the satellite receiving antenna gain is 12.6 (or 11.0 dB).   What is the power received at the satellite?   There are 5 dB of propagation loss on the path.

Solution:

10 KW	G_T	free-space loss	G_R	propagation losses	power received
70 dBm	+ 31.5 dB	+ (−203 dB) +	11.0 dB +	(−5.0) dB =	−95.5 dBm

Thus, the received power level is −95.5 dBm or about 2.8184×10^{-13} watt of power into the receiver of the satellite.

In looking at the power budget or the Frii's transmission equation, it is interesting to notice how the various factors change with frequency.   Assume that the antenna apertures and the spacing between the transmitter and receiver are kept constant.   Then as the frequency increases (λ decreases), we find that the free-space factor $(\lambda/4\pi\ell)^2$ decreases but that the antenna gains $(4\pi A_e/\lambda^2)$ increases for a given aperture area.   Since there are two antennas involved, the product of the two antenna factors and the free-space factor will increase proportional to f^2, or the total power received varies proportional to f^2:

$$P_R = P_T \left[\frac{A_{eT} A_{eR}}{\lambda^2 \ell^2} \right] = P_T \left[\frac{A_{eT} A_{eR}}{\ell^2 c^2} f^2 \right] \text{ watts}$$

(6-23)

Thus, for a given antenna size, it would be desirable to operate at higher frequencies.

There is a practical upper limit to this, however, which is established by the ability to generate large quantities of power at high frequencies, the increased atmospheric and ionospheric losses (H_p getting smaller) at high microwave frequencies, and the difficulty of accurately constructing antennas for the higher frequencies (i.e., A_e will become much less than the physical cross section of the antenna). For this reason, the optimum frequencies for the long-range transmission of signals lie in the middle region of the microwave band (from about 4 to 10 GHz).

Earlier, it was shown that if an antenna radiated only over 2π steradians (a hemisphere), it would deliver twice the power density to a point on that hemisphere than it would if it were an isotropic radiator which radiated over all 4π steradians of the sphere. We said such an antenna had a gain G_T of 2. If the antenna radiated only over one-fourth of a sphere (π steradians), then its gain would be 4, and so on. However, we do not find antennas that radiate uniformly over some portion of a sphere (i.e. over some solid angle given in steradians); rather they radiate with different intensities in different angular directions, and the computation of the gain is a more complicated matter to calculate than simply dividing 4π by the solid angle over which the radiation occurs. This is discussed in the next section.

6.4 ANTENNA PATTERNS

In the preceding section it was noted that the gain of an antenna is a function of the angular direction from the antenna to the receiving point (considering that the antenna is a transmitting antenna). Normally, when we speak of "gain," it is the power gain to which we are referring. However, when we measure the variation of gain with angular direction, we may consider the "pattern" of either the gain or of the E or the H field of the antenna. In the following discussion, the patterns referred to will be the *power gain* patterns of the antenna under discussion. Figure 6-6 shows the power-gain pattern for a typical microwave parabolic reflector antenna. The scale indicated is the value of the power gain. Here it is assumed that the pattern is only a function of θ, the angle by which the receiving point is displaced from the center line of the antenna (axial symmetry). It is assumed that there is no variation with the angle ϕ, which is the azimuth angle about the centerline of the antenna symmetry. As can be seen from the drawing, there are several sections (or lobes) to the pattern, with the large center lobe being referred to as

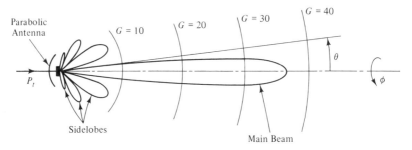

Figure 6-6 The power-gain pattern of a microwave parabolic antenna.

the "main beam" of the antenna. The smaller lobes are called the "sidelobes" of the antenna pattern. If the maximum gain as illustrated, for this example, is 37 (at the angle $\theta = 0$ degree), then the power *density* at a point ℓ from the transmitting antenna will be given by

$$\rho_\ell = P_T \left[\frac{G_{T_{max}}}{4 \pi \ell^2} \right] = P_T \left[\frac{37}{4 \pi \ell^2} \right] \text{Watts/m}^2 \tag{6-24}$$

while the power density received at a point at any angle θ from the centerline will be

$$\rho_\ell(\theta) = P_T \left[\frac{G_T(\theta)}{4 \pi \ell^2} \right] \text{Watts/m}^2 \tag{6-25}$$

Thus, the power density received at a point θ degrees off the axis of symmetry of this parabolic radiator will be $G_T(\theta)$ times that which would have been received from an isotropic radiator. At some values of θ, the received power density would be zero or near zero (these are called the nulls in the pattern). The sidelobes of the pattern illustrated in Figure 6-6 can be visualized as cone-shaped areas in space where the received power density will be substantial, but still much less than it would be if the receiving point were on the centerline where θ equals zero.

If the power density at a distance ℓ from the antenna is integrated over the entire 4π steradians of solid angle in the sphere surrounding the antenna, the total power received at the surface of this sphere of any radius ℓ must be equal to the power into the antenna, or

$$P_T = \int_{4\pi} \rho_\ell(\theta) \, d\Omega = \frac{P_T}{4 \pi \ell^2} \int_{4\pi} G_T(\theta) \, d\Omega \tag{6-26}$$

from which we have

$$\int_{A_{sphere}} G_T(\theta) \, dA = 4\pi\ell^2 \tag{6-27}$$

But the element of area dA subtended by the solid angle $d\Omega$ on the surface of the sphere of radius ℓ is equal to $dA = \ell^2 \, d\Omega$. Thus,

$$\int_{4\pi} G_T(\theta) \, d\Omega = 4\pi \tag{6-28}$$

In our case, where we have an antenna with axial symmetry, the solid angle $d\Omega$ is related to the plane angles $d\theta$ and $d\phi$ by the relation

$$d\Omega = \sin \theta \, d\theta \, d\phi \tag{6-29}$$

and, therefore, integrating this relationship for $d\Omega$ over the entire solid angle subtended from a point (4π steradians), we get

$$\int_0^{2\pi} \int_0^\pi G_T(\theta) \sin \theta \, d\theta \, d\phi = 2 \pi \int_0^\pi G_T(\theta) \sin \theta \, d\theta = 4\pi \tag{6-30}$$

or

$$\int_0^\pi G_T(\theta) \sin \theta \, d\theta = 2 \tag{6-31}$$

If we let the beamwidth be equal to $\Delta\Omega$ steradians (here letting this figure represent the average beamwidth—what the beamwidth would be if the gain suddenly dropped from its maximum value to zero rather than smoothly from maximum to zero, and if there were no sidelobes to be considered), then the maximum gain would be

$$G_{T_{\max}} = \frac{4\pi}{(\Delta\Omega)_{\text{beam}}} \tag{6-32}$$

Similarly, if we know what the maximum gain of an antenna is, we can approximate its beamwidth by the inverse relationship

$$(\Delta\Omega)_{\text{beam}} = 4\,\pi/G_{T_{\max}}$$
$$= \lambda^2/A_{eR} \text{ steradians} \tag{6-33}$$

Actual accurate specifications of the beamwidth are usually given in terms of the half-power beamwidth, or the point at which the value of the Poynting vector drops to one-half its maximum value, or the point where the E or H field magnitude drops to 0.707 of its value in the center of the beam. It is found in many cases that this half-power beamwidth is very close to the beamwidth given by Equation 6-33.

6.5 GAINS AND PATTERNS OF TYPICAL ANTENNAS

One type of antenna in use at the higher frequencies is the beam antenna. This may range from a half-wave dipole with director and reflector elements (see Figure 6-7) at the VHF frequencies (30 to 300 MHz) to a parabolic reflector element fed

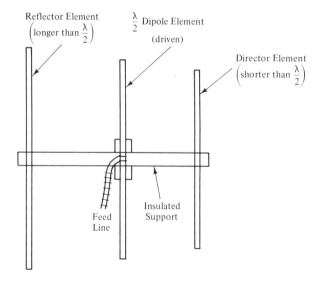

Figure 6-7 VHF 3-element beam antenna.

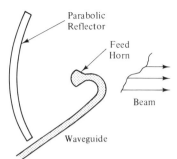

Figure 6-8 Microwave parabolic beam antenna.

by a horn from a waveguide (see Figure 6-8) at the microwave frequencies. The gain of a typical three-element VHF beam antenna may run from about 5 or 10 dB to 20 dB while the gain of a very large parabolic dish antenna may run as high as 70 dB or more. As indicated by Equation 6-19, the higher the frequency of operation of an antenna with a given aperture area, the higher the gain will be, with the gain being proportional to the square of the diameter of the antenna if it is a circular aperture and to the square of the frequency.

So far, we have discussed only the geometric or lossless gain of the antenna. The actual "gain"[2] of a given antenna is usually somewhat less than the amount which would be indicated by the foregoing discussion. This reduction in "gain" stems from two factors: construction inaccuracies and ohmic losses. The construction inaccuracies (referred to as surface irregularities in a dish antenna) become important only at the higher microwave frequencies, while ohmic losses may be important at all frequencies. The effect of construction inaccuracies is to lower the effective aperture area of a given antenna, thereby decreasing the gain of the antenna but also increasing its beamwidth. On the other hand, ohmic losses will lower the "gain" of an antenna without affecting its beamwidth. When the term "antenna efficiency" is used, the efficiency factor may include one or both of the preceding "losses"—it is important that the term antenna efficiency be defined whenever it is used.

The concept of half-power beamwidth was mentioned earlier. The beam limits for the half-power beamwidth may be viewed as the points where the gain factor $G(\theta, \phi)$ falls to half of the value of $G_{\max}$. For an antenna having axial symmetry, the solid angle of the antenna beam is related to the planar angle $\Delta\theta_{\text{beam}}$ by

$$\Delta\theta_{\text{beam}} = \sqrt{\left(\frac{4}{\pi}\right)\Delta\Omega} \text{ radians} \tag{6-34}$$

as shown by Figure 6-9.

[2] In the discussions of this text, antenna ohmic losses will be treated separately from the antenna gain. This is not always the case and the reader should be aware of this when reading other books or papers. In some cases D "directivity" is used for the "lossless" gain (i.e., $D > G$). It is also called the geometric gain.

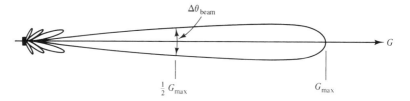

Figure 6-9   The half-power planar beamwidth angle (axial symmetry).

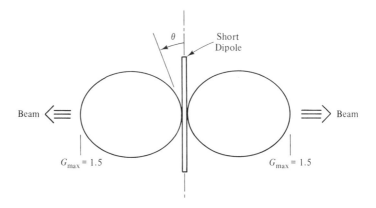

Figure 6-10   The pattern of the short dipole antenna.

The basic low-frequency antenna is the short dipole whose pattern is described by $G(\theta, \phi) = 2 \times \sin^2 \theta$ as illustrated in Figure 6-10. The maximum gain of the short dipole is equal to 1.5 and occurs at angles perpendicular to the direction of the dipole axis. The short dipole is not an efficient antenna for radiating power because of a low ratio of radiation resistance to the ohmic loss resistance (see Section 6.7). Consequently the longer $\frac{1}{2}$-wavelength dipole is usually the one which is used, although the $\frac{3}{2}$-wavelengths dipole is also sometimes used.[3] The pattern of the half-wavelength dipole (see Figure 6-11) very closely resembles that of the short dipole, the main difference between the two antennas being in the antenna efficiency and the driving-point (terminal) impedance. The half-wavelength antenna is widely used as the standard antenna at HF and VHF frequencies and is often referred to as the measurement standard for other types of HF and VHF antennas. Quite often the gain of a specific HF or VHF antenna will be given in decibels referred to the gain ($G = 1.64$ or $(G)_{dB} = 2.15$) of the standard half-wave dipole antenna rather than being referred to the gain ($G = 1$ or $(G)_{dB} = 0$) of the isotropic antenna. Again, it is important that when a gain figure is given for an antenna in the HF or VHF region, the person using this figure understand which reference is being used (the half-wave dipole or the isotropic antenna). This is usually not

---

[3] Although the pattern of a full-wave or $\frac{3}{2}$-wave dipole is considerably different from that of the short dipole or $\frac{1}{2}$-wave dipole.

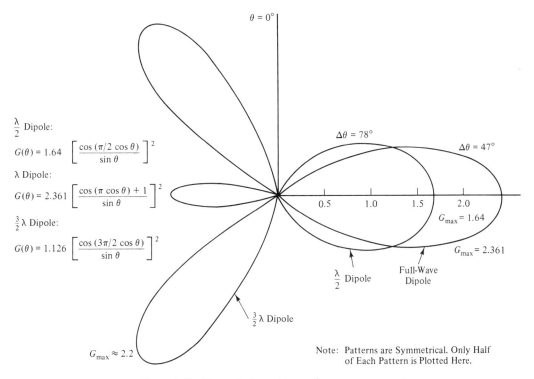

$\dfrac{\lambda}{2}$ Dipole:

$G(\theta) = 1.64 \left[\dfrac{\cos\,(\pi/2\,\cos\,\theta)}{\sin\,\theta} \right]^2$

λ Dipole:

$G(\theta) = 2.361 \left[\dfrac{\cos\,(\pi\,\cos\,\theta) + 1}{\sin\,\theta} \right]^2$

$\dfrac{3}{2}\lambda$ Dipole:

$G(\theta) = 1.126 \left[\dfrac{\cos\,(3\pi/2\,\cos\,\theta)}{\sin\,\theta} \right]^2$

Figure 6-11   The radiation patterns of some common antennas.

a problem in the microwave region for the isotropic radiator is almost always taken to be the reference at these frequencies.

6.6 ANTENNA POLARIZATION

When power is radiated from an antenna in a direction parallel to the earth's surface, the orientation of the antenna will determine whether the radiation propagates with the E vector of the EM radiation parallel to or perpendicular to the ground.   The first case is called horizontal polarization, while the second is called vertical polarization.   When receiving a wave with one polarization, if the receiving antenna is oriented in such a manner as to have the opposite polarization, no signal will enter into the receiving antenna.   Of course, it is possible to have the polarization of either the wave or the antenna such that the E-plane vector is neither horizontal nor vertical.   In this case, no signal will enter into the receiving antenna as long as the polarization of the incoming wave, and the antenna polarization are kept perpendicular to each other.   If the incoming wave and the receiving antenna have the same angle of polarization, then the entire signal will be intercepted by the receiving antenna, and if the polarization of the antenna and the wave are at some angle to each other which we denote by γ, then a fraction (cos γ) of the incoming E field will be intercepted by the antenna, or a fraction $(\cos \gamma)^2$ of the power in the incoming wave will be coupled into the receiving antenna and the receiver connected to it. The fact that if a polarization mismatch does occur (i.e., $\gamma \neq 0$) not all of the signal power density at the receiving point is coupled into the receiving system is accounted for by including a *polarization loss factor* in the system calculations. Again, as in the case of the free-space loss factor, this is not a loss in the ohmic sense, but merely an expression of the fact that some of the signal power is going elsewhere than into the receiving system. This polarization "loss" often becomes important, especially if the wave is transmitted through an ionized medium which may produce a change in the angle of polarization of the wave passing through it. Such a shift in the angle of polarization of the incoming wave at the receiving point may well be a function of time and medium conditions, so that it is not possible to fix the polarization angle of the receiving antenna to compensate for the shift that occurs in the polarization angle during the propagation of the signal.

Quite often the term *EFFECTIVE ISOTROPIC RADIATED POWER (EIRP)* is used in connection with a transmitting station. What this implies is the product of the transmitter power times the maximum gain of the transmitting antenna referred to an isotropic radiator. Sometimes it is more convenient to use the antenna gain of the subject antenna referred instead to the maximum gain of a standard half-wave dipole. In this case it is called only the *EFFECTIVE RADIATED POWER* or *ERP*. For the same subject antenna the EIRP value is 1.64 times (or 2.148 dB) greater than is the ERP value. In the case of a broadcast FM or TV station where the antenna is designed to have a doughnut-shaped pattern, with most of the power being transmitted out toward the horizon, we quite often hear the station announce that is transmitting "so many watts horizontal and so many watts vertical." This tells us that the antenna system of the station is designed to radiate out toward the horizon a wave whose polarization is tilted from the vertical at some angle γ. Since the power contained in the horizontally polarized portion of the propagating wave is equal to $ERP_{total} (\sin^2\gamma)$ and that contained in the vertical portion of the wave is equal to $ERP_{total} (\cos^2\gamma)$, it is easy for us to find both the station's total ERP and the polarization angle γ of its radiated signal level. The reason for dividing the transmitted power between horizontal and vertical polarization (or having the E-field polarization tipped at an angle) is to enable the station to serve both sets with horizontally polarized antennas (usually home sets) and those with vertically polarized antennas (automobiles). This information tells us something important if we live in a "fringe" area and wish to maximize the signal out of the receiving antenna terminals. We should tip the antenna on its side at the angle γ to minimize the polarization "loss" factor.

The preceding discussion has been concerned with what is called *linear polarization* where the E vector lies only in one direction in space (with the H vector perpendicular to it for propagation through free space). Another type of polarization which may be generated is called *circular polarization*. Circular polarization may be generated by having an antenna which produces a horizontally polarized wave located a quarter-wavelength in front of or behind a similar antenna which produces a vertically polarized wave. The two antennas are then fed with signals of equal magnitude and phase. The result is a composite wave composed of a horizontally polarized wave followed or preceded in space by a vertically polarized wave. When the E fields of the two waves are added vectorially at any given point along the propagation path, the resultant E-field vector has a constant amplitude at all values of time, but rotates in space. If the direction of rotation is counterclockwise as one looks toward the transmitter, the wave is called a right-hand circularly polarized wave (RHCP), and if it rotates clockwise, it is a left-hand circularly polarized wave.[4]

To intercept the full power from a circularly polarized (CP) wave, the receiving antenna must be designed to receive either RHCP or LHCP. If the incoming wave is LHCP and the antenna is designed to receive RHCP, no power will be transferred from the wave to the antenna output terminals. The same is true for a LHCP antenna receiving a RHCP wave. A CP antenna will, however, receive 50% of the power in a linearly polarized signal (and under special design conditions can be made to receive all of the power in a linearly polarized signal). Similarly, a linearly polarized antenna will receive 50% of the power in a CP wave which impinges upon it.

In some cases a wave may be generated which is similar to a CP wave where the horizontal and vertical components making up the CP wave are not of equal magnitude. Such a wave is called an elliptically polarized wave. Usually such waves are not generated by design, but result due to equipment malfunction or maladjustment, or they may result from a CP wave or a linearly polarized wave passing through a nonisotropic medium which modifies the polarization of the original wave. Thus, the elliptically polarized wave, if the ellipse is almost a circle, may be viewed as a modification of a CP wave, or if the ellipse is almost a straight line, it may be viewed as a modification of a linearly polarized wave.

6.7 ANTENNA IMPEDANCES

One way of considering an antenna is that it is a matching element which performs the function of matching the impedance of free space (the magnitude of the E-field vector to the magnitude of the H-field vector of a propagating wave, this ratio being equal to 377 Ohms) to the impedance of the transmitter, receiver, or transmission line attached to the terminals of the antenna. Different antennas will produce different matching ratios, depending upon the design of the antenna. For example, the terminal impedance of a half-wave dipole located free of all other surroundings is about 73 ohms. Thus, this antenna provides a match between a 73-ohm transmission line and free space. Other antennas have other terminal impedances. A short dipole, for example, may have less than 5 Ohms of terminal resistance. Usually it is desirable to design an antenna so that its terminal imped-

---

[4] Engineering definition; the RHCP and LHCP definitions of classical physics are just the opposite of this!

ance at the desired frequency of operation is a pure resistance. Sometimes, though, this is not possible, and there is a reactive component of the antenna terminal impedance. This effect is usually eliminated by the incorporation of lumped elements at or near the antenna terminals for the purpose of canceling out the reactive portion of the impedance, so that the transmission line attached to the antenna–plus–lumped-elements combination sees only a resistive termination. Such lumped-element circuits are called *antenna tuners*.

Another term which has been mentioned several times already is the antenna's *radiation resistance*. Mathematically, the radiation resistance is defined by Equation 6-35 where I_a is called the *antenna element current*. If the current value in this equation

$$P_T = \int_{A_{\text{sphere}}} \rho_\ell (\theta, \phi) \, dA = |I_a|^2 \, R_{\text{rad}} \qquad (6\text{-}35)$$

is the same as the current at the antenna terminals, then the radiation resistance would equal the terminal resistance for the situation where we had a lossless antenna with no reactance in the terminal impedance. In the case of a lossy antenna, the terminal resistance would be larger than the radiation resistance; the difference in the two can be thought of as a loss resistance in the antenna.

If an antenna is not completely isolated from all other surroundings (i.e., is not in "free space"), then its terminal impedance will be affected by its surroundings. Figure 6-12 shows how the terminal impedance of a half-wave dipole changes as it is brought closer to a completely reflecting parallel plane of infinite extent. With the antenna right next to the plane, its terminal impedance will be almost zero, while at about 0.35 wavelength away from the plane its terminal impedance will be about 95 ohms. It is not until the antenna is more than about three wavelengths away from the plane that its terminal impedance settles down to the nominal free-space value of about 73 ohms.

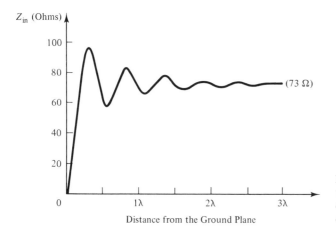

Figure 6-12 The variation of the terminal impedance of a half-wave dipole with its distance from an infinite conducting plane.

6.8 ANTENNA ARRAY FACTORS

Thus far we have been looking at only one particular antenna (which happens to have axial symmetry). In an actual antenna system we may have to consider an array composed of several antenna elements (which themselves may not have axial symmetry). In these cases we must go to describing the $\overline{\mathbf{E}}$ and $\overline{\mathbf{H}}$ radiation vector patterns for each element and for the array of the elements.

At a large distance from the antenna system (far-field case) each element will have its fields described by $\overline{\mathbf{E}}_i(r, \theta, \phi)$ and $\overline{\mathbf{H}}_i(r, \theta, \phi)$, which are called the $\overline{\mathbf{E}}$ and $\overline{\mathbf{H}}$ *patterns* of the element. We may also hear the term $\overline{\mathbf{E}}$ *plane* or $\overline{\mathbf{H}}$ *plane* applied as an adjective to either the $\overline{\mathbf{E}}$ *field* or the $\overline{\mathbf{H}}$ *field* patterns. This means that the pattern of the specified field is being *measured* in the *plane* in which the $\overline{\mathbf{E}}$ vector or the $\overline{\mathbf{H}}$ vector lies.

When several elements are arrayed together, the field of the array may be described by

$$\overline{\mathbf{E}}_a(r, \theta, \phi) = \sum_i \overline{\mathbf{E}}_i(r, \theta, \phi) \tag{6-36}$$

and

$$\overline{\mathbf{H}}_a(r, \theta, \phi) = \sum_i \overline{\mathbf{H}}_i(r, \theta, \phi) \tag{6-37}$$

In many cases the elements of the array are identical and have the same patterns, so that instead of a summation, the fields of the array may be given by equations that are easier to work with:

$$\overline{\mathbf{E}}_a(r, \theta, \phi) = \overline{\mathbf{E}}_i(r, \theta, \phi) \times A_a(\theta, \phi) \tag{6-38}$$

and

$$\overline{\mathbf{H}}_a(r, \theta, \phi) = \overline{\mathbf{H}}_i(r, \theta, \phi) \times A_a(\theta, \phi) \tag{6-39}$$

where $A_a(\theta, \phi)$ is known as the *array factor* for the antenna system being considered and is dependent upon the geometrical spacing of the elements and the relative magnitude and phase of the RF feed to each element. Thus, the total pattern of the array could be changed by either changing the elements and their pattern or by changing the relative position and/or the method of feeding the elements. Quite often, the pattern of an antenna system is changed during operation by changing the feeding of the various elements in the array.

Sometimes an array will include unfed (parasitic) elements that receive energy from the fed element through the induction of an RF voltage in the parasitic element. If the parasitic element is either longer or shorter than resonance length, the current in this element due to the induced voltage will suffer a phase shift. This induced current then radiates a field of its own, which, when combined with the field from the driven element, will produce a total pattern that is a modification of the pattern of the fed element. This type of technique is commonly used in HF, VHF, and UHF beam antennas.

Whenever a ground plane or other reflecting surface is in the proximity of an antenna, the pattern which results may quite often be obtained by considering the virtual image that results from "seeing" the reflection of the original antenna behind the reflecting plane.   The pattern factor of the original antenna and its *image* is then calculated and used to multiply the pattern of the original antenna to get the pattern which results from placing the ground plane near the antenna.

6.9 PROPAGATION

If an antenna or antenna system were located in free space away from all other surroundings, all propagation at any frequency would be in direct paths (rays) in what is commonly called "line-of-sight" propagation.   In this case the $\bar{E}$ and $\bar{H}$ vectors would always be perpendicular to each other and also perpendicular to the Poynting vector $\bar{P} = \bar{E} \times \bar{H}$ (which points in the direction of the propagation of the energy in the field).

However, as soon as other factors (such as the earth's surface, the ionosphere, etc.) enter into the picture, we have what is commonly called a "propagation" problem.   For instance, the presence of the ionosphere causes a bending of the electromagnetic radiation entering it at an angle from the underside so that quite often this radiation is "reflected" back down to the surface of the earth.   This effect is very dependent upon the frequency of the wave and upon the condition of the ionosphere (its level of ionization), which in turn is dependent upon the density of cosmic particles and radiation received from the sun.   For instance, just the diurnal change from day to night causes marked changes in both the density and the thickness of an ionospheric layer, as well as changes in its mean height above the earth, and this changes the point at which a beam of given frequency returns to the earth after being bent back by the ionosphere (see Figure 6-13).   If the frequency of the signal is high enough, if the density of the ionosphere is low enough, and if the ionosphere is thin enough at the time, the signal will pass through the ionosphere and out into space, although it may suffer bending and polarization rotation in the process.   It is only when the signal frequency is raised into the Gigahertz region that the bending becomes negligible, and even here the polarization rotation effect may still be present.

At the low-frequency end of the radio spectrum (up to about 2 to 5 MHz), the phenomenon of *ground-wave* propagation takes effect (see Figure 6-14).   Here the signal propagating from the antenna along the surface of the imperfectly con-

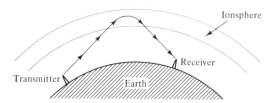

Figure 6-13   The refraction of HF signals back to the earth by the ionosphere.

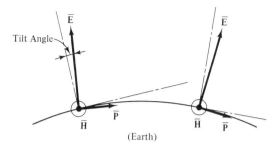

Figure 6-14 The propagation of a "ground-wave" signal along the earth's surface.

ducting earth has its Poynting vector $\bar{\mathbf{P}}$, tipped downward because the finite conductivity of the earth's surface causes a horizontal $\bar{\mathbf{E}}$-field component in the propagating wave. The result is that the wave is "trapped" to some extent to the surface of the earth and will thus follow the contour of the earth, making it possible to receive these low-frequency signals at distances beyond the line-of-sight distance from the transmitter (beyond the horizon, behind hills, etc.). The propagation of these ground-wave signals will depend upon the conductivity of the earth's surface and subsurface. The fact that the Poynting vector is tipped slightly toward the earth means that some of the power in the wave goes into the earth and is absorbed there. Thus, the wave is attenuated more than if it were propagating through free space.

Other factors may also have to be taken into account when considering the propagation of RF signals. For example, when a VHF, UHF, or microwave signal is beamed at the top of a sharp mountain ridge, the ridge acts like the edge of a razor blade when light is beamed at it, and by the process of *diffraction* the ridge causes some of the signal to be directed down into the valley on the other side of the ridge. Also, in urban areas where many tall buildings are present, especially ones with metal structures, the problems of signal reflection from the buildings (VHF and up) as well as possible parasitic excitation of a building or part of a building by an RF signal (VHF and down) may have to be taken into account.

The subject of signal propagation, like the subject of antennas themselves, is one that could easily fill many textbooks, and here it has only been possible to point out some of the main points of concern which the communications systems engineer has to be aware of in this area. In brief summary, the most important factors at the various frequencies might be listed as (1) in the low-frequency regions (up to about 2 to 5 MHz), the main phenomenon is ground-wave propagation, which is influenced primarily by the earth's surface under the path of propagation; (2) in the HF region (3 to 40 MHz), the main type of propagation is long-distance communications via waves "reflected" back to the earth by the ionosphere (what is commonly called "skip" phenomenon or "sky-wave" signals), and this is greatly dependent upon the condition of the ionosphere and the signal frequency; (3) in the VHF and UHF regions (20 to several hundred Megahertz, there is a transition from the skip phenomenon of the HF regions to the line-of-sight propagation of the microwave region, and depending upon the signal frequency and the path conditions one set of considerations or the other (or sometimes a combination of both) must be taken into account; and finally (4) in the microwave region (above several hundred Megahertz), the propagation is primarily line of sight, and the main considerations are the absorption of the wave by the medium through which it passes, as well as effects such as the rotation of the polarization vector, and so on.

In addition to considering merely what happens to the desired signal itself as it propagates from the transmitting antenna to the receiving antenna, we must also consider extraneous signals that are injected along the propagation path and that find their way into the receiving antenna. Such signals may be undesired information signals from other sources, or they may be electrical interference noise signals (automotive ignition systems, electric switches being operated, SCR dimmers, fluorescent lights, certain types of motors, etc.), or in some cases they may be delayed versions of the desired signal which have reached the receiving antenna via some alternate path of propagation and which interfere with the desired signal received via the desired path of propagation (what is called "multipath reception").

It is this latter phenomenon that is largely responsible for what is normally called "fading" of HF radio signals. It will not be possible to discuss all these interference and noise signals in detail. Some of them can largely be reduced or eliminated by proper antenna design, or by attacking the problem at its source (as in the case of impulse noise caused by automotive ignition systems or other electrical apparatus). However, one source of noise which it is not possible to eliminate is what is called *thermal noise*. This was discussed earlier, in Chapter 3, in connection with the noise from the different pieces of the actual equipment of the communications system (such as attenuators, resistors, amplifiers, etc.). Thermal noise also may enter into a system via the propagation path and be "picked up" by the receiving antenna. In the following section, we will be discussing this aspect of system noise. In the LF and HF regions, noise other than thermal noise is usually more important or predominant, and thermal noise, therefore, is of major concern only when working at VHF and higher frequencies.

6.10 ANTENNA NOISE TEMPERATURE

If we consider the terminals of the receiving antenna of a communications system, we find at these terminals not only the desired signal at a certain level but also various other noise and interference signals. The types of signals with which we will be concerned in this section are those due to the thermal noise which originates outside the antenna and which enters the antenna aperture and the thermal noise which is generated in the ohmic losses in the antenna itself. The total of these two noise powers at the antenna terminals may be expressed either as a noise power (in watts or in dBm) or as a noise temperature. The latter is often preferred because it is not dependent upon the bandwidth under consideration.

Consider now the geometry shown in Figure 6-15 where we have a small area dA_p at a temperature T_s which is radiating thermal radiation into a receiving antenna located a distance ℓ away from the small area. A_{eR} is the effective area of the antenna receiving the radiation from the area dA_p. If $d\rho_{NR}$ is the density

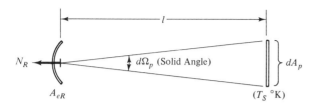

Figure 6-15 A small heated surface radiating into a receiving antenna.

in watts per square meter of the radiation from dA_p at the receiving point, then the noise power received by the antenna area A_{eR} will be

$$dN_R = \left[\left(\frac{\lambda^2}{4 \times \pi} \right) G_R \right] d\rho_{NR} = [A_{eR}] \, d\rho_{NR} \tag{6-40}$$

where at lower RF frequencies (microwave and below) and fairly large values of the source temperatures T_s (several tens of degrees Kelvin and higher),

$$d\rho_{NR} = \frac{2kBT_s dA_p}{\ell^2 \lambda^2} \text{ Watts/m}^2 \tag{6-41}$$

For a derivation of this equation, see Appendix A. Putting this result into Equation 6-40 gives

$$dN_R = \left(\frac{\lambda^2}{4\pi} \right) G_R \frac{2kBT_s dA_p}{\ell^2 \lambda^2} \tag{6-42}$$

But since $dA_p = \ell^2 d\Omega_p$, where $d\Omega_p$ is the solid angle that dA_p subtends when viewed from the receiving point, we have

$$dN_R = \frac{G_R kBT_s}{2\pi} d\Omega_p \tag{6-43}$$

These equations assume that all the power radiated by the area dA_p which reaches the area A_{eR} appears at the antenna output terminals. This is not true for radio antennas, since the preceding equations predict the *total* thermal radiation which includes all polarizations, whereas a radio antenna will intercept only one type of radiation polarization unless specially constructed with special matching networks. A radio antenna will thus only receive one-half of the thermal radiation incident upon it, and, therefore, a factor of one-half should appear in the equations when they are used for ordinary RF radiation. Sometimes the thermal radiation itself will exhibit some degree of polarization and the portion of the radiation received may be even less than one-half. We will account for this in the following equations by the factor U_N which can be viewed as the *noise polarization loss factor*. Its maximum value for a normal RF antenna will be one-half.

It is now convenient to put the foregoing in terms of noise temperature rather than noise power. To do this, we use the relationship that the noise power available at the antenna terminals is given by $N_A = kBT_A$ (where the A subscript refers to "antenna") and integrate equation 6-43 to get (including the U_N factor)

$$T_A = \frac{N_A}{kB} = \frac{U_N}{2\pi} \int_{4\pi} G_R(\theta, \phi) \, [T_{\text{background}}(\theta, \phi)] \, d\Omega \tag{6-44}$$

where T_A is the equivalent noise temperature of the thermal noise from the antenna terminals *if the antenna itself has no ohmic losses*. The background temperature which is a function of θ and ϕ is an expression of the fact that at different angles from the receiving antenna the sources producing the thermal radiation will not all be at the same temperature T_s, but that T_s will be a function of direction.

An interesting situation occurs when T_s is a constant over the entire beam-width of the antenna. Because U_N can have a maximum value of only 0.5, at first we might suspect that T_A would be at most only one-half of T_s. However, when we work out the mathematics, we find this is not the case:

$$T_A = \frac{U_N}{2\pi} \int_{4\pi} G_R(\theta, \phi) T_s d\Omega = \frac{0.5}{2\pi} T_s \int_{4\pi} G_R(\theta, \phi) d\Omega$$

$$= \frac{T_s}{4\pi}(4\pi) = T_s \qquad (6\text{-}45)$$

Thus, if a radio antenna is pointed at a source of temperature T_s, the noise temperature at the antenna terminals is also T_s. Noisewise, this is equivalent to putting a resistor (of a value equal to the antenna terminal resistance) whose temperature is equal to T_s in place of the antenna itself.

So far, we have assumed that T_s is the actual thermal temperature of the noise source. This is true only for an ideal black-body radiator. Actually, we may find that the *radio noise temperature* in the vicinity of the frequency of interest may be either greater or smaller than the actual physical temperature of the noise-producing object or area (a surface with low emissitivity may have $T_s \ll \text{T}_{\text{actual}}$ while a gas-discharge plasma may have $T_s \gg \text{T}_{\text{actual}}$ at some frequencies).

Next, we consider what happens when an antenna is looking at a target which is behind an attenuating medium (such as a microwave antenna pointed vertically at a target outside of the earth's atmosphere). This is shown in Figure 6-16. Here we consider that the portion of the signal which is transmitted through the lossy atmosphere along the line of propagation shown in Figure 6-16 is given by H_R while H_R' indicates the slant-path transmission through the atmosphere to the first main

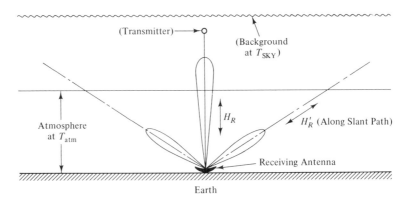

Figure 6-16 The effect of atmospheric loss on the antenna noise temperature.

sidelobe. Let us assume that all the antenna pattern is contained in the main beam and in the first sidelobe structure and that, if used as a transmitting antenna, the fraction of the total transmitter power P_t in the main beam is $K \times P_t$. The signal received from the transmitter out in space will be (from Equation 6-21)

$$S_R = G_{R\max} [P_T G_{T\max}] \left(\frac{\lambda}{4\pi\ell} \right)^2 H_R U_S \qquad (6\text{-}46)$$

where the quantity $P_T G_{T\max}$ is the *effective isotropic radiated power* (*EIRP*) of the transmitting station. The factor $[\lambda/4\pi\ell]^2$ is the free-space loss factor discussed earlier, and U_S is the signal polarization loss factor. The thermal noise received by the antenna will come both from the background behind the transmitter and from the atmosphere through which this background signal must pass.

Equation 6-44 in this instance reduces to

$$
\begin{aligned}
T_A =\ & \frac{U_{N,\text{beam}}}{2\pi} \int_{\text{main beam}} G_R(\theta, \phi) \{(H_R T_{\text{SKY}}) + [(1 - H_R) T_{ATM}]\} \, d\Omega \\[6pt]
& + \frac{U_{N,\text{sidelobe}}}{2\pi} \int_{\text{sidelobe}} G_R(\theta, \phi) \{(H'_R T_{\text{SKY}}) + [(1 - H'_R) T_{ATM}]\} \, d\Omega \\[6pt]
=\ & \left[\frac{2K U_{N,\text{beam}}}{(1)} \right] \times \{(H_R T_{\text{SKY}}) + [(1 - H_R) T_{ATM}]\} \\[6pt]
& + \left[\frac{2(1 - K) U_{N,\text{sidelobe}}}{(1)} \right] \times \{(H'R T_{\text{SKY}}) + [(1 - H'_R) T_{ATM}]\}
\end{aligned}
\qquad (6\text{-}47)
$$

If the antenna is not looking vertically at a target, but is at some elevation angle less than 90°, then H_R will be smaller and H'_R will be a function of the angle ϕ. If the elevation angle is small enough to permit a portion of the conical sidelobe to intercept the earth, then T_{SKY} must be replaced by T_{GROUND} for a part of the arc of ϕ. Both these conditions require a return to the integral form of Equation 6-44, where even U_N is usually now also a function of θ and ϕ.

There has been much information published on the values of T_{SKY}, since this is the primary area of concern of the science of radio astronomy. Areas of the sky that do not contain prominent radio sources will have a very low (50°K or less) radio noise temperatures at most wavelengths, whereas an area of the sky which contains one of the prominent radio stars or other sources (quasars, etc.) may have a noise temperature of several thousand degrees at some particular frequency. Many of these data are published either as sky "maps" or in tabular form in reference books on the subject and are of practical value to the engineer designing an extraterrestrial communications system. Similarly, much research has been done on the effects of radio wave propagation through the atmosphere and the ionosphere, and these data are also of interest to the communications engineer. Here the effects of weather conditions, solar activity, time of year, and so on must also be taken into account.

As can be seen from this chapter, the propagation portion of a communications link is probably the most complex portion insofar as the calculations required and insofar as the inability to predict what might occur with a high degree of accuracy. There are many factors on this portion of the link which are not well documented and which may in fact be quantities that change with time. Thus, the normal practice is to make generous allowances for the losses that might occur and the noise that might be injected on this portion of the communication system.

6.11 SPECIAL ANTENNA SYSTEMS

6.11.1 Tracking Antennas

There are several ways of causing an antenna to track a moving source. One technique is to monitor the level of the received signal while the antenna is moved slightly in azimuth or elevation. The antenna may then be manually pointed in the direction which gives the maximum signal. The process can also be automated with equipment which "dithers" the antenna and compares the AM modulation which is produced on the signal with the dithering signal and adjusts the nominal positioning of the antenna to minimize the phase shift between these two pairs of signals (one pair in azimuth, one pair in elevation).

Another type of tracking method is called *monopulse tracking*. In this scheme a dish antenna has four feed horns arranged as shown in Figure 6-17. The output of each horn is amplified by a separated amplifier (see Figure 6-18) and all four outputs are summed to create the signal output (summing may occur prior to or following the detection process). To generate the error signals fed to the servo drives, the sum of one pair of horns is compared with the sum of the other pair. For example, if the antenna pointing is slightly off in elevation the sum signal from horns 1 and 2 will be a little larger or smaller than will be the sum signal from 3 and 4. This difference is amplified in a differential amplifier and fed to the elevation (θ angle) servo drive which moves the antenna until the two sum signals are equal. Similarly, by comparing the sum of the 1–3 pair with the sum of the 2–4 pair, azimuth pointing errors can be corrected.

6.11.2 Steerable Phased Arrays

Figure 6-19 shows a digitally controlled steerable phased-array antenna system employing dipole antennas mounted above a ground plane. Each dipole is fed an equal portion of the transmitter power P_t via an individual phase-shift network.

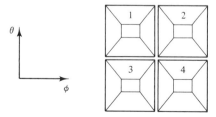

Figure 6-17 Arrangement of the four feed horns in a monopulse tracking antenna.

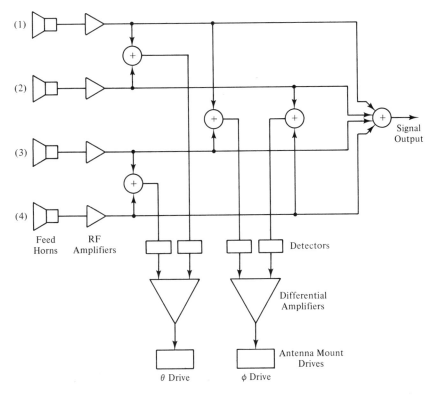

Figure 6-18   The electronics of a monopulse antenna tracking system.

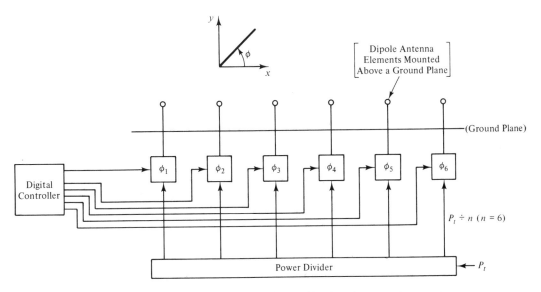

Figure 6-19   Digital controlled steerable phased-array antenna system.

The amount of phase shift introduced in each network is determined by the digital signal applied to it, these digital signals being derived from some type of digital control unit. By introducing different combinations of phase shift, the center of the beam pattern generated by the dipole array can be pointed in different ϕ directions (0 degrees $< \phi < 180$ degrees) and/or the beamwidth (in the ϕ plane) can be changed. Such antennas can scan in direction quite rapidly and are very reliable since there are no moving parts involved. Although an array of dipoles is shown in Figure 6-19, this technique can be applied to other types of antenna elements such as the slot antenna shown in Figure 6-20 (very often the slot antenna is filled with a solid dielectric so that a smooth surface is formed at the ground plane).

6.11.3 Physically Small Antennas

As mentioned back in the section on "open" antennas (i.e., not aperture-type antennas), it is important to keep the ohmic losses in the antenna, represented by R_{loss}, small in comparison with the antenna's radiation resistance R_{rad} in order to keep the antenna efficiency high. This represents the primary area of concerns which occurs when we try to make the physical size of the antenna small compared to a half-wavelength (as we would often like to do when $\lambda/2$ becomes more than a couple of feet). In transmitting antennas it means that a considerable portion of the power out of the transmitter may be dissipated in R_{loss} rather than being radiated from the antenna. In receiving antennas, it means that considerably less than one-half of the voltage induced in the antenna will appear across the load R_L (usually the receiver input resistance). (In the case where the antenna is a half-wavelength or larger in size and R_{loss} is much less than R_{rad}, approximately one-half of the induced voltage appears across R_L.) Also, the inclusion of a large amount of R_{loss} compared to the antenna's capture aperture (proportional to $R_{rad} + R_{loss} + R_L$) means that the noise contribution from R_{loss} may become significant compared to the received signal level. However, at the frequencies where physically small antennas are usually used (less than about 50 MH$_z$), the propagated noise received along with the signal (impulse noise, etc.) is usually greater than the thermal noise from R_{loss}.

One type of physically small antenna is the loaded dipole shown in various forms in Figure 6-21. The various forms of loading exhibit different characteristics, insofar as radiation resistance, loss resistance, and resonance sharpness are concerned. The combination of inductive plus capacitive loading gives the best results

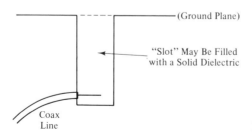

(Ground Plane)

"Slot" May Be Filled with a Solid Dielectric

Coax Line

Figure 6-20 A slot antenna element.

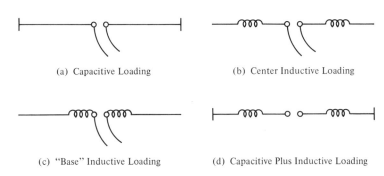

(a) Capacitive Loading   (b) Center Inductive Loading

(c) "Base" Inductive Loading   (d) Capacitive Plus Inductive Loading

Figure 6-21   Loaded short dipoles ($1 < \lambda/2$).

but is not always the easiest to implement mechanically.   The "base" inductive loading exhibits the highest loss resistance.   Figure 6-22 shows the shortened "quarter-wave" stub antennas (often used in mobile communication systems) corresponding to the four cases shown in Figure 6-21.

Another type of physically small antenna is the loop antenna, shown in various forms in Figure 6-23.   The low-impedance loop antenna consists of a loop constructed of heavy tubing included as part of a resonance circuit so that the large resonance-circuit currents flow through the loop.   By keeping the ohmic loss low in the loop and the resonance circuit, it is possible to produce a system which can be used for transmitting, although its efficiency is not too good.   The high-impedance loop shown in Figure 6-23b was used for years in table model broadcast-band receiver design.   Here the loop essentially forms the inductive portion of the resonance circuit.   A variation of the high-impedance loop is the "loopstick"

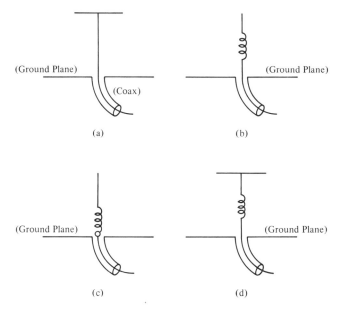

(Ground Plane)   (Coax)   (Ground Plane)

(a)   (b)

(Ground Plane)   (Ground Plane)

(c)   (d)

Figure 6-22   Loaded "quarter-wave" stub antennas.

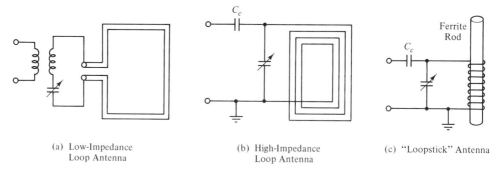

(a) Low-Impedance   (b) High-Impedance   (c) "Loopstick" Antenna
    Loop Antenna   Loop Antenna

Figure 6-23 Loop-type antennas.

antenna which consists of a coil wound on a ferrite rod. Its physical size is smaller, but its capture area is larger. It is the standard antenna now used in portable and table model broadcast-band receiver design. It can also be used in portable short-wave (1.7–20 MHz) receivers. The use of the high-Z loop or the loopstick for transmitting is not practical because of the high loss resistance and, in the case of the loopstick, saturation of the ferrite rod's magnetic curve.

6.12 RADAR SYSTEMS AND ANTENNAS

Figure 6-24 shows an airborne radar set which has an antenna which rotates in the azimuth direction around a 360-degree angle once every couple of seconds. This antenna transmits a fan-shaped beam which is narrow in the azimuth direction (usually a few degrees) but which is wide in the elevation direction (indicated by the angle α in the drawing). Usually the beam extends from α equal to 20 to 30 degrees up to α equal to about 50 to 60 degrees. The radar transmits a short pulse of radiation and then turns on its receiver. Radiation reflected from point A only has to transverse a distance $2s_A$ and thus arrives at the receiver first, while radiation reflected from B has to traverse the longer path $2s_B$ and thus arrives at the receiver

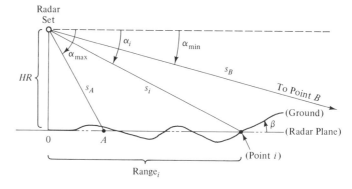

Figure 6-24 Airborne radar geometry.

last, after which the receiver is turned off and the transmitter transmits another pulse. This process continues as the radar antenna continues to turn.

In the radar set the electronic beam is deflected from the center of the CRT (cathrode-ray-tube) display to the outer edge at a nonlinear rate so that the point which the beam is at at any point in time corresponds to the point on the ground map where the reflection of energy would produce a receiver signal at that point in time. By using the receiver signal to intensity modulate the CRT beam, a radar picture can be drawn of the ground area under the radar. Bright areas in the picture indicate areas on the ground which reflect a large amount of the incident radar signal energy while dark areas on the CRT indicate ground areas which reflect little or no radar signal back to the radar set. This type of radar map display is called a PPI (plan-position-indicator) display and is only one of the many types of radar systems and displays which are in use. Our main interest in the present discussion will be the radar antenna, the radar signal path, and the reflection of radar energy from the ground.

Let us now analyze the situation shown in Figure 6-24. The angle α is called the depression angle, ranging between $\alpha_{\min}$ and $\alpha_{\max}$, and s is the slant distance from the radar to the reflecting point (a small area ΔA_i). HR is the altitude of the radar set about the nominal ground reference elevation (called the radar plane). Since the ground is hardly ever level and since the average slope of the ground in the small area ΔA_i determines how much energy is reflected back to the radar set, we need to also include the ground slope angle, indicated by β, in our calculations. We begin by analyzing the power out of the radar antenna terminals when the radar is receiving energy reflected from the small area ΔA_i. This is

$$P_{R,i} = G_R \left(\frac{\lambda^2}{4\pi} \right) \left(\frac{1}{4\pi s_i^2} \right) (\gamma_i)(\rho_{t,i})(\Delta A_i) \qquad (6\text{-}48)$$

where G_R is the gain of the receiving antenna in the direction of ΔA_i, $\rho_{t,i}$ is the radiation density received at ΔA_i from the radar transmitter, and γ_i is a scattering "gain" indicating how much of the received transmitter power is scattered back in the direction of the radar set by the area ΔA_i. The received power density is given by

$$\rho_{t,i} = \frac{G_T P_T}{4\pi s_i^2} \qquad (6\text{-}49)$$

and the incremental area covered by one pulse is

$$\Delta A_i = (s_i)(\text{BW})(\text{PLG}) \qquad (6\text{-}50)$$

where BW is the azimuth beamwidth in radians and PLG is the distance covered on the ground by one radar pulse. PLG is calculated as follows:

$$\text{PLG} = \left(\frac{c\tau_p}{2 \cos{(\alpha_i + \beta_i)}} \right) \qquad (6\text{-}51)$$

where τ_p is the radar pulse duration time and c is the speed of light. Making this substitution and also the one that $s_i = HR/\sin \alpha_i$, Equation 6-48 becomes

$$P_{R,i} = G_R \left[\frac{\lambda^2}{4\pi} \right] \left[\frac{\sin^2 \alpha_i}{(4\pi)(HR^2)} \right] [\gamma_i] \left[\frac{G_T P_T \sin^2 (\alpha_i)}{(4\pi)(HR^2)} \right]$$
$$\left\{ \left(\frac{HR}{\sin \alpha_i} \right) (BW) \left[\frac{c\tau_p}{2 \cos (\alpha_i + \beta_i)} \right] \right\} \qquad (6\text{-}52)$$

However, the same antenna is used for both transmitting and receiving and its gain is a function of α. Thus, substituting $G_R = G_T = G(\alpha_i)$ and collecting like terms we get

$$P_{R,i} = G^2(\alpha_i) \left(\frac{\lambda^2 \, P_T \gamma_i \, BW c\tau_p}{(4\pi HR)^3} \right) \left(\frac{\sin^4 (\alpha_i)}{\sin (\alpha_i) \, 2 \cos (\alpha_i + \beta_i)} \right) \qquad (6\text{-}53)$$

If we assume a rough scattering surface, it has been found that γ_i is given approximately by γ_{oi} times $\sin (\alpha_i + \beta_i)$. Making this substitution and letting the $\sin \alpha_i$ in the denominator of Equation 6-53 be equal to $\cos \alpha_i$ times $\tan \alpha_i$, we get

$$P_{R,i} = \left(\frac{\lambda^2 \, P_T c\tau_p \, BW}{2 \, (4\pi HR)^3} \right) \left(G^2(\alpha_i) \left(\frac{\sin^4 (\alpha_i)}{\cos (\alpha_i)} \right) \left(\frac{\tan (\alpha_i + \beta_i)}{\tan (\alpha_i)} \right) \gamma_{oi} \right) \qquad (6\text{-}54)$$

If now we *make* the antenna vertical gain function[5]

$$G(\alpha) = K \frac{\sqrt{\cos \alpha}}{\sin^2 \alpha} = G_o \left(\frac{\sin^2 (\alpha_{min})}{\sqrt{\cos (\alpha_{min})}} \right) \left[\frac{\sqrt{\cos (\alpha)}}{\sin^2 (\alpha)} \right] \qquad (6\text{-}55)$$

where $G_o = G(\alpha_{min})$ we find that many terms cancel, and we get

$$P_{R,i} = \left(\frac{\lambda^2 \, P_T \, c\tau_p BW G_o^2 \sin^4 (\alpha_{min})}{2(4\pi HR)^3 \cos \alpha_{min}} \right) \left(\frac{\gamma_{oi} \tan (\alpha_i + \beta_i)}{\tan (\alpha_i)} \right) \qquad (6\text{-}56)$$

In Equation 6-56, the factor γ_{oi} is related to the reflectivity of the incremental area ΔA_i and will depend upon the type of material present, the roughness of the surface, and perhaps the orientation of any reflecting planes or internal corners. The $\csc^2 (\alpha)$ times $\sqrt{\cos (\alpha)}$ function for the antenna vertical gain has resulted in creating an overall function for the received power which does not vary directly with depression angle or range for flat terrain where β equals zero. This produces a uniform level of signal into the radar receiver that varies only as a function of ground slope β_i and/or the reflectivity factor γ_{oi}.

In Figure 6-25 is illustrated one of the problems which occurs with a radar of the type just described. Here a point X at a ground range of R_x is also elevated by an amount HX above the radar plane. The signal reflected from this point appears to come from a point X' lying on the radar plane and hence is mapped on the CRT radar display at a range $R_{X'} < R_X$ instead of at its correct range R_X. This moving in of the point toward the origin is called the "layover" effect. A similar negative layover effect (point moving out away from the origin) occurs if the reflecting point lies below the radar plane instead of above it. The apparent range

---

[5] See Figure 6-26.

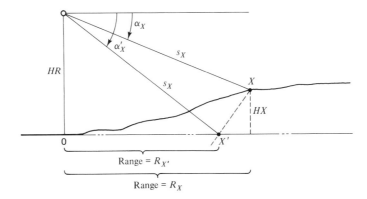

Figure 6-25 Shift of apparent reflector location due to height above or below radar plane (layover effect).

$R_{X'}$ is calculated as follows:

$$s_X = \sqrt{R_X^2 + (HR - HX)^2} \tag{6-57a}$$

$$R_{X'} = \sqrt{s_X^2 - HR^2}$$

$$= \sqrt{R_x^2 + \cancel{HR^2} - (2\ HR\ HX) + HX^2 - \cancel{HR^2}}$$

$$R_{X'} = \sqrt{R_X^2 - (2\ HR\ HX) + HX^2} \tag{6-57b}$$

The radar example given here is an airborne ground-mapping radar. The converse case of a ground-base PPI radar for displaying aircraft in flight or weather conditions is very similar to this. These are all examples of what is called *primary* radar (using just the reflected signal energy). Another type of radar is *secondary* radar in which a signal is transmitted on one frequency from the radar set and a transponder receiver-transmitter unit at the target point(s) sends back a reply on a different frequency. This type of radar is currently the type employed for most air-traffic-control (ATC) operations. It has the advantage that the return signal is not dependent upon the aircraft size, construction material or design, or its

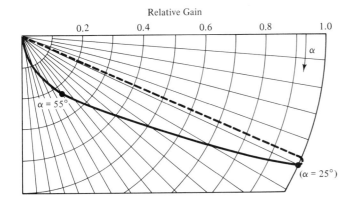

Figure 6-26 Plot of $(\sqrt{\cos\alpha}\ \csc^2\alpha)/(\sqrt{\cos 25°}\ \csc^2 25°)$ for $25 \le \alpha \le 55°$ (solid line) (dashed line is postulated beam roll-off).

relative orientation to the ground radar set.   Also, the reply sent back from the aircraft to the ground ATC radar may be encoded with PCM digital data giving the aircraft's identification, barometric altitude, or other pertinent information.

PROBLEMS

6-1. A 7.85-GHz microwave link has antennas 9.7 statute miles apart pointed at each other. One antenna has a gain of 23 dB and the other 28.5 dB.

  a. What is the effective aperture area of each antenna?

  b. If a signal of -65 dBm is required at the output of the receiving antenna, what must be the power input (in Watts) at the transmitting antenna?

  c. If the link were rebuilt to operate at 10.3 GHz using antennas of the same effective aperture (i.e., the same dishes as originally, but with feeds designed for the higher frequency), what would P_t have to be to get the -65 dBm received signal level? What happens to the respective gains of the two antennas when the frequency is raised from 7.85 to 10.3 GHz?

6-2. If two antennas are 25 statute miles apart, what is the free-space loss factor at 2.1 GHz?   at 6.3 GHz?

6-3. An antenna pattern is given by

$$G(\theta, \phi) = G_{max}[\cos^2 (5 \times \theta)][\epsilon^{-\theta/20°}]$$

  a. What is G_{max}?

  b. What is the planar $\frac{1}{2}$-power beamwidth of the main beam?

  c. What is the solid-angle $\frac{1}{2}$-power beamwidth?

  d. What is θ for the first null in the pattern?

  e. What portion of the total radiated power is in the main beam?

6-4. Assume that the $\overline{\mathbf{E}}$ field right at a dipole antenna is given by $\overline{\mathbf{E}} = K_1 i(t)$.

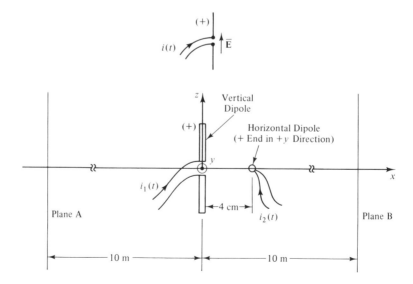

a. Using a piece of rectilinear graph paper, plot the **E**-field vector in the y–z plane A at $t = 0, 0.25 \times 10^{-10}, 0.5 \times 10^{-10}, \dots, 3 \times 10^{-10}$ (i.e., every 30° electrical degrees) if

$$i_1(t) = 2 \cos (2.0943951 \times 10^{10}\, t + 15°)$$

$$i_2(t) = 2 \cos (2.0943951 \times 10^{10}t - 95°)$$

b. Repeat for plane B.

6-5. If an antenna has a circular beamwidth with a planar half-power beamwidth of 3 degrees, what is the approximate theoretical beamwidth in solid angle units? What is the approximate gain of the antenna?

6-6. A parabolic-dish antenna with a gain of 30 dB operates at a frequency of 10 GHz. What is the effective aperture area of the antenna (in square meters)? If the 30 dB includes surface-roughness losses in the antenna such that the "efficiency" of the antenna is 60%, what would be the approximate physical aperture of the antenna? What would its diameter be?

6-7. If a 10-GHz antenna has a gain of 80 dB (*not* including any ohmic losses), what is the approximate area and solid-angle beamwidth of the antenna? If the beam is circular in cross section, what is the planar beamwidth angle in degrees? If the 80 dB *were* to include ohmic losses in the antenna, how would this affect the beamwidth?

6-8. If a transmitter at 3 GHz delivers 10 KW to an antenna with an effective gain of 10 dB, what would be the power density (in Watts per square meter) at a distance of 1 Km from the antenna?

6-9. If the signal transmitted by the station in Problem 6-8 is received by a station 10 Km away where the receiving antenna has an effective gain of 17 dB, and if the atmospheric absorption loss along the path is 1 dB, what signal power is available at the receiving antenna output terminals?

6-10. Given the following information, find the power (in watts) delivered to the receiver in each case:
a. Frequency = 8 GHz, distance = 18,000 *nautical* miles, P_t = 1000 Watts, G_t = 3000, $(G_r)_{dB}$ = 16 dB, absorption loss = 1.5 dB, transmitter feed line loss = 0.7 dB, receiving feed line loss = 0.5 dB.
b. Frequency = 2 GHz, distance = 21,000 statute miles, P_t = 36 dBm, G_t = 10 dB, G_r = 1000, and miscellaneous losses = 2 dB.

6-11. Using the standard spherical coordinate system where $\phi = 0$ degrees, $\theta = 90$ degrees $\Rightarrow x$ axis and $\theta = 0$ degrees $\Rightarrow z$ axis, and the far-field patterns for the $\lambda/2$ dipole,

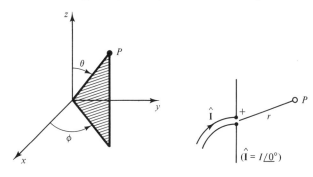

where $\hat{\mathbf{E}}_{\theta_d}(\theta, \phi) = K_1[\cos(\pi/2 \cos \theta)/\sin \theta]\, \varepsilon^{j(\omega t + \gamma_d)}$, where $\gamma_d = [-2\pi(r/\lambda)]$, where r = distance from the dipole center to the point of far-field measurement ($\hat{\mathbf{E}}_{r_d}$ and $\hat{\mathbf{E}}_{\phi_d} = 0$ in the far field), find $E(\theta, \phi)$ and $G(\theta, \phi)$ for the far field ($r \gg \lambda$) for

a. A *co-linear* array of $\lambda/2$ dipoles where $\hat{\mathbf{I}}_a = \hat{\mathbf{I}}_b = \hat{\mathbf{I}}_c = I\angle 0°$ and where the center of the center dipole is at the origin.

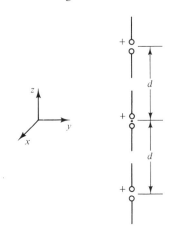

Note: The phase shift (in radians) at an angle (θ, ϕ) coming from an antenna displaced (x_i, y_i, z_i) from the array "phase center" is given by $\Delta\beta_i = (2\pi/\lambda)\Delta d_i = (2\pi/\lambda)\,[x_i \sin\theta \cos\phi + y_i \sin\theta \sin\phi + z_i \cos\theta]$.

b. A *broadside* array where $\hat{\mathbf{I}}_a = \hat{\mathbf{I}}_b = \hat{\mathbf{I}}_c = I\angle 0°$ and where the center of the center dipole is at the origin. The three dipoles lie in the x–z plane (i.e., where $\phi = 0$ degrees).

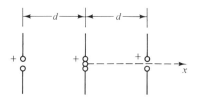

c. Both (a) and (b) if $\hat{I}_a = I\angle\alpha$, $\hat{I}_b = I\angle 0°$, $\hat{I}_c = I\angle-\alpha$.

6-12. Let a ground station be transmitting a signal at the 15-KW level through an antenna which has an actual gain of 55 dB.[6] Losses in the feed line of the antenna are 2 dB. The frequency of transmission is 8 GHz. What signal power will be available at the output of a receiver (gain = 28 dB) located in a satellite which is 24,000 statute miles distant from the earth transmitter if the actual gain of the satellite receiving antenna is 31.5 dB,[6] if there is 1 dB of loss in the feed line between the antenna and the receiver ($T_{ambient} = 200°$K), and if there is 0.5 dB of absorption loss between the transmitter and the receiver sites ($T = 275°$K)? If the receiver noise figure is 7 dB and its bandwidth is 5 MHz, what is the noise power output of the receiver? Assume

---

[6,7]* Not including ohmic losses in the antennas—these are included in the feed line losses.

that the earth's effective noise temperature is 275°K. What is the S_{out}/N_{out} at the receiver output (in decibels)?

6-13. Consider the earth's atmosphere to be 10 km high with an average temperature of 250°K. An antenna pointed at a communications satellite is pointed upward at an elevation angle of 50 degrees, and the background noise temperature of the sky at that elevation angle is 40°K. The satellite transmits on 8 GHz and is 24,000 statute miles above the earth's surface at the equator. The effective aperture area of the ground receiving antenna is 100 m²,[7] and the ohmic losses in the antenna and its feed are 2 dB at an ambient temperature of 300°K. The loss at 8 GHz looking *vertically* through the atmosphere is 1 dB. The power output of the satellite transponder is 15 Watts and its transmitting antenna has a gain of 25 dB.

    a. What is the noise temperature at the receiving antenna output terminals?

    b. What is the distance from the ground station to the satellite? (Use 24,000 miles as the circumference of the earth.)

    c. What is the signal level at the output of the receiving antenna?

6.14. What is meant by the ERP of a transmitting station? How is it calculated? What information do you need about the station to calculate its ERP?

6-15. Consider a radio-astronomy antenna system where the noise temperature at the antenna terminals is 40°K when the antenna is pointed at some radio-stellar source. The antenna feedline has a loss of 0.2 dB at an ambient temperature of 290°K and goes to a maser amplifier with a standard noise figure of $F_{o1} = 0.9$ dB and a gain of $G_1 = 27$ (ratio). The following stage in the radio-astronomy receiver has a standard noise figure of $F_{o2} = 8$ dB and a gain ratio $G_2 = 1000$. What will be the equivalent noise temperature of the entire system as referred to the input of the maser amplifier. (Note that the gains here are *not* given in decibel units.)

6-16. A satellite communication ground-station antenna is looking straight upward at the satellite as it passes overhead at an altitude of 10,000 statute miles.

    The satellite transmitter is at 9.8 GHz and has a power output of 20 Watts. The gain of the satellite transmitting antenna is 26 dB (without considering ohmic losses), and the ohmic losses in the antenna and feedline are 1.5 dB.

    The ground-station antenna has an effective aperture area (not including ohmic losses) of 100 m². Ohmic losses in the antenna and feed line amount to 2.0 dB, and the ambient temperature of the antenna and the feed line is 300°K.

    The transmission loss (due to scattering and absorption) through the atmosphere is 4.0 dB, and the mean ambient temperature of the atmosphere is 250°K. The sky-background temperature behind the satellite is 50°K.

    The ground-station receiving antenna feeds a receiver consisting of a preamplifier ($G = 20$ dB, $F_o = 2$ dB), a mixer stage ($G = 5$ dB, $F_o = 8$ dB), and an IF amplifier ($G = 40$ dB, $F_o = 7$ dB). The IF amplifier bandwidth is 3.5 MHz, and the RF preamplifier is tuned to the desired signal frequency and effectively rejects the image signal frequency.

    a. What is the solid-angle beamwidth of the satellite antenna?

    b. What is the power density (in Watts per square meter) received at the ground station from the satellite?

    c. What is the gain of the ground-station antenna (not including ohmic losses)?

    d. Assuming a circular beamshape for the ground station antenna, what is its planar beamwidth?

    e. What is the EIRP (including losses) of the satellite transmitting system?

f. What is the received power out of the terminals of the ground stations' receiving antenna and feedline?

g. What is the noise temperature at the same point?

h. What is the *excess equivalent noise temperature* of the receiver referred to the receiver input terminals?

i. What is the overall *standard noise figure* of the receiver?

j. What is the total noise power output at the output of the receiver's IF amplifier (including contributions from the antenna noise as well as the receiver's excess noise)?

k. What is the signal power output at the receiver's IF output?

l. What is the maximum information rate (in bits per second) which could be transmitted over this channel from the satellite to the ground station?

7

BROADCASTING AND VIDEO SYSTEMS

In North America commercial RF broadcasting is normally thought of as encompassing three main services: the AM broadcast band (540 to 1600 KHz), the FM broadcast band (88 to 108 MHz), and the VHF and UHF television bands. In recent years we have seen the concept of wired broadcasting (*cable TV*) primarily using frequencies in the VHF region and consumer interception of satellite-relay signals in the microwave region (*intentional* actual *broadcasting* from satellites to consumers is in the very near future). Also included within this chapter are the topics of shortwave AM broadcasting and audio and video recording and reproduction techniques.

7.1 AM BROADCASTING

This mode of broadcasting voice and music has been with us since the 1920 decade. At the present time it covers the range of 540 KHz to 1600 KHz, with stations being assigned transmitting frequencies which are integer multiples of 10 KHz. Carrier power levels of up to 50 KW are permitted, with this maximum limit being permitted only for "clear channel" stations—stations that do not share their frequency with other stations during the hours from sunset to sunrise when the maximum propagation distance occurs. Other stations of lesser power share their operating frequency with stations in other parts of the country. Since maximum propagation occurs at night, it is possible for more stations to share a frequency during the daytime hours than during the sunset-to-sunrise period. Hence, there are a large number of local-service AM broadcasting stations designated as daytime only (which may transmit only during the hours from sunrise to sunset).

The normal range of audio modulation frequencies which is employed is usually up to about 10 KHz, and a station must maintain its carrier frequency within ± 10 Hz. Peak modulation may not exceed $+125\%$ and -100% under any condition, and an audio limiting amplifier is usually employed to insure this.

AM broadcast transmitting antennas are normally towers (either free-standing or guyed) a quarter-wavelength in height and are fed at the base from one side of a transmission line. The other side of the line goes into an earth ground and/or counterpoise system. The counterpoise system, consisting of wire extending outward along the ground, can be buried in the ground, laid on the ground or elevated a foot or two above the ground. The latter type of counterpoise construction is preferred in areas where winter snow cover is expected so as to establish a ground plane which is relatively unaffected by the amount of snow cover present at any given time. In some cases, either to maximize the use of the transmitter power within the desired coverage area or to minimize interference with another station on the same frequency, a nonomnidirectional (in azimuth angle) radiation pattern is desired. This can be accomplished with an array of towers (usually two to six) with a definite relative spacing and orientation and fed at definite relative RF current phases and/or magnitudes. The propagation in different azimuth directions at AM broadcast-band frequencies is also greatly influenced by the geographical and cultural features (terrain, ground conductivity, large metallic structures, etc.), and one of the engineering tasks involved in applying for an AM station license is predicting the proposed station's coverage area as influenced by the surrounding geography.

Transmitters for AM broadcast may utilize either high-level or low-level modulation techniques (see Section 5.3.1.1). The "better quality" AM stations keep their average modulation percentage low, utilizing the audio limiting characteristics only on occasional signal peaks, while some AM station operators prefer to make extensive use of the audio limiter in order to increase the average sideband or "talk" power of the station, although when this is done, the available dynamic range of the transmitting system is reduced and objectionable background noise may be transmitted during pauses in conversation and music.

In an effort to get around the problem of the large modulation transformer in high-level plate modulation in high-power AM transmitters, some of the manufacturers of AM broadcast transmitters have come up with some unique modulation techniques. One of these involves using two RF amplifier tubes only one of which is normally supplying RF power to the output during unmodulated conditions. On negative modulation, the output of this tube is reduced by changing its screen-grid voltage, while on positive modulation, the second tube is essentially "turned on" and begins delivering power into the RF load. At the same time it causes the output impedance which the first tube "sees" to be lowered, and the first tube begins to deliver more RF power into the load until at the peak-positive-modulation point both tubes are contributing equal power to the load.

Another scheme utilizes a conventional class C stage, but instead of a modulation transformer for varying the high voltage applied to the stage, a pulse-type regulated power supply is used wherein the power suply voltage is controlled by the audio signal. The power supply's filter network must have a time constant long enough to filter out the pulse harmonics but short enough to permit the power supply voltage to follow the modulating signal. One system of this type (manufactured by Gates Co.) utilizes a 70-KHz pulse frequency with the pulse-duration modulation being obtained by adding the audio signal to a 70-KHz triangular waveform and using the sum signal to control a binary threshold detector which turns the pulses on and off. This is often referred to as "class D" modulation.

The rules and regulations of the Federal Communications Commission (Title 47 of the National Code), Section 73, Subpart A, give a vast amount of regulatory information concerning commercial AM broadcasting, including operating procedures, licensing, and exhaustive technical standard information, formulas, and charts. Subparts B and D of this same section include information on FM and TV broadcasting. Copies of the National Code are usually available at most of the larger university and public libraries throughout the United States.

7.1.1 Stereophonic AM Broadcasting

Stereophonic audio usually implies two separate audio channels originating from two microphones placed in the same environment, and it is intended to simulate as accurately as this permits the effect of sound arriving from different directions. In the case of a live broadcast, the stereo audio signal is usually generated from two such microphones or by mixing together a group of microphones located toward the left of the pickup area to produce one signal, while the other stereo signal is generated by mixing together a group of microphones located toward the right of the pickup area. However, in some instances, especially in preparing stereophonic recordings, the stereo effect may be simulated or enhanced by mixing together sounds from different sources into the two channels (designated "left" and "right") although there may be no actual left-right relationship involved.

Whatever the means by which the two stereo channels are generated, since the signal which is to be broadcast will be received by monaural (nonstereo) as well as stereo receivers, there must be some limitations placed upon the type of dual-channel signal which is broadcast as well as upon the techniques employed to encode and modulate this pair of signals upon the transmitted carrier. Obviously there must be some close relationship between the two (L and R channels) signals or their reception by a monaural receiver would just produce a meaningless jumble at the receiver audio output. Thus, while the two signals may be somewhat different, their sum must also be a useful audio signal. This is usually true of most stereophonic signals and is generally true of artificially produced stereophonic signals as well. Since the two signals then do somewhat resemble each other, the system of broadcasting them (as well as most systems of recording stereo) permits some degree of mutual interference or "crosstalk" between the two signals. The amount by which the two signals are kept distinct during the process of transmission through the broadcast channel (up to and including the audio amplifiers at the receiver) is called the stereo *separation*, and although 100% separation is theoretically attainable in the broadcasting system to be discussed, on a practical basis there is always some degree of mixing or crosstalk which occurs.

The technique for broadcasting the two stereophonic channels is to first form a sum channel $(L + R)$ and a difference channel $(L - R)$ from the original two channels $(L$ and $R)$. The sum channel is then used to modulate the carrier in such a manner as to enable existing AM broadcast receivers to detect it without any modifications to the receivers and without any noticeable level of audio distortion or noise being present due to the presence of the $(L - R)$ information on the same carrier. There have been a number of AM stereo broadcasting methods proposed,[1] all of which employ different versions of basic quadrature modulation techniques which produce the effects of amplitude and phase modulation (see Section 4.4). At this writing it appears as if a system somewhat like that proposed by the Motorola Corporation will become the standard for stereo AM in the United States. This system produces an RF output

$$v_o(t) = V_{cm} \{1 + [L(t) + R(t)]\} \cos(\omega_c t + \phi) \tag{7-1}$$

where

$$\phi(t) = \arctan \left[\frac{L(t) - R(t)}{1 + L(t) + R(t)} \right] \tag{7-2}$$

where the peak amplitude of $L(t) + R(t)$ is limited to being less than unity. Standard envelope detection of this signal by a monaural AM receiver will recover a signal proportional to $L(t) + R(t)$. A stereo receiver uses a modified quadrature detection system in which the instantaneous RF amplitude is first multiplied by $1/\cos \phi$, which results in an in-phase component proportional to $1 + L(t) + R(t)$ and a quadrature component proportional to $L(t) - R(t)$. These are then demodulated by in-phase and quadrature product detectors to produce the stereo sum and difference signals which are fed into a resistive matrix to produce the left-channel and the right-channel audio signals. A low-amplitude 5- or 10-Hz signal may be added to the transmitted signal in some manner (either added to the sum audio channel or else used to directly apply additional PM or FM modulation to the carrier) and used at the receiver to activate the stereo detection circuitry whenever a stereophonic signal is being broadcast.

---

[1] In 1982 the Federal Communications Commission backed out of the problem of trying to decide which AM stereo system to adopt and said that they would let "the open competitive marketplace" make the decision.

7.2 *SHORTWAVE BROADCASTING*

Shortwave broadcasting falls into three major types of operation: government, commercial, and religious.   In the United States and its territories, all three types are present, with only the commercial and religious broadcasters being directly regulated by the FCC.   U.S. government stations operate without FCC supervision, although for the most part, they follow the operating practices set forth by the FCC.   Shortwave (SW) broadcasting is basically AM, but with a few differences from standard-broadcast-band AM.   For one thing the power levels employed may be much higher, and the antenna systems employed are much different, especially since the frequencies of operation are higher and since it is also frequently desired to form a beam to a specific spot on the earth's surface.   Some SW broadcast stations have extensive antenna "farms" for beaming different frequencies to different parts of the world at different times.   It is also common for an SW broadcast station to broadcast the same programming simultaneously on several frequencies in the same SW band or in different bands.   By general international agreement, certain bands of frequencies are set aside for SW broadcast, although in some parts of the world, these are shared by other radio services.   It is not uncommon for some governments to not abide by these band assignments, and SW broadcasts may be quite often heard outside of the internationally agreed-upon bands.   At the present time the FCC recognizes the following bands: 5950–6200, 9500–9775, 11,700–11,975, 15,100–15,450, 17,700–17,900, 21,450–21,750, and 25,600–26,100 KHz.   Commercial and religious stations may be licensed to operate within these bands on 5-KHz increments within the bands.   Each year a station requests which frequencies it wishes to utilize during which hours of the day and with what type of antenna system and power, and the FCC prepares what amounts to a schedule of frequency use for the various stations.

One interesting difference between standard broadcasting and SW broadcasting is in the modulation percentage employed.   Because selective fading (where one frequency fades down while a frequency a few KHz away does not at the same time) occurs and regularly causes the carrier to decrease while the sidebands remain at normal received power, the ratio of carrier to sideband power in SW broadcasting is often made larger than for standard broadcasting.   This is based upon the assumption that the SW receiver will use a diode or envelope detector which produces a distorted output any time that the modulation percentage as observed at the detector becomes greater than 100%.

Information on FCC regulation of SW commercial and religious broadcasting can be found in subpart F of the FCC Rules and Regulations.   The International Telecommunications Union publications contain information about international agreements concerning SW broadcasting.

7.3 *FM BROADCASTING*

In discussing FM broadcasting in the United States we will be discussing all the services which are provided within the FM broadcasting band from 88 to 108 MHz whether this includes broadcasting to the general public or special broadcasting, point-to-point communications or remote-control services. The latter type of operations ride along "piggyback" on an FM station's main broadcasting carrier. An FM station's main broadcast channel may be either monaural or stereophonic(two simultaneous related audio channels).

For monaural FM broadcasting with no other services on the channel, the FM signal is modulated with audio up to a peak deviation of ± 75 KHz. The audio signal is defined by the FCC to exist from 50 Hz to 15 KHz with a 75-μsec preemphasis network in the audio channel. The preemphasis network essentially provides a flat response from below 50 Hz out to a corner frequency of about 2125 Hz at which point a 20-dB-per-decade (about 6-dB-per-octave) increase in the signal level is begun. This preemphasis produces what can be considered to be a PM signal for modulating frequencies above 2125 Hz (see Section 4.3). The purpose for having this preemphasis is to reduce the high-frequency audio noise at the receiver output (i.e., it reduces the "hiss" on the audio signal). More information on the use of preemphasis in FM broadcasting can be found in Chapter 4.

Frequency assignments in the FM band are at 200-KHz intervals starting at 88.1 MHz (designated as channel 201) and extending up to 107.9 MHz (channel 300). Certain of these channels are set aside for Class A stations (ERP up to 3 KW), while the other channels from 222 through 300 can be used by class B (ERP up to 50 KW) and class C (ERP up to 100 KW) stations. Some of the lower-numbered channels are set aside for low-power educational FM stations operated by schools and universities. The carrier frequency of an FM station must be maintained within ± 200 Hz of the assigned center frequency, and the FCC specifies the means by which the frequency is measured and this regulation complied with. The FCC also specifies the maximum amount of noise permitted on the signal and the allowable limits on the audio frequency response of the transmitter and studio equipment. The FCC Rules and Regulations, Subpart D, contains the specifications for FM broadcasting.

7.3.1 Stereophonic FM Broadcasting

The concept of splitting the stereophonic audio signals (left, L, channel and right, R, channel) into sum and difference signals has already been discussed in Section 7.1.1 on AM stereo broadcasting. What we will be discussing here is how the difference $(L - R)$ signal is added onto the monaural FM signal. The stereo sum signal $(L + R)$ is used to modulate the FM carrier in the conventional manner, except that the modulation percentage due to the $(L + R)$ signal may be only 90% of the ± 75-KHz maximum deviation limits (the other 10% of the peak deviation is reserved for a stereo *pilot* subcarrier as will be discussed shortly), and this only when both the L and the R channels are carrying the same information (i.e., when the $(L - R)$ signal level is zero). The $(L - R)$ signal, when it does exist, is used to DSBSC modulate a 38-KHz subcarrier, and this resulting DSBSC signal is then added to the audio $(L + R)$ signal. If the incoming audio signals (L and R) have frequencies up to 15 KHz, then the resulting *modulating* signal consists of an audio signal—the $(L + R)$ signal—from about 0 to 15 KHz and the DSBSC signal from 23 to 53 KHz.

However, to demodulate the DSBSC signal properly at the FM receiver, it is necessary to have information about the phase of the 38-KHz subcarrier signal which was suppressed in the DSBSC $(L - R)$ modulation process. This information could be passed along by transmitting a small amount of the 38-KHz subcarrier at 38 KHz, but this small signal would be "buried" down between the two sidebands of the DSBSC signal, and it would be hard to extract it from this location. Therefore the phase and frequency information about the 38-KHz subcarrier is conveyed by placing a *pilot subcarrier signal* onto the modulating signal at one-half of the original subcarrier's frequency (or at 19 KHz). This 19-KHz signal can be added in between the top of the main audio signal (15 KHz) and the bottom of the DSBSC signal (23 KHz) without disturbing either of these latter signals. The 19-KHz subcarrier's amplitude is made such that it alone will cause 10% (of 75 KHz) deviation of the main FM carrier. Similarly, the DSBSC signal's amplitude is adjusted so that if *only* an L *or* an R signal is present, the peak deviation of the main carrier due to just the DSBSC signal alone would be $\pm 45\%$ (the total peak deviation due to only one channel (L *or* R) being present in the stereophonic signal would then be $\pm 45\%$ from the main $(L + R)$ audio signal, $\pm 45\%$ from the DSBSC signal, and $\pm 10\%$ from the pilot signal, making a total of $\pm 100\%$). Figures 7-1 and 7-2 show the block diagram and the modulating signal spectrum, respectively, for the commercial stereo FM system. These drawings also show the inclusion of an *SCA* (subsidiary communications authorization) *subcarrier* which will be discussed in the following section.

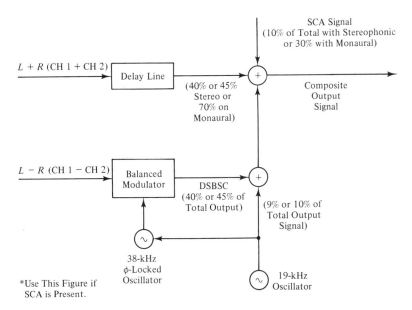

Figure 7-1   FM stereo broadcasting.

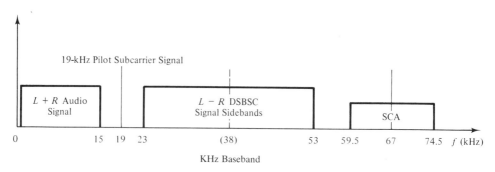

Figure 7-2   FM stereo/SCA broadcasting modulating signal.

Ignoring the presence of the SCA signal for the moment, the basic idea behind the stereo FM system is that a monaural receiver will receive and demodulate the entire signal as shown in Figure 7-2, but it will filter out everything above 15 KHz in the demodulated signal coming out of the FM discriminator detector and apply only the $L + R$ audio signal to the audio amplifier of the receiver. In a stereo receiver a low-pass filter following the discriminator allows the $L + R$ audio signal to pass to one input of an additive matrix. A bandpass filter permits the $L - R$ DSBSC signal to pass to an AM demodulator where it is combined with a 38-KHz *injection carrier* which is phase locked to the 19-KHz pilot subcarrier. The output of the AM demodulator is the $L - R$ audio signal, and it is fed to the second input of the additive matrix.

In the matrix the two signals $L' = A_1 (L + R) + A_2 (L - R)$ and $R' = A_1 (L + R) - A_2 (L - R)$ are formed. If $A_1 = A_2$ and if no other mixing or crosstalk has occurred in the transmission process, then $L' = L$ and $R' = R$. However, because of system limitations plus the fact that electronically, it is not possible to make A_1 *exactly* equal to A_2, there will be some amount of the L channel appearing in R' and some amount of R in L'. Crosstalk may be introduced at the point where the $(L + R)$ and $(L - R)$ signals are created from the L and R channel signals, in the audio amplifiers in the studio consoles (where the L and R channels are usually physically in the same cabinet and often operate from the same power supply), in the transmitter where the gains of the $(L + R)$ and the $(L - R)$ signal amplifiers and modulators may be slightly different, along the propagation path where the main audio channel $(L + R)$ and the subcarrier channel are not prop-agated equally (since they produce different sets of FM sidebands), in the receiver where the handling of the $(L + R)$ audio channel and the $(L - R)$ DSBSC channel may cause unequal changes in the two signals (such as in the AM detector where the phase of the 38-KHz injection carrier may not be exact), and so on.

Because of the crosstalk thus introduced, it will not be possible to obtain 100% separation of the L and R channel signals. However, enough separation (usually better than 20 dB) is obtained with a good receiver operating on a good antenna in the primary coverage area of an FM station to produce a very pro-nounced stereophonic effect.

Figure 7-3 shows block diagrams for a stereo FM transmitter and receiver (again with SCA included). In the transmitter the part represented by "FM carrier

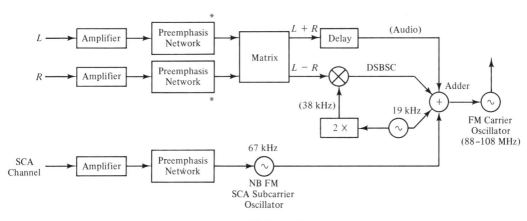

(a) Transmitter

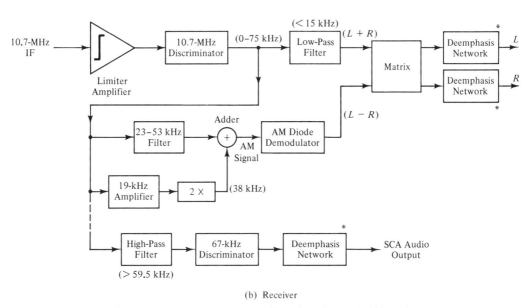

(b) Receiver

* = Preemphasis (Differentiation) and Deemphasis (Integration) Networks.

Figure 7-3 Stereo FM with SCA.

oscillator'' may actually be any of the FM modulation methods discussed in Chapters 4 and 5 (usually the modulation takes place at some subharmonic of the transmission frequency with frequency-multiplier stages employed in the transmitter). In the receiver, the demodulation process just discussed takes place and the $(L + R)$ and $(L - R)$ channels are obtained and fed into the decoding matrix. Table 7-1 is a listing of some of the specifications applying to stereo FM and SCA services along with some of the reasons why the 38-KHz subcarrier is not transmitted as such.

TABLE 7-1   STEREO/SCA FM DATA

Reasons for not transmitting the 38-KHz subcarrier

1.   Too much of the total composite modulating signal voltage (and, hence, of the total allowable 75-KHz station deviation) would be taken up by the subcarrier voltage.

2.   It could be transmitted at 38 KHz but at reduced power, but then it would be difficult to separate it out from the $L - R$ sidebands in order to use it to derive a phase reference at the receiver.

3.   Therefore, the subcarrier *information* (frequency and phase) is carried to the receiver via a 19-KHz pilot signal. Note that if a pilot signal at twice 38 KHz (76 KHz) had been used, it would not be possible to send the phase information without a 180 degree phase ambiguity.

The $L - R$ signal is AM modulated onto the missing carrier, producing a DSBSC $L - R$ signal centered at 38 KHz. The maximum audio frequency of the modulation applied to the 38-KHz subcarrier is 15 KHz (versus 10 KHz in a standard AM broadcast station).

SCA information

1.   Types of services
     Supplemental broadcasting—background music (MUZAK®, etc.); special-interest groups (the blind, special educational programming, weather reports, sports events, time signals, etc.).
     Remote control of FM repeater station from the master station, the telemetry of the main transmitter status back to the studio, etc.
     Digital data broadcasting links (being proposed).

2.   SCA subcarrier frequencies—stereo stations 53–75 KHz; mono stations also may have in 20–53 KHz range.

3.   Percentage of main carrier modulation signal—on stereo station SCA is limited to a total of 10%; on mono stations the SCA is limited to 30%.

4.   Common subcarrier frequencies in use—67 KHz (with a maximum NBFM deviation of 8 KHz and audio of 20 to 7500 Hz); on mono stations a second SCA carrier at 41 KHz is common.

Stereo pilot carrier

     19 KHz ± 2 Hz and between 8 and 10% of the total baseband signal magnitude.

7.3.2 SCA (*Subsidiary Communications Authorization*)

As has been mentioned several times thus far, Figures 7-1 through 7-3 and Table 7-1 show the presence of an auxilliary type of communications service which may ride along "piggyback" on the signal of either a monaural or a stereo broadcast FM station. This auxilliary service (SCA) will now be discussed.

SCA is a means whereby broadcasts to a select group of people (the blind, subscribers to background music programming, special education classes, etc.) may be made over an FM channel or whereby the FM channel may be used to send telemetry data from a transmitter site to the studio (in the case where the transmitter is remote-controlled from the studio), to send control signals to a remote low-power transmitter which picks up and rebroadcasts the main-channel programming from an FM station master transmitter (for example, to reach into a valley where the master transmitter's signal could not reach), to send communications to a remote-broadcast crew, and to send facsimile data (weather maps, etc.) and other special-purpose one-direction communication and control services.

The concept behind allowing SCA operations on an FM broadcast station is that since a maximum peak deviation of ± 75 KHz is allowed on the main FM carrier, the modulating signal could contain frequency components of up to 75 KHz without causing m_f to drop below unity. Since a monaural FM station's modulating signal has components only up to about 15 KHz and a stereo station up to about 53 KHz, room obviously exists for adding in additional information-carrying signal components. For a monaural station the FCC says these must fall into the range from 20 KHz to 75 KHz, while for a stereo station they must fall into the range of from 53+ to 75 KHz. Of course, when these additional signal components are added onto the modulating signal, the amplitude of the main-channel audio (up to 15 KHz) and the stereo $(L - R)$ and 19-KHz pilot subcarrier must be reduced so that the peak values of the composite modulating signal never exceeds the levels which causes a ± 75-KHz deviation of the main FM carrier.

For a monaural FM station where two SCA subcarriers are permitted, the combined peak value of the subcarriers must not exceed 30% of the permitted peak modulating signal. For a stereo station with one SCA subcarrier, this figure must be no more than 10%. SCA subcarriers are narrow-band FM (NBFM) modulated with a maximum deviation of ± 8 KHz and with the subcarrier modulating signal frequency range limited to 20 Hz to 7500 Hz (this keeps most of the modulated subcarrier frequency components within ± 7.5 KHz of the subcarrier center frequency). The two subcarrier frequencies used are 67 KHz for both monaural and stereo stations plus 41 KHz for monaural stations with two SCA subcarriers. These center frequencies must be maintained within ± 500 Hz. In stereo stations the normal maximum reduction of the stereo signal components is to between 40 and 41% each for the $(L + R)$ and $(L - R)$ signals, or the $(L + R)$ modulation reduced to between 80 or 82% when the L and R channels are identical and to between 10% or 8% respectively for the 19-KHz pilot subcarrier, although less reduction may be used with less than 10% SCA subcarrier injection. In any event the cross-modulation products which cause interference with the detected main-channel audio must be down by at least 60 dB from the main-channel audio signal at 100% modulation of both channels.

The FM station may use the SCA for its own purposes (remoted transmitter telemetry and control, control of a distant repeater, communications with a remote-pickup crew, etc.), or it may lease the SCA channel to some other service (such as Muzak, Inc.,® and others) which provides special services (such as background music to stores and offices, etc.) to a select group of receivers. Usually these outside-agency broadcasts are not intended to be in the public domain, and therefore the agency providing the service may charge for its reception and use.

7.3.3 Quadraphonic and Multichannel Audio Broadcasting

In recent years, the concept of being able to "surround" the listener with sound (at least on a two-dimensional basis) has produced the concept of "quadraphonic" or "four-track stereo" sound which ideally utilizes four separate channels to produce the effect (actually, as we shall see later, only three discrete channels are really needed). At the present time, the only popular method of recording and presenting the four discrete channels is via magnetic tape recording where four discrete tracks are recorded side by side on the magnetic tape. This will be discussed later in the section on recording techniques and formats. Most of the other techniques have involved matrixing (mixing) the four channels into two discrete channels to be recorded or broadcast.

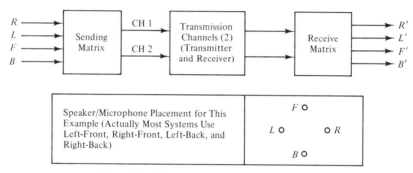

Transmission Matrix

CH 1 = $R + F - B$
CH 2 = $L + F + B$

Receiving Matrix

$F' = $ CH 1 + CH 2 = $2F + R + L$ (Monaural Signal)
$B' = $ CH 2 − CH 1 = $2B + L - R$
$L' = $ CH 2 = $L + F + B$ (Stereo Left)
$R' = $ CH 1 = $R + F - B$ (Stereo Right)

(a) A Possible Four-Channel "Matrix Method" (Two Discrete Channels or 4-2-4 Four-Channel Matrixing)

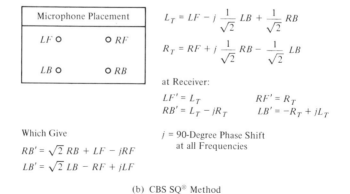

Which Give

$RB' = \sqrt{2}\, RB + LF - jRF$
$LB' = \sqrt{2}\, LB - RF + jLF$

$L_T = LF - j\,\dfrac{1}{\sqrt{2}}\, LB + \dfrac{1}{\sqrt{2}}\, RB$

$R_T = RF + j\,\dfrac{1}{\sqrt{2}}\, RB - \dfrac{1}{\sqrt{2}}\, LB$

at Receiver:

$LF' = L_T$ $RF' = R_T$
$RB' = L_T - jR_T$ $LB' = -R_T + jL_T$

$j = $ 90-Degree Phase Shift at all Frequencies

(b) CBS SQ® Method

Figure 7-4   Examples of 4-channel matrix-type quadraphonic systems.

Transmission Matrix

$L_T = (LF + jLB) \cos\theta + (RF + jRB) \sin\theta$
$R_T = (RF - jRB) \cos\theta + (LF - jLB) \sin\theta$

Receiving Matrix

$LF' = L_T \cos\theta + R_T \sin\theta$
$RF' = R_T \cos\theta + L_T \sin\theta$
$RB' = R_T \cos\theta - L_T \sin\theta$
$LB' = L_T \cos\theta - R_T \sin\theta$

LF ○	○ RF
LB ○	○ RB
Microphone Placement	

Provides Only 3-dB Separations Between Any Two Adjacent Channels but Complete Separation Between Diagonally Opposite Channels.

$$\tan\theta = \frac{F - B \text{ Spacing}}{L - R \text{ Spacing}}$$

(c) Sansui Four-Channel Matrix Method (Another 4-2-4 Method)

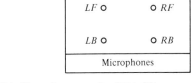

LF ○	○ RF
LB ○	○ RB
Microphones	

Two Main Channels: $R = RF + RB$ $L = LF + LB$
30-kHz FM Subcarrier*: $R_s = RF - RB$ $L_s = LF - LB$

At Playback: $RF' = \frac{1}{2}(R + R_s)$ $LF' = \frac{1}{2}(L + L_s)$

                        $RB' = \frac{1}{2}(R - R_s)$ $LB' = \frac{1}{2}(L - L_s)$

*Will Not Broadcast Through American FM Stereo Stations.

(d) CD-4® (4-4-4) Matrix Method of Quad Recording
(RCA and Japan–Victor)

Figure 7-4 (continued)

This latter technique is referred to as 4-2-4 quadraphonic as opposed to 4-4-4 quadraphonic wherein the four discrete channels are recorded or broadcast. There are several types of 4-2-4 matrixing methods. Figure 7-4a shows a straightforward matrixing method. The original four channels are front (F), back (B), left (L), and right (R), while the four channels re-created at the receiver are F', B', L', and R', respectively. As can be seen, the four channels out of the receiver matrix are actually mixtures of the original four channels. However, some of the original four-channel effect is still preserved, although it is somewhat diluted. Most of the commercial quadraphonic systems are based upon a square placement of the microphones rather than the diamond arrangement of Figure 7-4a. This square placement gives a left-front (LF), a right-front (RF), a left-back (LB), and a right-back (RB) designation for each channel. Figure 7-4b shows the CBS SQ method of 4-2-4 quadraphonic recording which involves introducing 90-degree phase shifts into the mixing matrix and compensating 90-degree shifts into the receiving matrix. This gives more separation than a straight resistive matrix mixing technique can. The Sansui 4-2-4 method shown in Figure 7-4c is similar, but has the advantage that adjustments can be made for the microphone-placement aspect ratio (the ratio of left-to-right spacing divided by front-to-back spacing). The inverse of the aspect ratio is equal to the tangent of an angle called theta. This system provides complete separation (theoretically) between the diagonally opposite microphones, but only 3-dB separation between adjacent channels.

In addition to the matrix methods of attempting to consolidate four channels of sound into two discrete channels, other techniques have been developed to try and actually send four (or three in some cases) discrete channels. One of these 4-4-4 techniques is the CD-4 method of RCA and Japan-Victor wherein 30-KHz subcarriers are utilized on each of the two stereo channels to carry the front-minus-back information for each side while the main signal on each stereo channel carries the front-plus-back signal from each side (see Figure 7-4d). Since these modulated subcarriers are actually mechanically inscribed on the track of the stereo record it is obvious that very light styli must be used in the playback process and that the record may be more subject to abuse and general deterioration than a conventional stereo recording. Also the four-channel effects can not be broadcast over standard stereo FM stations in the US, although the main-channel signals are stereo (two-channel) compatible.

One method of transmitting "all-around" sound is based upon the fact that four channels are not really needed, but that sound coming from any direction of the compass can be simulated with only three loudspeakers (perhaps using a front-left, L, a front-right, R, and a center-back, B, microphone placement as shown in Figure 7-5a. By matrixing these three signals as shown in Figure 7-5a, a mono-phonic signal (sum signal), a stereo (left-right) difference signal, and a back-to-front difference signal are obtained. By using the concept of quadrature modulation (see Chapter 4), the two difference signals can be placed upon the DSBSC 38-KHz stereo FM subcarrier as shown in Figure 7-5b. These three signals can then be separated out as shown in the receiver circuit of Figure 7-5c, and put back through a matrix to produce the original three input channels discretely. The only mixing that would occur would be due to crosstalk among the A, B, and C channels in the modulation-transmission-demodulation process. The system is compatible with monaural FM receivers (which receive only the A channel) and with some type of stereo receivers that detect only in-phase modulation of the 38-KHz sub-carrier (i.e., the $L - R$ difference signal).

7.4 AUDIO RECORDING TECHNIQUES

7.4.1 Mechanical Recording Techniques

The oldest form of audio recording was the purely mechanical method developed by Edison wherein a groove was cut into the surface of a cylinder or disk by a needle attached to a diaphragm at the neck of a "morning glory"–type horn. The

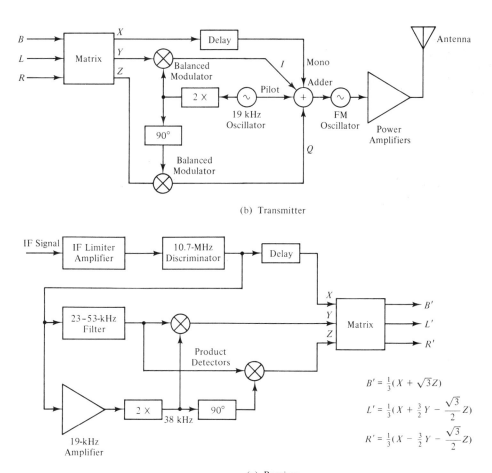

L
(Left)
○

R
(Right)
○

B
(Rear)
○

Speaker/Microphone
Placement

From B, L, and R Generate Three Signals:

Monophonic $X = B + L + R$
Stereo Difference ($L - R$) $Y = L - R$
Back-to-Front Difference $Z = \dfrac{2}{\sqrt{3}}\,(B - \tfrac{1}{2}L - \tfrac{1}{2}R)$

(a) Signal Relationships

(b) Transmitter

(c) Receiver

$B' = \tfrac{1}{3}(X + \sqrt{3}Z)$

$L' = \tfrac{1}{3}(X + \tfrac{3}{2}Y - \dfrac{\sqrt{3}}{2}Z)$

$R' = \tfrac{1}{3}(X - \tfrac{3}{2}Y - \dfrac{\sqrt{3}}{2}Z)$

Figure 7-5   Three-channel FM system (proposed).

reproduction or playback process was the reverse of the recording process. Later the cutting process was implemented using an electrodynamic cutting head wherein the needle was moved by an electromagnet whose coil was driven by the output of an audio amplifier. The reproduction process used a piece of piezoelectric quartz crystal attached to the needle. When the needle introduced strains in the quartz crystal, it produced a signal voltage which could then be amplified by an audio amplifier and used to drive a loudspeaker. In recent years the natural quartz crystal has been replaced either by a cartridge using an artificial piezoelectric ceramic material or by a magnetic cartridge wherein the needle is attached to an armature of magnetic material which moves in proximity to a coil of wire and produces a small voltage in the coil. This voltage is quite small and must be amplified by a preamplifier to bring its output up to the same level as is obtained directly from the quartz or ceramic crystal cartridges.

Stereo disk recording is accomplished by recording one channel on one side of the sloping walls of the record groove and the other channel on the opposite wall as shown in Figure 7-6. The stylus transmits these two directions of motion to an armature which moves with respect to two magnetic coils, varying the air-gap reluctance and thus causing a voltage to be induced in each coil in proportion to the amount of variation in the groove wall for that channel. Stereo pickup cartridges are also manufactured using two piezoelectric ceramic elements mounted at right angles and attached to a common stylus.

7.4.2 Basic Magnetic Recording Methods

Although mechanical disks are still popular for commercially recorded material (largely because they can be easily mass produced), much present-day audio recording is done via the use of magnetic-tape recording using various types of tapes and tape formats. Basically all audio magnetic-tape recording involves "permanently" magnetizing a highly retentive magnetic powder that has been deposited on the surface of a tape as the tape moves past a magnetic recording "head" consisting of an electromagnet with an air gap as shown in Figure 7-7. The tape

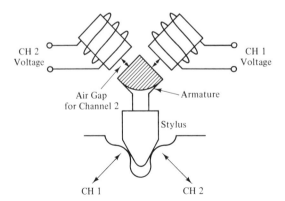

Figure 7-6 Stereo phonograph pickup.

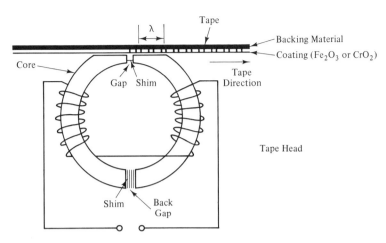

Figure 7-7 Tape recorder record/playback head.

material is commonly Mylar® or some form of polyester plastic (although paper tape was used in the early days of tape recording) and the coating is usually powdered iron oxide (Fe_2O_3) or chromium oxide (CrO_2). Typical magnetization curves for these two materials are shown in Figure 7-8a.

One of the problems encountered in magnetic tape recording is the nonlinearity of the oxide's magnetization curve (as shown in Figure 7-9) for low values of H. If the magnetizing signal were just a sine wave centered about zero (as shown by the dashed line) the resultant retained β in the magnetic oxide would be a highly distorted waveform as shown in the $\beta(t)$ plot. One way of overcoming this would be to offset the magnetizing function $H(t)$ by applying a d.c. bias sufficient to center $H(t)$ on a linear portion of the curve (point A in Figure 7-9). This method is used on some of the cheaper magnetic tape recorders but is not very satisfactory because it does not produce an optimum S/N ratio on playback. Another technique is to add a high-frequency (above the audio region, often at about 50 KHz) signal to the audio signal. The result is a total signal where the upper and lower envelope functions are accurate reproductions of the audio waveform (note that this is *not* AM modulation—it is merely *adding* the bias signal to the audio signal). When the tape is played back through a low-pass filter which rejects the bias signal frequency, the filter output is an accurate reproduction of the audio signal added to the bias signal at the time of recording.

One phenomenon of magnetic tape recording is the space-time convolution process which takes place as the tape is recorded. This results from the fact that the magnetic field which is produced across the air gap and which penetrates into the tape oxide has a finite width along the direction of the tape as is shown in Figure 7-10a for a value of $i_{head}(t) = 1.0$ units. The tape moving across the air gap in the $-v$ direction can be thought of as the air-gap function moving along the tape in the $+v$ direction. The "permanent" magnetic field which is "recorded" at the point $v = x$ on the tape is proportional to the integral of the current value

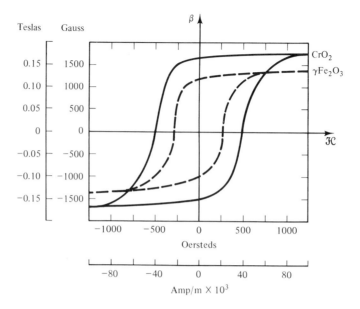

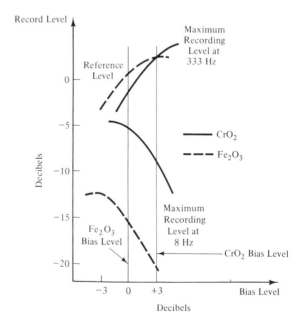

Figure 7-8 Magnetic recording materials characteristics.

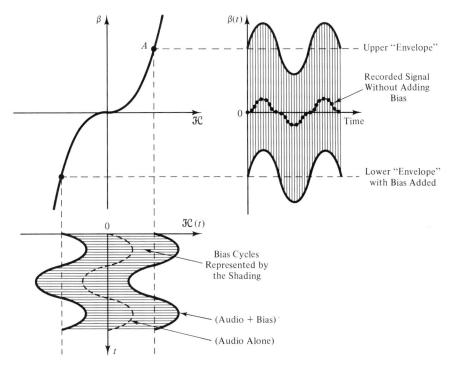

Figure 7-9 How adding a high-frequency bias signal to the audio signal reduces non-linear distortion.

of $\beta_g(x)$ as it slides past the point, or

$$\beta_r(x) \propto - \int_a^{-a} \beta_g(v)i_{\text{head}}(\tau - t)dv \tag{7-3}$$

where $\tau = t$ when $v = 0$ and i_{head} is the audio signal component of the current in the recording head.   Letting $v = v_{\text{tape}}t$ we then get

$$\beta_r(x) \propto - \int \beta_{g\tau}(\tau - t)i_{\text{head}}(\tau)d\tau \tag{7-4}$$

where $\beta_{g\tau}$ is just $\beta_g(v)$ mapped into the time domain (the larger the tape velocity is, the "thinner" $\beta_{g\tau}$ will be along the time axis).

The net result of all this is that the finite width of the $\beta_g(v)$ function causes a "smearing" of the recorded function so that $\beta_r(x)$, whre $x = v_{\text{tape}}t$ is *not* exactly proportional to $i_{\text{head}}(t)$.   The "wider" the function $\beta_{g\tau}$ is on the time axis due to $\beta_g(v)$ being wide on the v axis or v_{tape} being small, the more smeared the recorded function $\beta_r(x)$ on the tape will become and the more limited the high-frequency response of the system will be.   Thus, to enable the recording of the higher audio tones, the recording-head air gap must be kept quite small (about $1/1000$ inch or less) and the tape speed moderately high ($1\frac{7}{8}$, $3\frac{3}{4}$, $7\frac{1}{2}$, and 15 inches per second are the most common standard tape speeds for audio recording).

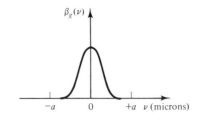

(a) Tape-Head Gap Function

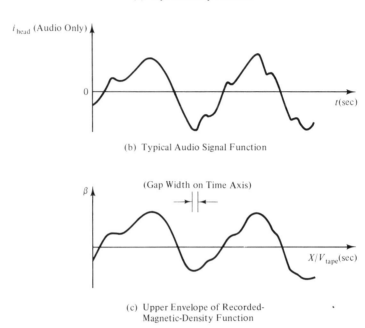

(b) Typical Audio Signal Function

(c) Upper Envelope of Recorded-
Magnetic-Density Function

Figure 7-10   The effects of a finite-width tape-head-gap function.

7.4.3 Magnetic Tape Recording Formats

The oldest popular form of magnetic recording was magnetic wire recording which was beset by many mechanical and handling problems and never became really popular. The next approach was reel-to-reel magnetic tape using $\frac{1}{4}$ inch-wide tape. Early formats used the entire width of the tape for one track (full-track format), but the two-track format was soon adopted wherein one-half of the tape width was used when recording with the tape traveling in one direction until the takeup reel was full. This reel was then turned over and placed in the feed reel position, and the now-empty feed reel became the new takeup reel. When the tape was re-started, the recording commenced on the unused half of the tape width. This type of format (half-track monaural) not only permitted recording twice the program material on the tape, but also eliminated the necessity to rewind the tape after each usage. It became a very popular form of magnetic tape recording in the 1950s and 1960s.

With the advent of stereo and quadraphonic recording, other reel-to-reel tape formats were adopted (see Figure 7-11a). The single-track stereo and the four-track quad are single-direction formats. The four-track stereo is a bidirectional format developed at a time when head technology was such that it was difficult to construct a head having the two stereo channels side by side, and as a result the format shown was adopted, resulting in noncompatibility between four-track stereo and half-track monaural.

Two other types of magnetic-tape recording that are popular are the cassette formats and the continuous-loop cartridges. The standard cassette is essentially

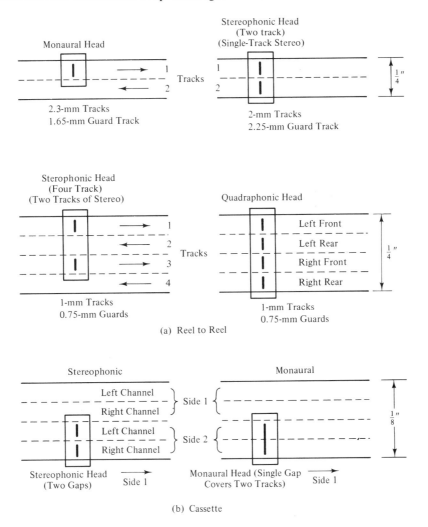

(a) Reel to Reel

(b) Cassette

Figure 7-11   Reel-to-reel and cassette recording formats.

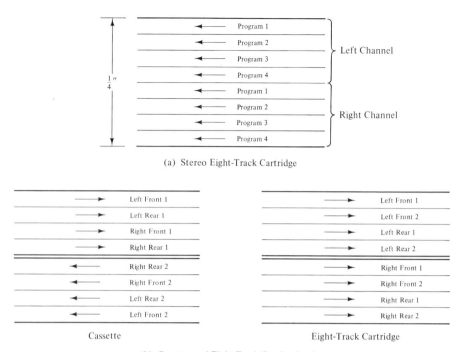

(a) Stereo Eight-Track Cartridge

Cassette   Eight-Track Cartridge

(b) Cassette and Eight-Track Quadraphonic

Figure 7-12 Stereo and quadraphonic tape formats.

a miniature reel-to-reel type of operation where the playback and takeup "reels" are enclosed in a plastic housing. The standard tape size is $\frac{1}{8}$-inch width with a four-track format (see Figure 7-11b). Fortunately, the mistake that was made in going from half-track monaural to bidirectional stereo was avoided in cassette technology, and thus we find that monaural and stereo cassettes are compatible.

In addition to the standard cassette size there are "microcassettes" intended for use in miniature personal tape recorders and large-format cassettes which have recently been introduced for hi-fidelity music recoding (some people think that these will eventually replace reel-to-reel tape recording and LP case recording). The eight-track cartridge is a continuous loop of $\frac{1}{4}$-inch tape with the ends spliced together with a short length of metal foil tape. The stereo format is as shown in Figure 7-12a. Whenever one "program" is completed and the metal foil tape passes through the tape mechanism, the heads are automatically shifted to the tracks for the next program. When all four programs (1 through 4) have been played, the heads shift back up to program 1 and the whole sequence is repeated. Thus this type of recording will play indefinitely and thus is very popular for continuous background music applications.

Another version of the continuous-loop cartridge is the two-track stereo loop cartridge used in broadcasting stations for spot announcements (such as program openings and closings, commercials, etc.) or for short program segments. Quite often the tape used in these cartridges is prepared on a reel-to-reel tape deck. Some broadcast stations have large machines which automatically play one cartridge after another all day long (i.e., no "live" material except for possibly some news broadcasts or weather and time reports).

Quadraphonic (four-channel) magnetic tape recording uses reel-to-reel, cassette (usually large format), or eight-track cartridges with all four tracks recorded side by side. Unfortunately, the decision as to which channel goes on which track has produced inconsistent formats for the quad cassettes and cartridges (see Figure 7-12b).

7.4.4 Reduction of Noise in Magnetic Tape Recording

One of the problems associated with magnetic tape recording is the fact that the high-frequency response is greatly dependent upon factors such as the width of the effective gap in the tape head, the speed with which the tape passes by the head, the level of the high-frequency bias signal, and in some cases the granularity of the oxide coating on the tape. As has been previously discussed, the smaller the head gap and the higher the tape speed, the better the frequency response. There has been considerable advance in tape-head technology in recent years, and consequently the newer tape recorders have higher frequency response and/or are capable of better performance at lower tape speeds than were the older machines. Many of the earlier machines used tape speeds of 15 inches per second (ips), 30 ips, or even 60 ips. With better head technology and tapes with finer powders, the popular reel-to-reel speeds today are $3\frac{3}{4}$ ips and $7\frac{1}{2}$ ips, with some machines also making $1\frac{7}{8}$ ips available. Eight-track cartridges run at $3\frac{3}{4}$ ips and standard cassettes at $1\frac{7}{8}$ ips.

The dependence of frequency response upon the level of the high-frequency bias can be seen in Figure 7-13. As can be seen, increasing the head bias current produces a large reduction in the third harmonic distortion produced, but also greatly decreases the output at high audio frequencies. Some compromise between harmonic distortion generation and loss of high-frequency response is normally made when selecting the bias level to be used for a given tape material. The bias level to be used is also affected by the tape oxide (Fe_2O_3 versus CrO_2), and many tape recorders now have a two-position selection switch for this purpose.

The loss of high-frequency response in magnetic tape recording can be compensated for by including a great deal of treble boost in the recording and playback amplifiers, but this tends to increase the level of high-frequency noise or "hiss" present in the audio output. Thus other more sophisticated techniques were developed. One of these techniques which is now widely used is the Dolby® technique, and it has almost become an industry standard.

The Dolby-B® system is based upon the use of a high-pass filter with a variable cutoff frequency used in a manner which in some respects resembles a frequency-dependent automatic gain control. Figure 7-14 shows a basic block diagram of the recording/playback process and the Dolby-B control network. When signal levels are low, the feedback loop in the network lowers the cutoff frequency of the filter, thus permitting most of the audio high-frequency range to be boosted

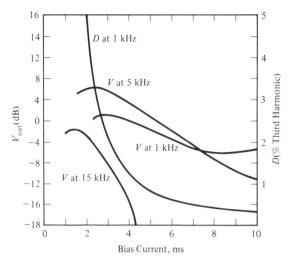

(a) Third Harmonic Distribution at
1 kHz and Output at 1 kHz,
5 kHz, and 15 kHz Versus High-
Bias Current (I_{signal} = 0.63 mA)

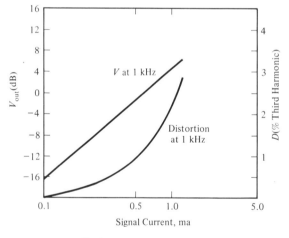

(b) Output and Distortion at 1 kHz

Figure 7-13 Effects of high-frequency
bias upon response characteristics.

for recording, thus bringing the low signals up above the noise level introduced in
the recording/playback process. At the receiving end, the effective gain is likewise
reduced over most of the audio range. As the level of the input signal to be
recorded increases, the cutoff frequency of the high-pass filter is moved upward,
resulting in signal boost being applied only to the very high audio frequencies.
Thus, at high signal levels most of the audio frequencies are not boosted prior to
being recorded, and on the playback end the gain is not reduced for these frequencies. The net result is better frequency response and larger dynamic range

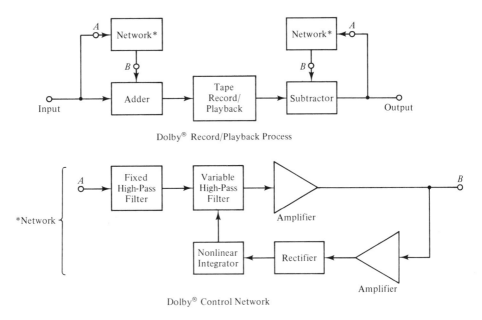

Figure 7-14   The Dolby-B® noise reduction system.

(80 dB with Dolby-B® versus about 70 dB for ordinary good-quality magnetic tape recordings).

The Dolby-C® system is somewhat the same as Dolby-B in its operation except that there are two boost channels having different characteristics. It provides about a 10-dB dynamic range improvement over Dolby-B. The dbx® noise-reduction system is a nonfrequency-dependent system which reduces the recording gain for high signal levels and allows lower-level audio signals to be recorded at higher S/N levels. In general the dbx system will produce a better S/N ratio upon playback for very quiet passages while the Dolby-C system will provide less tape "hiss" (high-frequency noise) during moderate-to-loud passages. It should be noted that although the term "companding" is often applied to the operation of these systems, they are really fast-acting *gain-control* systems and do not operate in the same manner as the *instantaneous* companding systems commonly used before digitizing analog audio signals (see Section 5.1.3.2).

Another means by which a greater dynamic range may be obtained while at the same time giving improved frequency response and less distortion is by digitizing the audio signal and recording the resulting PCM (pulse-code-modulation) signal on magnetic tape or optically on a plastic disk which is read by an electro-optical system employing a finely focused laser beam (this will be discussed in more detail in Section 7.8.1 on optical audio and video recording techniques). At the playback end each frame of the PCM signal is sent to a D/A converter where an analog audio signal is produced. Since the digitizing noise can be made arbitrarily small by choosing a sufficient number of bits per sample, the dynamic range of this type of system is limited only by the dynamic range of the amplifiers and other electronics

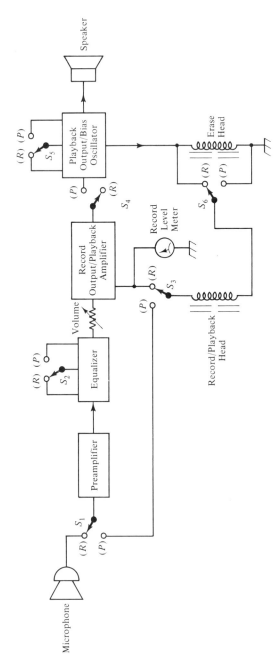

Figure 7-15 Block diagram of an integrated tape recorder.

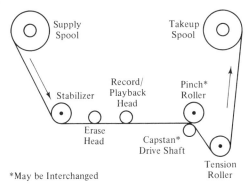

(a) Typical Reel-to-Reel Mechanism

*May be Interchanged

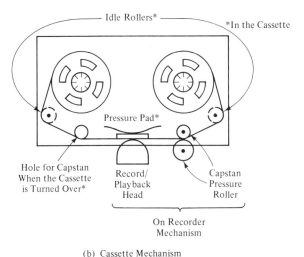

(b) Cassette Mechanism

Figure 7-16 Tape recorder mechanisms.

involved and not by the limitations of the recording process itself. A dynamic range of up to 120 dB is typical of this type of audio recording system.

7.5 TELEVISION AND VIDEO SYSTEMS

7.5.1 Basic Black-and-White TV

A very elementary television system could be constructed by applying sawtooth waveforms to the horizontal and vertical deflection plates of a TV camera tube and the same waveforms to the deflection plates of a TV picture tube. If the vertical sawtooth frequency was about 20 to 40 Hz and the horizontal sawtooth frequency about 500 times the vertical sweep frequency, a TV "raster" would be generated at both the camera tube and the picture tube. All that would be needed would be to amplify the camera-tube video output and apply it to the control grid (brightness control) of the picture tube. However, such a TV system would have several problems: it would be necessary to send three signals from the camera to the picture tube (the vertical sweep signal, the horizontal sweep signal, and the brightness signal), the "retrace" lines that occur during the rapid fall of the sawtooth waveforms might be visible, and the fact that the horizontal and vertical sweep oscillators are free running and not synchronized to each other or to anything else might cause some problems. Also, scanning the entire picture from top to bottom before starting over again at the top might cause some flicker problems.

To get around some of the problems, synchronized-sweep TV systems were

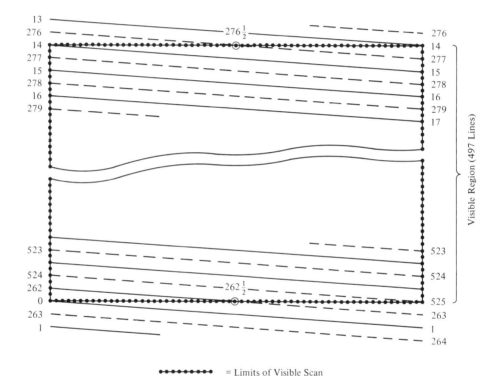

•••••••• = Limits of Visible Scan

Times shown are "line times"

Examples: Line 1 starts at $t = 0$ and ends at $2t = 1$.
Line 93 starts at $t = 92$ and ends at $t = 93$.
Line 525 starts at $t = 524$ and ends at $t = 525$.

Lines 1 to middle of line 263 are in field 1 and are shown solid; lines from midpoint of 263 to and including line 525 are in field 2 and are dashed.

Vertical blanking from $262\frac{1}{2}$ to $276\frac{1}{2}$ and from 0 to 14.

Figure 7-17   TV picture raster scan.

developed.   The original U.S. black-and-white TV system used a vertical-sweep frequency of approximately 60 Hz locked to the local 60-Hz power-grid frequency where the camera was located.   The horizontal-sweep frequency was exactly 262.5 times this, or 15,750 Hz.   Since the multiplier was not an integer, the sweep lines in alternate vertical-sweep "fields" were spaced between the sweep lines of the preceding and following fields.   Thus, there were effectively 262.5 times 2 or 525 sweep lines in the picture.   It required two 262.5-line "fields" to produce one complete picture or "frame."   The "frame rate" of this TV system was 30 frames per second (one-half of the field rate of 60 per second).   This process is known as "interlaced scanning" and does much to eliminate any frame flicker in the TV picture.

Figure 7-17 shows the raster that is produced.   Note that some lines are missing completely because they occur during the times at which the vertical retrace from bottom to top occurs.   Also, some of the lines that are shown are not part of the active picture.   Actually, there are only $248.5 \times 2 = 497$ lines inside the rectangle defining the active picture area.   Vertical "blanking" (the shutting off of the picture tube's electron beam) occurs from line times 0 to 14 and from $262\frac{1}{2}$ to $276\frac{1}{2}$.   The picture tube electron beam is also turned off during each horizontal retrace (horizontal blanking).   The blanking operations eliminate the problem of visible retrace lines.

To avoid having to send the horizontal and vertical sweep signals (deflection signals), these are generated locally at the TV receiver by horizontal and vertical sweep oscillators which are synchronized with the sweep oscillators in the TV camera.   The synchronization is accomplished by superimposing synchronizing pulses upon the camera tube video (brightness) signal.   Synchronization of the horizontal oscillator is accomplished by a single pulse which occurs after the horizontal blanking is turned on at the end of each horizontal line.   Synchronization of the vertical oscillator is more complex, especially since synchronization of the horizontal oscillator must be maintained during the vertical retrace.   It is accomplished with an integrator circuit which is first allowed to discharge to a minimum level by sending a series of very narrow pulses (equalizing pulses) as shown in Figure 7-18a or b.   Then a series of wide pulses is sent and the integrator circuit begins to charge up.   Near the end of this train of six wide pulses the integrator-circuit output reaches the level which initiates the vertical retrace.   A series of equalizing pulses and horizontal sync pulses gets the horizontal oscillator well established again in its waveform while the vertical retrace is taking place and the new vertical sweep is being initiated.   Note the difference in the drawings in (a) and (b) of Figure 7-18 and correlate these with the raster line numbers in Figure 7-17 (the line-number points are given in parenthesis in Figure 7-18).

Other drawings in Figure 7-18 detail the dimensions of the horizontal and vertical sync pulses and the equalizing pulses and where these occur on the blanking pulse.   Also shown is the "color burst" signal for synchronizing the color subcarrier oscillator in the color TV receiver.

At this point it should be noted that with the advent of color TV, the vertical-sweep frequency was changed to 59.940052 Hz and the horizontal-sweep frequency to 15,734.264 Hz (the latter is the color subcarrier oscillator frequency of 3.579545

(a) Start of add-numbered fields.

(b) Start if even-numbered fields.

Horizontal Dimensions Not to Scale is 1, 2, and 3

(c) Two complete scan lines

Detail Between 3-3 in (b)

234

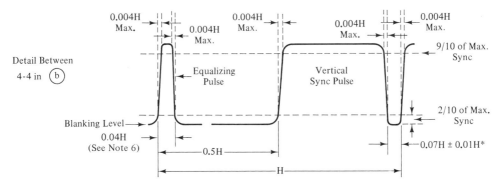

Detail Between
4-4 in (b)

0.004H Max.

0.004H Max.

0.004H Max.

0.004H Max.

9/10 of Max. Sync

Equalizing Pulse

Vertical Sync Pulse

Blanking Level

0.04H (See Note 6)

0.5H

2/10 of Max. Sync

0.07H ± 0.01H*

H

(d) Equalizing and vertical sync pulses.

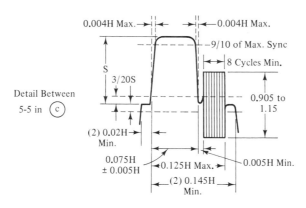

0.004H Max.

0.004H Max.

9/10 of Max. Sync

8 Cycles Min.

Detail Between
5-5 in (c)

S

3/20S

0.905 to 1.15

(2) 0.02H Min.

0.075H ± 0.005H

0.125H Max.

0.005H Min.

(2) 0.145H Min.

(e) Horizonital sync pulse and color burst.

NOTES:

1. H = Time from start of one line to start of next line.
2. V = Time from start of one field to start of next field.
3. Leading and trailing edges of vertical blanking should be complete in less than 0.1 H.
4. Leading and trailing slopes of horizontal blanking must be steep enough to preserve minimum and maximum values of $(x + y)$ and (x) under all conditions of picture content.
*5. Dimensions marked with asterisk indicate that tolerances given are permitted only for long time variations and not for successive cycles.
6. Equalizing pulse duration must be between 0.45 and 0.55 of the duration of the horizontal synchronizing the pulse duration.
7. Color burst follows each horizontal pulse, but is omitted following the equalizing pulses and during the broad vertical pulses.
8. Color bursts to be omitted during monochrome transmission.
9. The burst frequency shall be 3.579545 mc. The tolerance on the frequency shall be ± 10 cycles with a maximum rate of change of frequency not to exceed $^1/_{10}$ cycle per second per second.
10. The horizontal scanning frequency shall be $^2/_{433}$ times the burst frequency

11. The dimensions specified for the burst determine the times of starting and stopping the burst, but not its phase. The color burst consists of amplitude modulation of a continuous sine wave.
12. Dimension "P" represents the peak excursion of the luminance signal from blanking level, but does not include the chrominance signal. Dimension "S" is the sync amplitude above blanking level. Dimension "C" is the peak carrier amplitude.
13. Start of Field 1 is defined by a whole line between first equalizing pulse and preceding H sync pulses.
14. Start of Field 2 is defined by a half line between first equalizing pulse and preceding H sync pulses.
15. Field 1 line numbers start with first equalizing pulse in Field 1.
16. Field 2 line numbers start with second equalizing pulse in Field 2.
17. Refer to test for further explanations and tolerances.
18. During color transmissions, the chrominance component of the picture signal may penetrate the synchronizing region and the color burst penetrates the picture region.

Figure 7-18 NTSC Television waveforms for color transmission.

MHz divided by 227.5 as will be explained in the next section on color TV). In transmitting the composite TV signal, the maximum transmitter RF output occurs during the transmission of the synchronizing pulses (100% level). The blanking level is at 75% maximum RF voltage (black level is at 70%), and the maximum picture brightness occurs at 12.5% of the maximum RF voltage. Thus, this is a "negative modulation" system with respect to brightness.

Generation of the synchronizing and blanking pulses is usually accomplished by a sync generator such as is shown in Figure 7-19. Here the turning on or off of the sync-level pulses or the blanking-level pulses is controlled on a coarse basis by outputs from a ring counter setting or resetting a flip-flop for each type of pulse to be produced (blanking, vertical sync, horizontal sync, or equalizing) while the

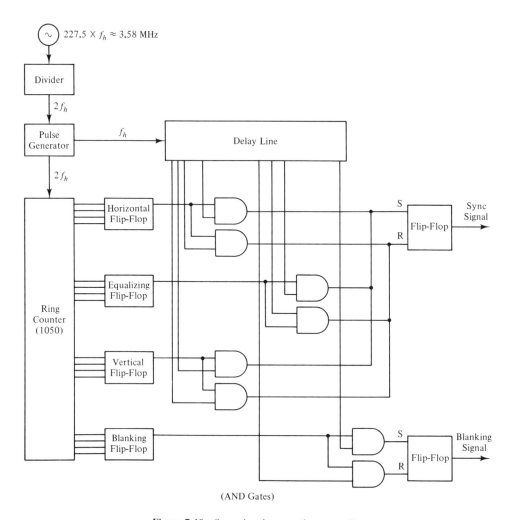

Figure 7-19 Sync signal generation at studio.

fine timing is controlled from outputs along a delay line driven by short duty-cycle pulses at a frequency of f_h.

Only about 82% of each horizontal-sweep period (T_h = 63.5556 μsec) or about 51.1156 μsec is active picture sweep time. If the aspect ratio of the picture is 4:3 (i.e., the width is $\frac{4}{3}$ times the height) and if the same resolution is to be maintained in the horizontal as in the vertical direction, then there would have to be 248.5 × $\frac{4}{3}$ complete cycles of the video brightness signal in 52.1156 μsec or about 6.3577 MHz as the highest frequency component in the video signal. Actually in the U.S. TV system the bandwidth is only slightly more than 4.2 MHz, so the horizontal resolution is not quite as good as the vertical resolution. The spectrum which the composite signal produces extends upward from 0 to 4.2 MHz but occurs in packets of energy centered around multiples of f_h = 15.734264 KHz with considerable "open space" between each packet. When the video signal is used to modulate an RF carrier a technique called "vestigial sideband AM" is used wherein the full range of upper sideband frequencies are transmitted (up to 4.2 MHz above the carrier) but only a portion of the lower set of sidebands (down to about 1.0 MHz below the carrier) are transmitted. This asymmetrical method of modulation is compensated for at the receiver by the shape of the video IF passband. This will be discussed in detail later in this chapter.

The TV sound channel is on a separate carrier located 4.5 MHz above the picture carrier and FM modulated with ±25 KHz of maximum peak deviation. The total video plus audio channel bandwidth is 6 KHz with the picture carrier frequency located 1.25 MHz inside the lower band edge and the sound carrier frequency located 0.25 MHz inside the upper band edge. There are currently 82 TV broadcast channels assigned in the United States (12 "VHF" and 70 "UHF" channels).

Table 7-2 lists these channel frequency assignments. Note that some of the channel assignments are on a shared basis with other radio services. Table 7-3 gives some of the specifications of the present-day U.S. TV system (the NTSC-compatible color/black-and-white system). Further standards specification for the NTSC system can be found in the FCC Rules and Regulations (Title 47, Section 73 of the U.S. National Code).

7.5.2 Color TV Techniques

Before beginning a discussion of how color TV works and the various systems that have been tried or adopted, it is first necessary to present some elementary information on color and how the human eye perceives it. Many good books have been written on color from different viewpoints (the physicist's, the engineer's, and the artist's, for example). One of the better ones for our purpose is *The Science of Color* by the Committee on Colorimetry of the Optical Society of America and published by them.[2]

Approaching the subject of color first from the physicist's point of view, electromagnetic radiation of from 380 to 800 mμ (millimicrons wavelength where

---

[2] Committee on Colorimetry of the Optical Society of America, 1963.

TABLE 7-2 U.S. TV CHANNEL FREQUENCIES

Channel no.	Frequency band (MHz)	Channel no.	Frequency band (MHz)	Channel no.	Frequency band (MHz)
2	54–60	30	566–572	57	728–734
3	60–66	31	572–578	58	734–740
4	66–72	32	578–584	59	740–746
5*	76–82	33	584–590	60	746–752
6*	82–88	34	590–596	61	752–758
7	174–180	35	596–602	62	758–764
8	180–186	36	602–608	63	764–770
9	186–192	37†	608–614	64	770–776
10	192–198	38	614–620	65	776–782
11	198–204	39	620–626	66	782–788
12	204–210	40	626–632	67	788–794
13	210–216	41	632–638	68	794–800
14	470–476	42	638–644	69	800–806
15	476–482	43	644–650	70	806–812
16	482–488	44	650–656	71	812–818
17	488–494	45	656–662	72	818–824
18	494–500	46	662–668	73	824–830
19	500–506	47	668–674	74	830–836
20	506–512	48	674–680	75	836–842
21	512–518	49	680–686	76	842–848
22	518–524	50	686–692	77	848–854
23	524–530	51	692–698	78	854–860
24	530–536	52	698–704	79	860–866
25	536–542	53	704–710	80	866–872
26	542–548	54	710–716	81	872–878
27	548–554	55	716–722	82	878–884
28	554–560	56	722–728	83	884–890
29	560–566				

*In Alaska and Hawaii, the frequency bands 76–82 Mc/s and 82–88 Mc/s are allocated for nonbroadcast use. These frequency bands (Channels 5 and 6) will not be assigned in Alaska or Hawaii for use by television broadcast stations.

†Channel 37, 608–614 MHz, is reserved exclusively for the radio astronomy service until the first Administrative Radio Conference after January 1, 1974, which is competent to review this provision.

Source: From Title 47, Section 73, Subpart A of the U.S. National Code.

1 millimicron = 10^{-9} meter) produces the sense of visible color in the human eye. If only one wavelength of radiation is present at a time, the color sensation is classed as *spectrally pure* color, with deep violet at the short-wavelength end of the visible spectrum and red at the long-wavelength (low-frequency) end of the spectrum. Radiation of a wavelength slightly shorter than visible purple is called ultraviolet (most children and some young adults can see somewhat into the ultraviolet region), and radiation with wavelengths longer than red is called infrared radiation. Infrared is commonly referred to as heat radiation, and the lower end of the infrared region merges with the very-short microwave region.

TABLE 7-3 NTSC COLOR TELEVISION STANDARDS

Frequencies	Sound carrier 4.5 MHz above the picture carrier
	Horizontal sweep frequency = 4.5 MHz/286 = 15,734.264 ($\pm$) 0.044 Hz
	Vertical sweep = f_h/262.5 = 59.94 Hz
	Chrominance subcarrier = $f_h \times 227.5$ = 3.57954506 MHz
	(The old black-and-white standards were a vertical-sweep frequency 60 Hz and a horizontal sweep of 60 $\times$ 262.5 or 15,750 Hz.)
	The sound carrier is FM modulated with a maximum deviation of ± 25 KHz and a total bandwidth of about 120 KHz (remember the figures for standard broadcast FM are ± 75 KHz and 200 KHz).
	There are up to (4.25 MHz/15.734264 KHz) = 270 sidebands present for the luminance (black-and-white) signal modulation, thereby permitting 270 cycles per each horizontal sweep or 270 *pairs* of alternating light and dark lines across the width of the picture. There are 525/2 = 262.5 *pairs* of light-dark lines in the vertical direction. (Neither of the foregoing takes into account the portion of the picture lost due to horizontal or vertical blanking.)
Modulation levels	Horizontal pulse peak amplitude = 100% V_{cm}
	Blanking pulse level amplitude = 75% V_{cm}
	Picture black level amplitude = 70% V_{cm}
	Picture white level amplitude = 12.5% V_{cm}
Color subcarrier synchronization	Color subcarrier information demodulation requires a locally generated 3.579545-MHz oscillator in the receiver which is synchronized with the camera-chain oscillator. To do this synchronization, an 8-cycle burst of the camera oscillator signal is placed in the composite signal following each horizontal sync pulse, before the horizontal blanking is removed. This area in the pulse train is commonly called the "back porch" of the horizontal sync pulse.

Within the visible region the spectrum from red to purple goes like this: red, orange, yellow, yellow-green, green, greenish-blue, blue, and violet. Spectrally pure colors are called *saturated* colors because they contain only radiation of one wavelength. To the human eye, they are very brilliant and must actually be seen to be appreciated. Most of the spectrally pure colors cannot be reproduced by any of the normal methods of reflective color mixing (called *subtractive* color mixing) as is done with paints and dyes. However, it is possible to produce an approximation of most of the spectrally pure colors by passing sunlight or other "white" light through a prism or a transmitting color filter.

Colors that are not spectrally pure are a combination of radiation at two or more visible wavelengths. Actually what is happening is that there are two or more wavelengths of spectrally pure color entering into the human eye at the same time and the physiology of the eye and brain is what produces the effect of different (nonspectral) hues or the effect of nonsaturated color (i.e., pink in place of red). There have been numerous attempts to quantify the manner in which the human eye perceives color hues and color saturation. The most accepted of these is the

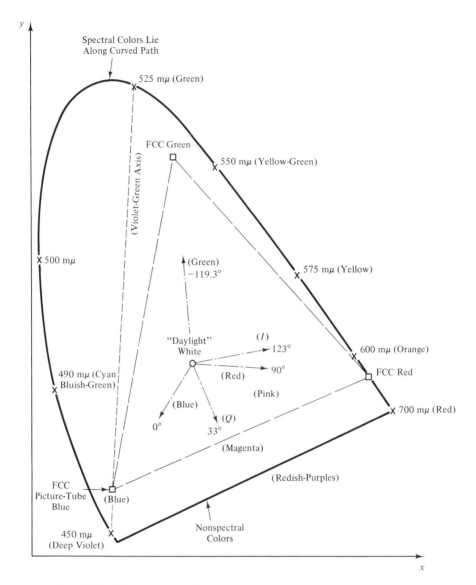

Figure 7-20   CIE chromaticity diagram (area enclosed by triangle is producible by color TV).

CIE[3] chromaticity diagram, a black-and-white version of which is shown in Figure 7-20.   It is a two-dimensional (x and y) plot on which the spectral color range forms a horseshoe-shaped pattern.   It is based upon average responses of how people perceive and match colors.   Different nonspectral colors can be produced

[3] Commission Internationale de l'Eclairage.

by adding together *radiation* of two or more colors on the horseshoe (spectrally pure colors). This is called *additive* color mixing and should not be confused with *subtractive* color mixing, which occurs when paints or dyes of two or more colors are physically mixed together to produce a paint or dye of another color.

For example, in Figure 7-20, it is possible by mixing deep violet light with green light to produce any of the color hues and saturations that lie along the *dotted* line between the 450-mμ point on the horseshoe and the 525-mμ point (i.e., along the violet-green axis). Similarly, by mixing any three spectral colors, it is possible to create any color effect that lies within the triangle defined by the three color-to-color axes.

If the three primary colors are properly chosen, then it is possible to include the point which is marked "daylight white" on the diagram and which represents the sensation produced outdoors in sunlight under certain weather atmospheric and geographic conditions. Similarly, it is possible to produce a somewhat more restricted range of color sensations to the human eye by mixing together three nonspectrally pure colors of light. This is what is done in the color TV system. The *dashed-line* triangle shown connects three points which are defined as FCC blue, FCC green, and FCC red and are three colors which are producible by the electron excitation of certain phosphor materials (Note: *phosphor* has no relation to *phosphorous*). It is quite evident that the color TV system using these three primary colors is not going to be able to reproduce all the colors that the human eye can perceive, but it does reproduce most of the colors that occur naturally in nature. It is primarily lacking in the saturated blue-green region. *Saturation* of any color can be measured by how far out the point representing the color is on the line drawn from the "daylight-white" point through the point representing the color to the outside of the diagram.

The CIE chromaticity diagram is a means of specifying by means of two coordinates, *x* and *y*, the hue and saturation of a color but does not indicate the brightness or intensity of the colored light reaching the eye from some part of a scene. Thus, a third variable, the *intensity* (or *brightness* or *luminance*) is needed. This will be discussed shortly, but first, consider the fact that if two variables, *x* and *y*, can specify hue and saturation, then on a color TV system, these two variables could be time functions which, when taken together at any point in time, would specify the point on the chromaticity diagram defining the hue and saturation of the color in the part of the scene being scanned at that point. However, from a practical sense, it would be better to work with signals which could have negative values as well as positive values and thus it would be better to choose a coordinate system which has its origin somewhere in the center of the chromaticity diagram rather than at the lower left-hand corner outside of the range of permissible hue/ saturation combinations. One good choice for the origin of the new coordinate system might be at the point marked "daylight white" since here the absence of the "chrominance" signal voltages would mean essentially "no color hue discernible" or zero color saturation (i.e., "white" light).

In almost all color TV systems, this is what is done. Also, the new coordinate system does not have to have the same axis directions as the CIE diagram, but

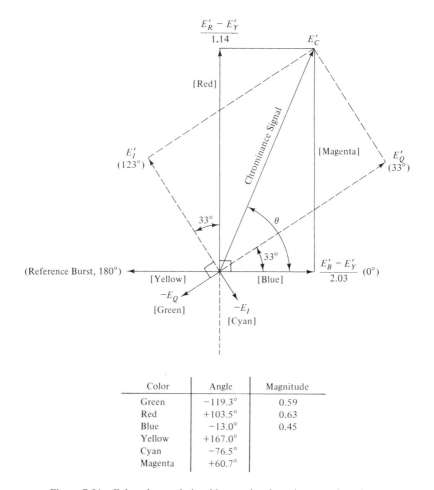

Color	Angle	Magnitude
Green	$-119.3°$	0.59
Red	$+103.5°$	0.63
Blue	$-13.0°$	0.45
Yellow	$+167.0°$	
Cyan	$-76.5°$	
Magenta	$+60.7°$	

Figure 7-21   Color phase relationships on the chrominance subcarrier.

instead can be mapped into a new rectangular coordinate system having in the NTSC[4] color TV system a "horizontal" axis Q and a "vertical" axis I (see Figure 7-21), which were chosen for reasons to be discussed later.   The mapping from one coordinate system (the x and y of the chromaticity diagram) to the other (the Q and I of the TV system) is nonlinear, and hence if the Q and I axes are shown on the CIE diagram, they do not appear to be orthogonal (see Figure 7-20, the dash-dot lines).   Also shown as dash-dot lines in Figure 7-20 are the directions in the new coordinate system to the picture-tube red, green, and blue points.   The angle values given alongside each of these lines and the angles given alongside the Q and I axes are the angles CCW from the horizontal axis in a *third* rectangular coordinate system where the Q axis is tipped upward at 33 degrees and where the

---

[4] National Television System Committee.

horizontal axis represents the blue saturation signal and the vertical axis represents the red saturation signal in the color TV system.

7.5.2.1 The NTSC color TV system.
In the American color TV system the scene being observed by the TV camera is split into three beams after passing through the lens of the camera. One beam goes through a red filter to the "red" camera tube, the second through a blue filter to the "blue" camera tube, and the third through a green filter to the "green" camera tube. Since the color TV system had to be compatible with the existing black-and-white TV system, one thing that is done with the three signals is to sum them together in the correct proportions to form a luminance or brightness signal which is transmitted as part of the original black-and-white TV signal. The proportions of the signals from the three camera tubes in the luminance signal are 59% for green, 30% for red, and 11% for blue because our eyes observe green as a rather brilliant color and blue as a dark color, with red somewhere in between.

To transmit the color information and still not have more than three channels of information, we *could* have simply transmitted two difference signals,[5] blue minus the luminance signal and red minus the luminance signal, and then at the receiver simply have re-created the blue and red signals by adding the luminance signal value to each of the difference signals. The green signal could then have been recovered by subtracting the proper proportions of red and blue signal from the luminance signal. However, rather than do this, it was decided to create two new signals Q and I by adding together different proportions of blue minus luminance and red minus luminance in each case (see the formulae in Figure 7-22). The advantage of creating these two new signals is that the Q signal axis represents hue changes for which the human eye cannot distinguish great detail (i.e., alternate bars of magenta and blueish-green tend to merge together when seen from a distance) while the I axis represents changes in hue (from blue to yellow) which the eye can see in more detail. Thus, it is not necessary to transmit color information in as much detail for objects having Q axis colors as it is for objects having I axis colors, or in other words, the bandwidth of the Q axis signal does not need to be as great as that of the I axis signal.

Neither the Q axis nor the I axis signal needs to transmit as much detail as does the luminance signal. The reason for this is that for very fine detail, the eye does not distinguish color at all, but only bright versus dark. Thus, the luminance signal needs a much wider bandwidth than does either of the chrominance signals.

In the American (NTSC) color TV system, the luminance signal has a bandwidth of more than 4.2 MHz, while the I signal has a bandwidth of about 1.5 MHz and the Q signal a bandwidth of about 500 KHz (however, many of the less expensive color receivers limit the usable luminance-signal bandwidth to about 3.8 or 3.9 MHz or less and the bandwidth of both the I and the Q chrominance axes to less than 500 KHz). Figure 7-22 shows how these signals are created by matrixing. Figure 7-23 shows the bandwidths of the I and Q signals after modulation on the subcarrier.

---

[5] This is done in the European SECAM system.

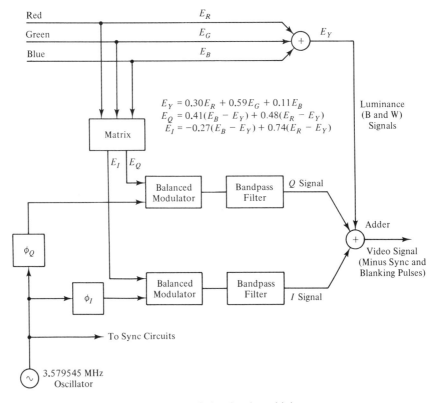

$$E_Y = 0.30E_R + 0.59E_G + 0.11E_B$$
$$E_Q = 0.41(E_B - E_Y) + 0.48(E_R - E_Y)$$
$$E_I = -0.27(E_B - E_Y) + 0.74(E_R - E_Y)$$

Figure 7-22 Color signal combining.

After obtaining the Q and I signals the next question is how to transmit them along with the original TV signal (i.e., the luminance signal and the synchronization and blanking information which makes up the black-and-white portion of the TV signal). Recalling from the previous section that we said that the spectrum of the black-and-white TV signal consisted of packets of spectral power grouped around multiples of the horizontal line frequency with a lot of open spectral space in

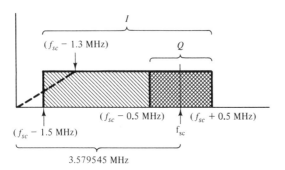

Figure 7-23 Spectrum of the chrominance subcarrier signal.

between these packets, we realize that there might be a way in which we could sandwich the chrominance information into these open spaces. This is exactly what the group that designed the NTSC system did. Realizing that the chrominance signal would also consist of packets of power spaced at multiples of the horizontal line frequency, they decided to put a subcarrier halfway between harmonics 227 and 228 of f_h, setting this frequency at 3.579545 MHz and readjusting the horizontal line frequency to 3.579545/227.5 or about 15.734264 KHz. The choice of these frequencies was made to avoid creating a beat note under certain conditions with the sound carrier at 4.5000 MHz removed from the main video carrier. The composite *video* signal now consisted of the original black-and-white signal (luminance plus blanking and synchronization) plus a 3.579545-MHz DSBSC quadrature-modulated subcarrier signal, where a portion of the upper sideband resulting from modulating the subcarrier with the I signal has been filtered out so that the entire modulated subcarrier signal extends from about 1.5 MHz to about 4.1 MHz, which keeps it within the region already occupied by the black-and-white TV video signal.

   One problem still exists: DSBSC modulation of the subcarrier is used, and since quadrature modulation is involved, it is necessary to transmit information concerning the frequency and phase of the subcarrier signal itself to the color TV receiver in order for the receiver to be able to demodulate the chrominance signal. This problem is solved by incorporating in the color TV receiver a very stable 3.579545-MHz oscillator which needs to be updated in frequency and phase only on an occasional basis ("occasional" in this case refers to once every horizontal line period or every 63.55555 microseconds). This updating is done by putting a short burst of the subcarrier oscillator signal on the "back porch" of each horizontal blanking pulse, immediately following the horizontal sync pulse. This is shown in Figure 7-18e.

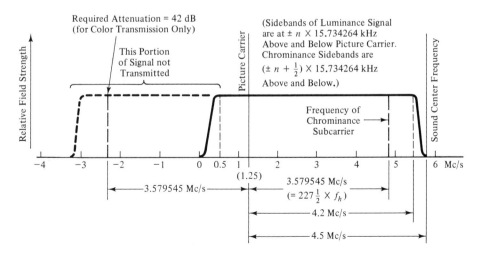

Figure 7-24   Idealized picture transmission amplitude characteristic.

Other special types of signals may also be added to the composite TV video signal before it is used to AM vestigial-sideband modulate the TV transmitter. For example, in Figure 7-18 we see that there are several horizontal-line periods preceding entry into the active picture region. These time periods may be used for the transmission of special-purpose test signals or network cue signals (alphanumeric digitally coded data). Also, they can be used to transmit information concerning program light level and color hue (many of the more expensive new color TV receivers make use of these signals to adjust the picture brightness and hue automatically), or they can be used to convey information which can be used to produce printed information at the bottom of the screen for the hearing-impaired viewer. An example of how this information is placed on the TV signal is shown in Figure 7-25 for the chrominance-adjust signal. Some of the other types of signals are binary-digitally encoded during the vertical-blanking interval.

After the composite TV video signal is formed, it is used to modulate the TV transmitter, resulting in an RF spectrum as shown in Figure 7-24 where 0 is at the lower edge of the allotted RF TV channel. It must be realized at this point that a very complicated signal is present (i.e., the video portion is vestigial-sideband AM for the black-and-white and blanking and synchronizing information; it is DSBSC-quadrature modulation, with partial SB suppression, AM for the chrominance information, and it contains a separate FM-modulated carrier for the sound information). In addition, the FCC permits stereo sound and two SCA-type subcarriers on the sound channel for the purposes of second-language audio trans-

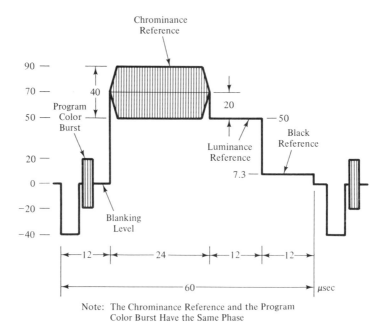

Figure 7-25   Vertical interval reference (VIR) signal.

mission and/or transmitter telemetry or other station purposes.[6] Such a signal must be handled quite carefully when it is received and amplified or demodulated, because any amplitude or phase-frequency distortion which takes place will seriously degrade one or more of the functions which the signal has to perform in the color receiver.

Figure 7-26 shows the block diagram of a typical color TV receiver (without any automatic brightness or color-hue control). The incoming signal is first amplified and down-converted in the receiver RF stages. For VHF TV signals, there is usually one stage of superheterodyne conversion, while for TV signals in the UHF band, there are usually two superheterodyne conversions. The signal then goes into the main (or "video") IF amplifier. At this point, correction is made for the fact that the lower sideband was partially suppressed when the main TV carrier was AM modulated. This is done by having the IF amplifier amplify the frequencies above $f_{\text{CIF}} + 0.8$ MHz twice as much as it does the carrier signal. There is a linear rolloff from $f_{\text{CIF}} + 0.8$ MHz down to $f_{\text{CIF}} - 0.8$ MHz (see Figure 7-27, which shows the IF passband and also the spectrum of the I and Q chrominance signals within the IF passband). The signal out of the main IF goes into two detectors: the sound detector, which is a diode detector followed by a tuned circuit at 4.5 MHz, and the video detector, which is a diode envelope detector followed by a video amplifier with a passband from about 30 Hz up to more than 4.2 MHz. There is also a d.c. output from the video detector which is used for AGC purposes.

In this type of receiver design the diode sound detector actually operates as a nonlinear additive RF mixer and uses the video carrier signal at f_{cIF} in the manner of a local oscillator signal to down-convert the FM sound signal from its frequency of $f_{\text{CIF}} + 4.5$ MHz to 4.5 MHz. This 4.5-MHz signal then passes through a sound IF amplifier with a center frequency at 4.5 MHz and a bandwidth of about 67 KHz to an FM limiter-discriminator where the audio signal is detected. In an alternative type of receiver design the sound carrier enters a sound-IF amplifier at 41.25-MHz center frequency after passing through the first main IF amplifier stage.

Once the TV video signal has been recovered from the video diode (envelope) detector, it goes several places: without any further processing (except perhaps for a delay line which delays it until the corresponding portion of the chrominance signal is processed), it goes to the cathodes of the picture tube (along with a d.c. signal for adjusting the picture brightness). This is the luminance (or $-E_Y$) signal. The detected video signal also goes to the I/Q amplifier which amplifies the chrominance signal and feeds it to the color quadrature demodulators which do not produce the I and Q signals but, rather, the $E_R - E_Y$ and $E_B - E_Y$ signals since the phase of the reference signals fed into the demodulators is not the phase of the I and Q signals but rather are signals 33 degrees behind the Q signal and 33 degrees behind the I signal (see Figure 7-21).

The reference signals for the color demodulators are obtained from a 3.579545-

---

[6] If stereo sound is broadcast, the subcarrier is at $f_{sc} = 2 f_h$ with ± 50-KHz peak deviation of the main sound carrier being permitted due to the DSBSC $(L - R)$ sidebands. The stereo pilot subcarrier is equal to f_h, the *second audio program* (*SAP*) subcarrier is at $5f_h$, and the station-control SCA-type subcarrier is at $6.5f_h$.

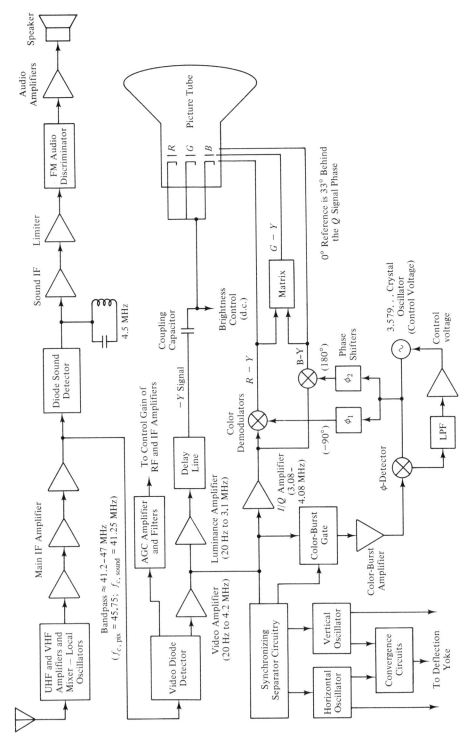

Figure 7-26 A typical basic color TV receiver.

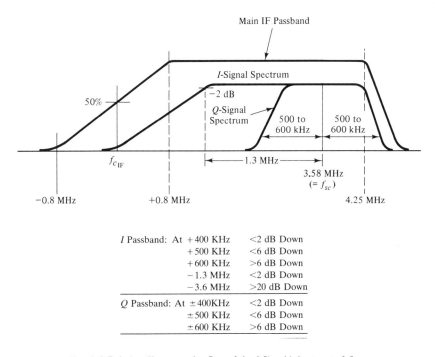

I Passband: At $+400$ KHz	<2 dB Down
$+500$ KHz	<6 dB Down
$+600$ KHz	>6 dB Down
-1.3 MHz	<2 dB Down
-3.6 MHz	>20 dB Down
Q Passband: At ± 400KHz	<2 dB Down
± 500 KHz	<6 dB Down
± 600 KHz	>6 dB Down

(I and Q Relation Chosen so that Part of the I Signal is kept out of Q
Channel and the Q Channel is Associated with Low-Visibility Signals.)

Figure 7-27   IF passbands of color TV receivers and spectrums of the Q and I
signals.

MHz crystal oscillator which is phase-locked-loop synchronized to the color-burst
signal which is present once every 63.55555 μsec for eight cycles of the color
subcarrier signal.   This color-burst signal is passed to the PLL via a color-burst
gate which is turned on whenever the color-burst signal is due to arrive. The color-
burst gate is controlled by a part of the receiver's synchronizing circuits.   These
latter circuits detect the portion of the video signal which is above the blanking
level and provide the shaping of the horizontal sync pulses for the synchronizing
of the horizontal-sweep oscillator as well as the integration of the vertical sync
pulse train to produce a pulse output for synchronization of the vertical-sweep
oscillator.   The sweep oscillators provide the sweep voltages to the horizontal and
vertical deflection coils on the picture tube as well as signals to the convergence-
correction coils which make sure that the electron beams from all three guns in
the picture tube hit the same point on the screen for all deflections of the beams.

The horizontal deflection circuit in all large-screen TV sets also incorporates
a high-voltage boost circuit which allows some of the energy which has to be dumped
or dissipated during each horizontal return (or "flyback") to be put to use instead
of being converted to heat.   The horizontal-deflection-signal transformer also usually incorporates a high-voltage winding of many turns which produces the very

high voltages required by the picture-tube anode (up to 30,000 volts in a large-screen color TV set).

Returning to the color demodulator circuits, the $E_R - E_Y$ and $E_B - E_Y$ signals are fed to a resistive matrix where the $E_G - E_Y$ signal is produced. These three difference signals are then fed to the three electron guns in the color picture tube. Since they are fed to the respective grids and since the cathodes all have the $-E_Y$ signal, the difference signal in each case is equal to each of the three color signals (i.e., E_R, E_G and E_B). These signals then decrease the negative bias on the respective grids and cause more beam current to flow in the respective guns, thus producing a more intense amount of light from the dot of phosphor which the beam from each gun is hitting at the time.

Although picture tubes are constructed in many different manners, the color-mask tube is one of the most common in present usage. It consists of the three electron guns in the neck of the picture tube and a picture-tube face which is a matrix of tiny dots of phosphor of the three primary colors with three such dots in each group. Behind each dot group there is a mask which is a plate with holes in it, one hole for each color-dot group. The selection of which dot is hit by the electron beam is controlled by the relative positioning of the center of the color-dot group, the hole in the mask, and the position of the gun from which the beam is arriving. Thus one gun in the neck of the tube is aligned such that it only can excite the red dots, gun 2 can only excite the green dots, and gun 3 can only excite the blue dots. Since all guns are operating simultaneously, the three dots are excited at the same time in intensities determined by the color hue, brightness, and saturation characteristic of that part of the TV picture. This type of color presentation is termed a *simultaneous color TV system* (i.e., all three colors are displayed simultaneously). Other types of color TV systems or displays may display only one color at a time and flicker very rapidly from one color to another (i.e., from red to green to blue). These are called *sequential color systems*.

One type of sequential color system is the field-sequential system. This was one of the systems proposed for the United States during the period when color TV was being developed. Basically it proposed sending one $262\frac{1}{2}$ line field in red, the next in green and the next in blue, and so on. Using the old black-and-white field rate of 60 per second, this would mean that the three colors of the picture would be sent every $\frac{1}{20}$ second and a complete picture (all three colors over all lines) once every $\frac{1}{10}$ second. One of the big arguments for the adoption of this system was that a black-and-white TV (picture-tube sizes at this time were about 10 or 12 inches across) could be converted to color by placing a "color wheel"consisting of colored gelatins mounted on a synchronously rotating wheel in front of the black-and-white screen. However, when it was shown that a simultaneous color system could be devised which was also black-and-white compatible, the field-sequential system was voted down and the simultaneous system (developed largely by RCA) was adopted by the NTSC as the U.S. standard. Additional information on the NTSC system standards, as now set by the FCC, as well as standard definitions of the technical terms can be found in Title 47, Section 73 of the U.S. National Code.

At this point it is informative to take a look at the amplitude of the NTSC RF signal as it is transmitted. Figure 7-28 shows a portion of one line scan beginning with the horizontal sync pulse and the eight-cycle color-synchronizing burst. It is assumed that on this particular scan, the following colors are encountered: gray, white (a high-intensity gray), saturated bright green, saturated bright blue, lower-intensity (dimmer) saturated green, "light" (lowly saturated) green, and saturated red. For gray and white, the chrominance subcarrier modulation level is zero, and thus we have only the main (luminance) video signal. Where highly saturated colors are present, there is a large ($+$) and ($-$) variation around the luminance-signal level produced by the sidebands caused by the heavy modulation of the chrominance subcarrier (this rapid variation at approximately 3.58 . . .-MHz is represented by the shading shown in Figure 7-28). When the color being scanned at a particular point on the line is not saturated (i.e., see the "light" green region in Figure 7-28), the small amplitude of the approximately 3.58 . . .-MHz sidebands produces a smaller variation about the luminance-signal level than when the area being scanned is a highly saturated color. Also, it should be remembered that the

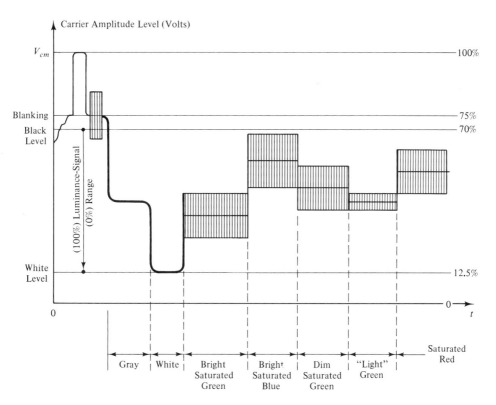

Figure 7-28 Part of one line scan showing NTSC signal levels for various brightness/hue/saturation combinations.

level of the luminance signal depends upon what hue is being scanned as well as upon the intensity of the hue (to compensate for the apparent "brightness" of different colors to the human eye—59% of maximum luminance-signal level for maximum-brightness saturated green, etc.). The *phase* of the subcarrier sidebands is determined by what hue is being currently scanned, but this is not indicated in Figure 7-28 (due to using shading to represent the approximately 3.58 . . .-MHz subcarrier sidebands signal).

7.5.2.2 Foreign color TV systems (PAL and SECAM).

These are two European systems developed at about the same time as the American NTSC system. The PAL system in particular bears considerable resemblance to the NTSC system. Both PAL and SECAM use the basic chromatic luminance relationship that the luminance signal is made up of 59% green signal, 30% red signal, and 11% blue signal. In the PAL system, the red difference signal is $0.877 (E_R - E_Y) = E_v$, and the blue difference signal is $0.493 (E_B - E_Y) = E_u$. Both these are within a fraction of a percent of the red and blue difference signals in the NTSC system. However, instead of forming an I and Q signal as in the NTSC system, the E_v and E_u signals are used directly to modulate the color subcarrier using quadrature modulating as in NTSC. The sidebands generated are permitted to extend downward to about 1.6 MHz below the subcarrier and upward by 0.57, 1.07, or 1.6 MHz, depending upon whether the TV system has the sound carrier 5.5, 6.0, or 6.5 MHz above the picture carrier. The PAL subcarrier is located at 4.3361875 MHz above the picture carrier between the black-and-white sidebands 277 and 278, giving a horizontal line frequency of 15,625.901 Hz and a vertical frequency of 50.003 Hz (European TV has 625 lines in two interlaced fields and European power systems are nominally 50 Hz).

The synchronization of the receiver subcarrier injection oscillator is accomplished by sending a 13-cycle synchronizing burst *on the top of* the horizontal sync pulses (instead of on the back porch as in the NTSC system). Also, an added feature of this system is that the phase of the 13-cycle burst is alternated every line. This reduces the color signal's phase error in the receiver, but adds to the receiver complexity.

The SECAM III system is somewhat different from the NTSC and PAL systems in its method of sending the chrominance information. Its two color signals are a red difference signal $D_r = -1.9(E_R - E_Y)$ and $D_b = 1.5(E_B - E_Y)$ which are not in the same ratio as in the NTSC or PAL systems. However, the same ratio is not necessary here since the signals are used differently. Rather than generating a quadrature-modulated signal where the amplitude represents color saturation and the phase the color hue, these signals are used directly, making use of the fact that it is not necessary to have as much vertical color detail as vertical luminance detail. Thus, the two color signals are transmitted alternately, or each color signal is transmitted for a *pair* of horizontal scan lines. Thus the red difference signal is transmitted on one horizontal scan and the blue difference signal on the next horizontal scan. These two signals alternately FM modulate a subcarrier at 4.4375 MHz above the picture carrier. This subcarrier is *on* horizontal line harmonic 284.

It is interesting to note that since the modulation is for alternate horizontal lines, the basic harmonic frequency present in the signal which modulates the subcarrier is $f_h/2$ rather than f_h as in the PAL and NTSC systems. This, coupled with the fact that FM rather than quadrature AM modulation is used, permits the subcarrier to be on a multiple of f_h and still have the chrominance sidebands in between the sidebands of the black-and-white signal. In the SECAM III systems, the lower sideband extends downward 1.40 MHz and upward 0.57, 1.07, or 1.40 MHz, depending upon the picture-sound carrier spacing used at the transmitter. The maximum frequency deviation of the FM-modulated chrominance subcarrier permitted in the SECAM III system is 500 KHz.

7.6 TV CAMERA AND STUDIO EQUIPMENT

7.6.1 TV Camera Imaging Devices

During the course of TV development a number of different types of camera tubes and other imaging devices have been developed. One of the early (1939) practical camera tubes was the *iconoscope* (see Figure 7-29). It consisted of an image area (the plane where the image formed by the camera's lens system is projected) made up of an array of small mosaic elements of silver oxide or antimony cesium (photoelectric materials) deposited upon a mica substrate about 10 × 12.5 cm in size with a platinum coating on the other side of the mica substrate. An on-axis lens system focused the image on the mosaic side of this "target area" and an off-axis electron gun and its associated deflection coils scanned a beam of electrons in a raster pattern across the mosaic.

Because of the off-axis mounting of the electron gun, the deflection circuitry had to compensate for the "keystone" pattern which would otherwise be produced by linear raster scanning. A positive potential was applied to the platinum electrode on the back of the mica substrate. As the electron beam scanned over a particular element in the photoelectric mosaic, that element would temporarily change its potential by several volts because of the secondary emission of electrons from the element's material each time it was hit by the high-energy beam of electrons.[7] The presence or absence of light falling upon that particular element determined the exact change of voltage which took place, causing it to vary by about 0.2 volts from a "dark" element to one which was fully illuminated. This change in potential due to the scanning (and the modulation in the change caused by the light pattern (image) on the mosaic surface) was capacitively coupled into the platinum electrode on the other side of the mica substrate. The 0.2-volt modulation due to the illumination variations was what comprised the video signal out of the tube.

The iconoscope was followed by the *orthicon* and the *image orthicon* tubes (see Figures 7-30 and 7-31). Both these tubes eliminated the electron-scanning

---

[7] A grounded conductive coating on the inside of the glass envelope collected these secondary electrons.

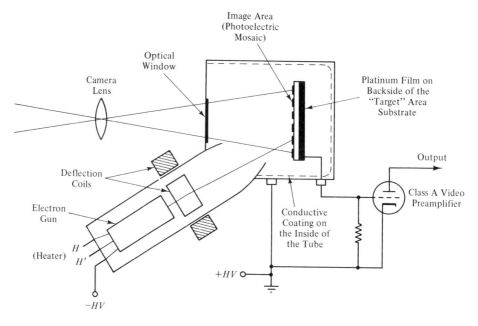

Figure 7-29   The video iconoscope.

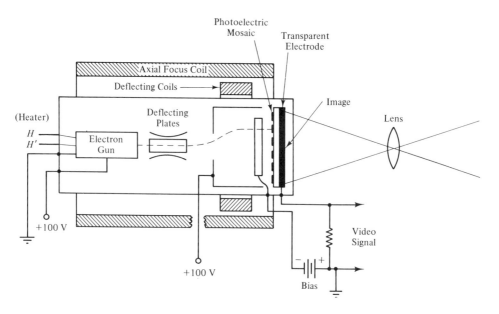

Figure 7-30   The orthicon tube.

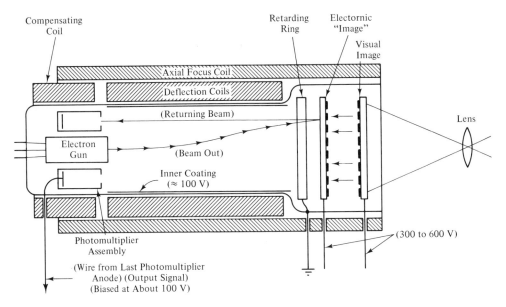

Figure 7-31   The image orthicon tube.

keystoning problem of the iconoscope by using a transparent target substrate which allowed the lens to be placed on the other side of the target mosaic, thereby allowing the electron gun to be placed on axis on the active side.   Two other modifications which differentiated the orthicon tube from the iconoscope were the presence of a decelerating electric field near the target area, which slowed the incoming electrons in the beam and which also removed the secondary-emission electrons from the vicinity of the target, and a beam-deflection system, which caused the electron beam to arrive perpendicularly to the plane of the target.   Both these features insured a more even video output signal over the entire scanned area.   In the *image* orthicon, the optical image was focused on a transparent photocathode wherein the presence of light photons falling upon a photocathode element caused electrons to be emitted.   A uniform electric field caused these electrons to be accelerated toward the main target area where they formed an "electronic image" of the visible scene projected upon the photocathode.   The presence of this electronic image on the main target modulated the number of electrons removed from a low-energy beam of electrons scanning the target area.   The electrons *not* removed from this beam were collected by a photomultiplier anode assembly.   The current into the final anode of this assembly was what comprised the video output signal.

The *vidicon* tube (introduced by RCA in about 1950) employs the principle of photoconduction rather than photoemissitivity as its operating principle.   Antimony trisulphide is used as the photoconductive material and is used as the center of a "sandwich" (see Figure 7-32) consisting of a transparent conductive coating on the optical lens side of the sandwich and a conductive mosaic on the electron

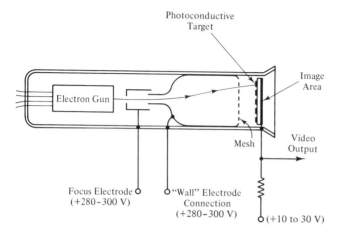

Figure 7-32 The vidicon tube (deflection coils not shown).

gun side.   As the electron beam is scanned over an element which is "in the dark," no current flows through the photoconductive material under that element into the transparent electrode on the other side of the sandwich, while when the beam scans over an element which is illuminated, there will be a flow of current from the electron beam through the photoconductive material into the transparent electrode. Thus, as the target area is scanned, the current into the transparent electrode varies according to the amount of light falling upon the various elements as they are scanned by the electron beam.

One of the big advantages of the vidicon tube as compared with earlier camera tubes is its much smaller size.   A number of variations on the vidicon have been introduced.   The *plumbicon* tube, which was introduced by the Phillips Company in 1962, uses lead oxide as the photoconductive material which allows the tube to have a "dark" current of almost zero.   The silicon-diode vidicon (each element in the mosaic behaves as a separate photodiode) was put out by Bell Labs in 1958 and has the characteristic of possessing a very high low-light-level sensitivity, which makes it quite useful in a number of applications such as industrial TV, in-field TV news coverage, security, and so on.   A number of other materials such as zinc selenide, zinc cadmium telluride, cadmium selenide, and arsenic sulphide have been used in fabricating multilayered photoconductive targets.

Among recent developments in producing a video signal from an optical image are the various types of solid-state imaging devices.   Most of these are of the *charge-coupled-device* (*CCD*) type wherein an array of photoelectric elements is arranged in such a manner that any one row (or column) can be made to operate like an *analog* shift register (as opposed to conventional *binary digital* shift registers).   Thus, after photoexposure of all the elements in a row (one shift direction), the data stored in the elements of this row can be "read out" by clocking the shift register, producing at the serial output of the shift register an analog signal that is equivalent to "scanning" the image line represented by that particular row of elements in the mosaic.   One of the interesting aspects of this device is the possibility of scanning a number (or even all) of the rows of the mosaic simultaneously, producing a number of parallel (simultaneous) scan-line analog signals.

7.6.2 Color TV Cameras

The first types of nonsequential color cameras developed employed three separate camera tubes with the same image being focused on the targets of all three of these tubes as shown in Figure 7-33 by means of a single lens and a set of dichroic mirrors or prisms.   These are optical devices which have surfaces coated with optical interference layers which provide color-selective reflection of the light passing through them.   Two of these mirrors are used to reflect the blue (short-wavelength) end and the red (long-wavelength) end of the visible spectrum into a "blue" and a "red" camera tube, respectively.   The light not reflected by either of these mirrors has wavelengths in the middle of the spectrum and is directed to the "green" camera tube.

Another method for producing a color TV signal utilizes one camera tube with a color filter consisting of very narrow alternating vertical stripes of green, cyan, and white (clear) in front of the tube.   The image falling upon the tube's target area then consists of stripes of green-signal, green-plus-blue-signal, and total-luminance-signal colors.   Thus, during a horizontal scan, there is a rapid effective switching of the tube's video signal output among these three colors.   If this signal

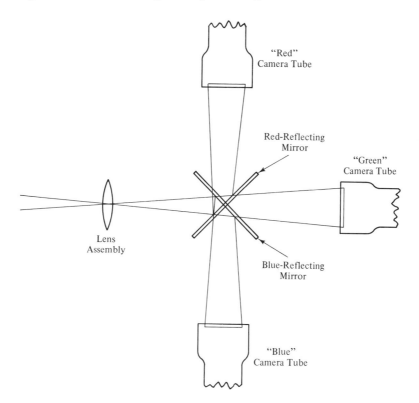

Figure 7-33   The use of selective-wavelength-reflection (dichroic) mirrors in a three-tube color TV camera.

is passed through a low-pass filter so that the three signals are effectively averaged, the result is just the luminance signal. If this same signal is passed through a band-pass filter and the upper and lower envelope functions of the resulting filter output signal are detected, the red and the blue color signals may be extracted by a matrixing process. These can then be combined with the luminance signal out of the low-pass filter to produce the red, blue, and green video signals.

This technique is one that can be easily applied to solid-state CCD-type imaging devices because the fabrication of the color-stripe optical filter can easily be included into the fabrication process for the CCD device itself.

7.6.3 Motion-Picture to TV Conversion

One problem that occurs when conventional 24 frames-per-second motion-picture filmage is to be carried via TV is the difference in the frame rates between the motion-picture films and the field-rate standards of either European or North American television. This is compensated for in the European system by running the film at 25 frames per second (two 50-Hz TV fields per motion-picture frame), which causes everything in the picture to move 4.17% faster than it should (a 100-minute "movie" lasts only about 96 minutes). For North American TV, alternate motion-picture frames are scanned for two and three 60-Hz TV fields. Thus, two motion-picture frames are converted into five TV fields, resulting in a 0.10% slowdown in the picture.

In either case there are two ways of generating the TV signal from motion-picture film. The first involves using a modified projector which flashes the film image onto a camera tube during the vertical retrace period of the TV scan. The tube's target stores this image, and it is scanned during the next active TV field period. Mechanical movement of the film (which occurs once every two or three TV fields in the North American system) occurs while the scanning of the camera-tube target is taking place. The second film to TV conversion technique is the *flying-spot scanner* method. In this method the film moves on a continuous (non-stop) basis while a spot of light derived from the face of a special high-intensity CRT and a special lens system is projected onto and scanned over the motion-picture film frame as it moves past. The light transmitted through the film is split up by dichroic mirrors and is sent to three photomultiplier tubes to produce the red, green, and blue video signals. It should be noted that the motion of the light spot on the face of the high-intensity CRT must not only be kept in synchronization with the TV sweep pattern, but must also track the motion of the film as it moves past. Such flying-spot systems have very excellent image-conversion properties and quality, but they are quite complicated and quite expensive.

7.6.4 TV Studio Equipment

In a typical network TV outlet the program material will normally, at various times, come from the network, from locally originated "live" cameras, from several mag-netic-tape machines, from a motion-picture projector and so on, sometimes with several shifts between sources occurring in the matter of a few minutes. It is important, therefore, that all these sources are in synchronization with each other to prevent vertical rolling of the picture at the time of transition from one source to another. This generally implies having a studio master-sync oscillator which is kept in step with a network synchronization signal. All the studio's live cameras are synchronized with the studio oscillator so that it is possible to do a *dissolve* (gradual fade-out/fade-in) or a *wipe* (intrusion of the new picture into the old) from one camera to another, to insert the image from one camera into the corner of the image from another camera, to overlay text or captions, and so on, at any point on the screen, and to achieve other studio "special effects."

Also, the master-sync signal is used to bring a mechanical device such as a videotape recorder, a projector, or a flying-spot scanner into synchronization during its start-up period prior to switching to its video output. In the event that it is necessary to switch to a nonsynchronized program source (such as a portable field-pack "live" camera for instance), the studio sync may have to be temporarily shifted to that of the portable source (the studio video-output signal should remain at the blanking level while this is done so that the sync readjustment is not visible to the viewers). Another alternative is to process the signal from the camera (usually digitally) into an equivalent signal which *is* in sync with the studio master oscillator. This digital technique can also be used to correct for time-base errors, chrominance-phase errors, and other problems occurring between the incoming analog signal and the studio's master synchronization timing. There are many other interesting and complex problems and solutions to be found within the confines of a modern TV studio, but they are too numerous and too complex to be discussed at the present time.

7.7 VIDEO MAGNETIC TAPE RECORDING

As mentioned in Section 7.4 on audio magnetic tape recording, the maximum signal frequency which can be recorded on a magnetic tape is proportional to the tape speed and is inversely proportional to the length of the head gap. While great progress has been made in reducing the length of the head gap and permitting slower tape speeds to be used for audio tape recording, it is still necessary to use a tape speed in inches per second (ips), which is at least numerically equal to about one-fourth of the highest frequency to be recorded expressed in KiloHertz. Thus to record directly a TV signal with a maximum frequency of about 4000 KHz (4 MHz) would require a tape speed of about 1000 ips (or about 57 miles per hour), which would empty a 2400-foot reel of tape in less than 30 seconds. Obviously, some better technique is needed!

7.7.1 Transverse-Scan Videotape Recorders

In 1956 the Ampex Corporation introduced a recording machine that produces a writing speed of about 1500 ips by writing 64 slightly tilted tracks per linear inch of tape across the width of a 2-inch-wide magnetic tape which moves forward at a rate of 15 ips (see Figures 7-34 and 7-35). It does this with a transversely rotating

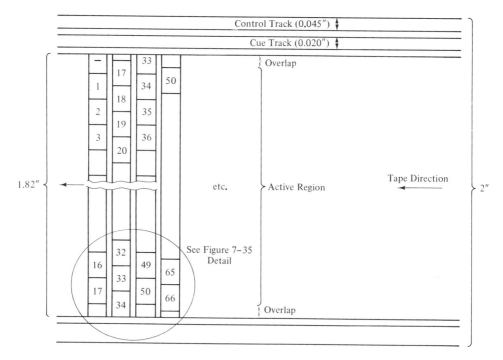

Figure 7-34 Format of a 2-inch professional videotape format.

drum which contains four recording heads evenly spaced about its circumference and which rotates at 240 revolutions per second (or 14,400 rpm!), thereby moving one of the four heads across the width of the tape once every $\frac{1}{960}$ of a second (see Figure 7-36) at the approximate 1500 ips of writing speed.

To synchronize the rotation of the head drum with the forward motion of the tape during playback, it is necessary to record a control track of synchronizing pulses by conventional recording techniques (i.e., a stationary head) along one edge of the tape (see Figures 7-34 and 7-36). Also recorded longitudinally along the tape edges are a cueing track and one or more audio tracks. Upon playback the synchronizing track is used to control the speed and the mechanical "phase" of the drum rotation (so that each of the transverse tracks is scanned by one of the four heads on the circumference of the drum) as well as to control the electrical

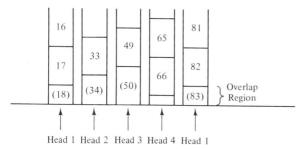

Figure 7-35 Detail of track-ending shown in Figure 7-34.

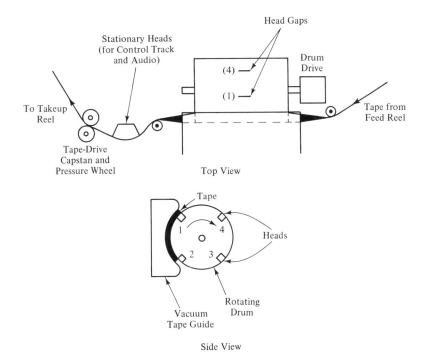

Figure 7-36   Studio videotape recorder (transverse tracks).

switching among the heads so that there is a smooth transition from the signal from a head which is just finishing the reading of one of the transverse tracks to the signal from the head which is just beginning to read the next track (note that in Figure 7-35 there is an overlap in the recording of line scans at the bottom end of one track and the top end of the next track—line 17 appears at the end of the first track and the start of the second track, etc.).   The synchronization track is also used to control the speed of the entire recorder mechanism so that when the operator begins "rolling" a tape, the frame-synchronizing indication on the synchronization track of the tape is brought into agreement with the studio's master clock oscillator.   This prevents the loss of vertical synchronization when switching from other program material to the taped program.   It also permits electronic or actual cut-and-splice editing of the video tape without the loss of vertical synchronization in the finished tape.

Although the effective writing speed along the transverse video tracks is sufficient to allow the direct writing of the video signal, the nonlinearities of the tape material's hysteresis loop would prevent the direct linear recording of the video amplitude upon the tape (see the discussion of this same problem in audio magnetic tape recording in Section 7.4.2 and Figure 7-9).   To solve this problem by the use of a high-frequency bias (as is done in the audio case) would necessitate a bias frequency of 10 to 20 MHz.   Also, the S/N level obtainable would not be very good.

To get around these problems the actual signal which is recorded upon the tape is an FM signal which is produced by using the complete composite color video signal to frequency modulate a carrier oscillator. The resulting FM signal is then clipped to a square wave that is then applied directly to the recording heads to be recorded onto the tape. The carrier is modulated in such a manner as to make the "white" video signal level correspond to 10.0 MHz, the blanking level to 7.90 MHz, and the synchronization level to 7.06 MHz in what is called the *standard high band* (or HB) system (several other black-and-white and color standards have also existed at various times).

Although the transverse-track quad-head (or *quadruplex method*) videotape recorder (or VTR) has been a broadcast industry standard since its introduction in the late 1950s, it has a number of disadvantages: the switching among the four heads splits a frame up into 32 parts. If the heads are not perfectly matched in performance, the chrominance-signal amplitude changes, which results in a variation of the color-saturation level among different parts of the picture. Also, since the tape must move forward continuously at a fixed speed, there is no possibility for stop-frame or slow-motion playback modes.

7.7.2 Helical-Scan Videotape Recorders

To solve the problems mentioned in the preceding paragraph, a number of different types of machines called *helical-scan* VTRs were developed during the 1970s. All these systems record the video signal on a track that is only slightly tilted from the longitudinal (or along-the-tape) direction as opposed to the type of tracks previously discussed which are only slightly tilted from the transverse (or across-the-tape) direction. This permits tracks which are long enough so that a full field of video can be recorded along one track with only one active recording head (this permits recording a complete video frame on two adjacent tracks). This is done by wrapping the tape around the circumference of a drum whose axis of rotation is roughly perpendicular to the direction of tape motion. The wrapping is done in a slightly helical manner as is shown in Figure 7-37. The amount of tape wraparound is approximately 360 degrees so that the same head can be used to write successive fields. Although there are a number of standards and therefore slight variations from standard to standard, typically the tilt of the tracks is about 4 to 8 degrees, the tape speed is about 10 ips in the normal mode, and the drum speed is about 50 to 60 revolutions per second for a 6-inch diameter drum, with about a 16-inch-long track for one field with a track-to-track spacing of about 0.16 inch in a direction along the tape. A tape width of 1 inch has been standard for these "professional" types of helical-scan recorders.

Figure 7-37 shows another common feature of most models of this type of VTR called AST (or *automatic scan tracking*) which is a method by which the head is properly positioned vertically to keep it on the specified track which it is currently reading. This also makes it possible for the tape to be played in "stopped-frame" or "slow-motion" modes. This is done by stopping the tape (for the stopped-frame mode) at a point where the complete tracks for two adjacent fields are available, increasing the rotation speed of the drum to compensate for the loss of

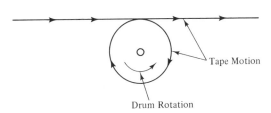

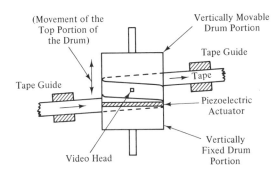

Figure 7-37 Helical-scan videotape mechanism.

the forward tape speed, and then moving the head up and down as the drum rotates, reading one of the two tracks on one rotation and reading the adjacent track on the next rotation, then reading the first track on the next rotation, and so on.

One method of achieving the vertical (across-tape) motion of the tape head is by splitting the drum into two parts as is shown in Figure 7-37. A piezoelectric-crystal driver is placed between the lower (vertically fixed) part of the drum and the upper part. The amount of voltage applied to the crystal determines the amount of spacing between the two parts of the drum and hence the vertical position of the heads which are on the upper part of the drum. Although only one head is shown in Figure 7-37, there are usually six heads occurring in pairs spaced 120 degrees apart with the first pair being designated erase heads, the second pair record heads, and the third pair playback heads. The first head in each pair is used to record the equalizing pulses and any digital information occurring during

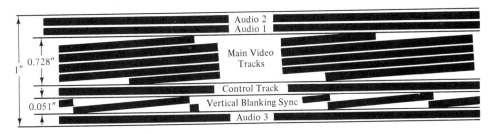

Figure 7-38 Helical-scan tape format (track angles are exaggerated—tracks actually overlap more than shown here; the main video tracks are about 16 inches long each).

the vertical blanking interval on a separate set of sloped tracks near one edge of the tape (see Figure 7-38), while the second head records the actual analog video signal data in form of a frequency-modulated 7.06- to 10.0-MHz carrier (using the HB standard mentioned in the previous section) on the main sloped tracks. The longitudinal tracks along the edges of the tape are used in the same manner as in the quadruplex method for recording audio, cueing, and synchronization signals.

An important part of almost any type of video recording/playback system is the inclusion in the playback-mode electronics of circuitry that allows the substitution of a previously "read" line of video for the line of video currently being read whenever it is found that the latter is missing or otherwise unsuitable for display. This is usually accomplished by feeding the currently displayed "good" video line into the input of a one-horizontal-line-period delay line as that video line is scanned. If the state of the next line of video is then detected as missing, noisy, or otherwise unusable, an immediate switch is made to the delay line's output, and the previous line of video, which was put into the input of the delay line and is now available at its output, is then used as the output from the playback unit.

7.7.3 "Home" Videotape Recorders

"Home" VTRs use variations of the helical-scan principle just described with one important difference: the chrominance information is extracted from the signal before recording, limited to a 500-KHz bandwidth and modulated onto (or in most cases heterodyned down from the 3.579. . .-MHz subcarrier frequency to) a subcarrier frequency whose frequency is $40f_h = 629.37$. . . KHz in the VHS recording format (or about 688 KHz in the Betamax format), producing frequencies in the band from about 129.37 to 1129.37 KHz (in the VHS system). The remaining video signal (luminance plus synchronizing information) is low-pass filtered to about a 3-MHz bandwidth (which eliminates most of the original chrominance information from the signal as well as some of the fine detail in the luminance signal). This signal is then used to FM modulate a carrier in such a manner that the white level corresponds to 4.4 MHz and the sync level to 3.4 MHz in the VHS system (4.8 and 3.5 MHz in the Betamax system). The clipped FM signal thus produced is then bandpass filtered to a range from about 1.1 MHz to about 7.5 MHz. The DSBSC chrominance-information signal just discussed (whose bandwidth is approximately 0.11 to 1.1 MHz) is then added onto this essentially black-and-white signal, and the sum of these two signals is used to drive the recording head. It is interesting to note at this point that the FM signal derived from the black-and-white signal essentially fulfills the purpose served by the high-frequency bias signal in audio magnetic tape recording, thus permitting the linear recording of the lower-frequency analog DSBSC chrominance signal. This technique of separating the chrominance from the luminance and sync information and putting it on a lower-frequency subcarrier is called the *color-under heterodyning* system of videotape recording.

Both popular home VTR systems (the *VHS* and the *Betamax* systems) use the color-under signal-processing technique and the helical-scan method of recording the information onto the tape utilizing two recording heads on the drum with approximately 180 degrees of tape wraparound the drum. Both use a $\frac{1}{2}$-inch cassette-type tape cartridge but have different tape formats and different tape-loading mechanisms. At the normal recording speed, the VHS system uses a 33.4-millimeters (mm)/sec normal tape speed (note the shift to metric specifications on the newer systems) around a 62-millimeters diameter drum rotating at about 30 revolutions per second,[8] giving a tape-writing speed of about 5.8 m/sec, whereas the Betamax system has a 40-mm/sec normal tape speed and a 74.49-mm-diameter drum rotating at about 30 revolutions per second, giving a tape-writing speed of about 6.9 m/sec. Both systems offer slower-tape-speed options which allow the user additional recording time per cassette (2 and 3 times) but at the expense of some loss of picture detail because of the slower ($\frac{1}{2}$ or $\frac{1}{3}$ times) tape speed. Like their big brothers in the broadcast field, these small VTRs also have longitudinal synchronizing and audio tracks recorded by stationary heads along the edges of the tape.

Even more miniaturized VTR formats and machines are being designed, tested, and introduced for consumer use at the time of this writing. One of the most promising formats uses a small cassette of 8-mm-wide (0.315-inch) tape. When coupled with a solid-state CCD-type camera "tube" and microelectronics, this lends itself to the production of a portable one-unit TV camera/recorder not much larger than the 8-mm film/magnetic strip audio/visual cameras introduced a couple of decades ago. It will probably make any type of photographic film "home movie" systems completely obsolete.

7.8 OPTICAL AND OTHER TYPES OF AUDIO AND VIDEO RECORDING SYSTEMS

Although the primary types of audio and video recording systems used up to this point in time have been either mechanical or magnetic, other types of techniques have been used and/or tried. One of the most promising at the present time is the use of a highly focused optical laser beam to read data recorded as pits or spots on the inner reflective surface of two transparent disks cemented together back to back. The reflected light from the laser beam is on/off modulated by the absence/presence of a pit in the reflective coating. This binary signal can be used in one of two basic ways: its fundamental component can represent a frequency-modulated carrier signal carrying analog modulation data, or alternatively, it can represent a binary digital PCM signal. Both these techniques will be illustrated in the examples to be discussed. A somewhat related recording technique, the RCA CED (capacitance electronic disk) or *Selectravision*® technique will also be discussed.

---

[8]Actually 29.97 revolutions per second in the stopped-frame mode and slightly less when the tape is moving.

7.8.1 Audio Digital Optical Disk Recording

As has been discussed in previous chapters and in Section 7.4.4, by digitizing an analog signal and then transmitting and receiving (or, alternatively, by recording and "playing back") the digital signal and then reconstructing an analog signal from the "received" digital information, it is possible virtually to eliminate all the noise normally associated with transmitting and receiving (or recording and playing back) the analog signal directly in analog form. The only noise added by this method, when it is done correctly with appropriate digital error-correcting methods, is the small amount of *digitizing noise* associated with the fact that the voltage interval between digitizing levels is finite (this can be made small enough to be negligible if enough bits per sample are used).

The "standard" digital audio optical disk (often called the *compact disk*) uses a recording format of 44,056 16-bit data samples per channel per second, recorded on an outward-spiraling track moving at a linear velocity of about 1.3 m/sec. This requires a disk speed of from 500 (when on the innermost track) to 200 (when on the outermost track) rpm. The total recording time is about 74 minutes on a disk which is 12.0 cm (about 4.72 inches) in diameter and 2.5 mm thick. Each side has a 1.2-mm-thick transparent coating onto which the laser beam is focused in a spot about 0.8 mm in diameter. Because of the disk's refractive index of 1.5 and the focus angle of the laser beam onto the surface of the disk (a ± 30-degree cone), additional focusing of the beam occurs within the plastic coating to a spot size at the reflective surface of only about 0.0017-mm (1.7 microns) diameter. This permits using a spiral track pitch of 1.6 microns and a spot width of about 0.5 micron. One electromechanical servo system must be used to keep the laser beam spot properly in focus as the disk rotates, and a second servo system is needed to keep the spot positioned transversely on the track.

The left and right channel analog signals are each sampled at a rate of 44,056 times per second and digitized using a 16-bit 2's-complement digitizing code. Each 16-bit data word is split into two 8-bit words, and each of these 8-bit words is transformed into a 14-bit word using an *EFM (Eight-to-Fourteen Modulation) Code* transformation to make the binary bit stream self-clocking (i.e., so that there are no long intervals of 0's or 1's in the NRZ binary code). Six samples from each of the two stereo channels (or $6 \times 2 \times 2 \times 14 = 336$ bits), 24 synchronization bits, 14 control bits, 112 error-correction bits,[9] and 102 merging and LF-suppression bits are combined into one 588-bit digital frame which is recorded onto the optical disk at a bit rate of 4.3218 Megabits per second (Mbps). There are 7342.67 such data frames recorded every second, which gives an *average* data rate of 4.3175 Mbps. The digital error-correcting method used will correct for a bit error rate (BER) of about 10^{-4} (the BER due to "reading" a disk which has not been excessively abused is expected to be about 10^{-5}). If so many bits *are* in error as

---

[9]See Chapter 11 for a discussion of error-correcting codes. The particular code used in this case is known as the *Cross Interleave Reed-Solomon Code* (CIRC), which is a combination of the *Cross Interleave Code* and the *Reed-Solomon Code*.

the disk is read that the error-correcting scheme fails for the reconstruction of a given sample point, the analog value of that sample is then obtained by interpolating from adjacent correct (or corrected) sample values. If the BER gets as bad as 10^{-3}, about 1000 samples per minute will have to be interpolated (there are about 2.643 million samples per minute per channel).

Since the standard procedure used to produce audio digital optical disks is via a pressing technique somewhat similar to that used to produce ordinary phonograph records, this is basically a playback-only system as far as the user is concerned, although some machines capable of producing the digital disks directly from recordable blank disks have recently been introduced on the market. These use a high-powered laser beam which can actually burn off the reflective coating on the backside of the transparent material to produce the "off" spots needed to record the binary signal. The chief disadvantage compared to either the analog or digital recording of audio onto magnetic tape is the lack of editing and medium-erase capabilities.

7.8.2 Optical Video Disks

Another application of optical disks read via a laser beam is to store prerecorded video programs for sale for playback use by the consumer. Unlike the audio digital optical disk, the video optical disk uses analog FM modulation of a recorded carrier signal rather than recording a PCM digital signal. In this respect the technique is somewhat similar to that used on the "professional" videotape recorders. The FM frequencies used for the video signal are 9.2 MHz for the white level, 8.1 MHz for the black level, and 7.5 MHz for the synchronizing level. The chrominance information is included as part of the composite video signal used to FM modulate the carrier. The stereo audio signals are used to modulate the carrier at about 2.3 MHz ($146 \times f_h$) for the left channel and about 2.8 MHz ($178 \times f_h$) for the right channel. A peak FM deviation of ± 100 KHz is used for both audio channels.

The optical system is almost identical to that of the digital audio disk system except that the disk diameter is much larger (300 mm or about 12 inches) and it spins much faster (1800 rpm in the constant-angular-velocity mode and 600 to 1800 rpm in the extended-play constant-linear-velocity mode). The linear velocity varies from about 173 to 455 mm/sec in the constant-angular-velocity mode and is about 150 mm/sec in the constant-linear-velocity mode.

As in the case of the videotape recorders, an important feature of the video electronics in some of the playback units manufactured is circuitry which allows the replacement of a bad line of video (as read from the disk) with the previous line of video received from the disk. In addition to the standard video signal and the stereo audio signal, a vertical-retrace-interval PCM digital signal can also be encoded on the disk and used for frame-number identification, thereby giving the unit the capability of accessing any single frame which is stored. This makes the unit quite valuable for use as a random access, computer-controlled, video storage/presentation medium for such uses as computer-assisted instruction (CAI) and similar applications.

7.8.3 CED Disks

The RCA *Capacitance Electronic Disk®* system uses a disk with mechanical grooves which guide a diamond pickup stylus which has a metal film on its back edge. A capacitance effect exists between this metal film and the conductive material of the disk. By controlling the depth of the groove during the manufacture (pressing) of the disk, this capacitance is varied as the disk rotates. This capacitance variation is detected by applying a small voltage between the metal film and the conductive disk and using the current that flows into and out of the varying capacitance as the signal to be processed. The 12-inch CED disk rotates at about 450 RPM, which allows about 7.5 video fields to be recorded per revolution. As in the case of the *home* videotape recorders, a *chrominance-under luminance* process is employed with the black-and-white signal FM modulating the carrier such that the peak sync level corresponds to 4.3 MHz and the white level to 6.3 MHz. The recorded FM luminance signal extends from about 2.1 to about 9.3 MHz, and the chrominance information, limited to a 500-KHz bandwidth, is quadrature modulated onto a subcarrier at about 1.53 MHz. Provisions are made for compatible (with older playback units) stereo by FM modulating the $(L + R)$ signal onto a 716-KHz carrier and the $(L - R)$ signal onto a 905-KHz carrier using an audio noise-reduction technique (type CX® developed by CBS laboratories) prior to generating the ± 50-KHz peak-deviation FM signal for each audio channel.

7.9 CABLE TV AND SATELLITE TV SYSTEMS

7.9.1 CATV Systems

Community antenna television systems have been around for quite some time, originally appearing in small isolated communities as a means for bringing off-air TV signals into areas where reception would not have been possible with small home-type TV antennas or else would have required the installation of a tall tower supporting a large rotatable high-gain TV antenna at each home or apartment site. These early systems essentially were little more than what their name implied: a number of high-gain TV antennas, usually mounted on a tall tower on top of a nearby hill, with each antenna being pointed at some distant VHF TV station, an amplifier for each of those channels, and a combining/distributing system with a network of cables going out to the various homes and establishments in the community. In some cases the FM broadcast band (which falls within the range of the VHF TV bands) was also broadband-amplified and put onto the cable. Later, when UHF TV stations began to appear, some of the CATV systems added equipment to translate the receivable UHF TV signals down to a locally unused VHF channel for distribution over the cable network. Some of the larger CATV systems added a small studio and began originating local programming for a few hours each day, utilizing another of the unused VHF channels. Since most of the sets then in use could only receive the VHF TV channels, this is all that these early systems distributed, and that is where the situation remained for a number of years.

Developments in other areas of the TV industry in the 1970s resulted in many new possibilities for CATV systems, and a whole new industry began to develop. The possibility of distributing a large number of new UHF off-air receivable signals as well as TV signals obtainable directly from communication-relay satellites meant that many additional channels of regular broadcast-type programming could be distributed as well as special-purpose TV fare such as continuous video news and weather channels, special channels carrying "children's" and "adult" programming, and the like.   Also, the advent of computer-generated video made possible the use of several channel slots for local news, weather, public service, and advertising "billboards"; and low-cost TV cameras made possible economical local programming.   Although most receivers were by now equipped with UHF tuners, sending additional channels over the cable at these high UHF frequencies was not desirable for a number of reasons, the chief one being the high cable losses which would be experienced.

It was at this point that a whole new approach was taken toward *CAble TV systems* (same acronym, but a new meaning and a new technology).

The present CATV systems are not merely antenna/amplifier/distribution systems, but rather are systems that receive TV signals from a wide variety of sources (off-air signals, microwave and cable-relayed TV signals, TV and audio signals from relay satellites, locally generated program material, etc.), process it to produce high-quality baseband video signals, and then "broadcast" these signals on RF carriers into a complex cable distribution system.   In some cases, one *headend* location may relay (via microwave, trunk cable, or fiber optics) the video signals to several distribution-system *hub* locations.   Each hub then "broadcasts" the signals into the cables of its local distribution network.   The trunk cable links used between a headend and its hubs usually involve using each video signal to FM modulate a carrier with deviations of from ± 1 MHz to ± 4 MHz and RF transmission bandwidths of 12 to 14 or more MHz per channel.   This results in virtually noise-free distribution of the video signals to the system hubs.

The frequencies used for the actual distribution of the signals to the consumer depend upon how many channels are to be put onto the cable (usually anywhere from 20 to 54 separate channels) and the exact scheme to be used.   These frequencies usually start at about 54 MHz and go up into the region of 400 MHz maximum for a 54-channel system.   Several severe problems can occur in such a wideband system which is simultaneously handling so many complex signals.

The biggest "internal" problem is the severe intermodulation that can result from the mixing of two, three, or more of these signals if any system nonlinearities are present.   The intermodulation products which result from two or three signals thus combining may very well cause interference with one or more of the other signals on the cable.   In addition to minimizing the production of the intermodulation products themselves, there are several techniques which can be used to reduce the perceived effects of such spurious signals when they do occur.   One such technique is phase-locking all the carrier frequencies to one common oscillator. This makes any interference pattern produced on the TV screen stationary and therefore less perceived and less objectionable to the viewer.   One such scheme places all the frequencies at multiples of 6 MHz (the HRC or *Harmonically Related*

Carrier scheme).   Another scheme is the *Constant Interval Channeling plan* (sometimes referred to as the IRC or ICC scheme), which places the carriers at $(6n + 1.25)$ MHz.   With some type of phase-locked scheme (or *SyncLock* as it is often called), the interfering carrier signal must still be 41 to 45 dB below the level of the desired signal for the interference not to be objectionable.   However, when SyncLock is *not* used, the interfering signal must be at least 53 dB below the desired signal level.   In some cable systems the interference between the TV audio and video signals may be reduced by placing all the TV sound channels on the cable at the lower end of the cable spectrum (54 to 60 MHz, for example).

Another problem with cable TV systems is that although they are supposed to be self-contained "closed" systems (and therefore can utilize frequencies assigned for other uses in the "outside world"), they do, in fact, "leak" radiation.   By this we mean that some of the signals within the cable system may escape into the outside world and cause all kinds of problems to other users of the RF spectrum (for example, a cable system in a given geographical area may be restricted from utilizing frequencies within the aircraft communication and navigation bands if leakage problems might present a hazard to aircraft operation in that area).   On the other hand, leakage *into* the cable can also occur when there are strong "outside world" RF signals present in the area in which the cable system operates, and this may make some of the channel slots essentially unusable in that area.

The cable distribution system itself usually consists of 75-Ohm rigid metallic coaxial cable, often with a dry-nitrogen dielectric with the inner conductor being supported by plastic beads.   A 1-inch-diameter cable has a loss of about 19 dB/mile at 100 MHz and 45 dB/mile at 400 MHz, while for 0.412-inch diameter cable the figures are 45 and 100 dB, respectively.   Repeater amplifiers are located about every 15 to 20 dB apart (at the highest frequency on the cable).   Since the RF loss of a cable section increases with frequency, the gain of the amplifiers must be higher at the high end of the cable band than at the low end.   Cable drops into individual homes are usually via more flexible solid-dielectric 75-Ohm cable, which has a higher loss per foot than does the rigid distribution cable, but which nonetheless is acceptable for short runs.   The tap-off must be designed so as to minimize any signal absorption from or reflection into the main cable regardless of what is done to the drop cable at the consumer's end.   Since few TV sets are made to accept cable signals directly, a converter must be installed between the end of the cable and the TV set.   This converter translates a group of the cable signals to the set's range of VHF channels.   Since most of the cable systems offer a selection of extra-cost optional channels in addition to a "standard" service, the converter is "programmed" to provide those services contracted for by the customer and is then sealed to (it is hoped) prevent tampering and the reception of the unauthorized channels.   In some cases a descrambler device may be installed and activated for a given period of time by coded signals addressing a particular converter box to permit the reception of certain charged-for programs.

Many interesting possibilities are being explored for future cable TV systems, including such things as videotext signals (essentially low-bandwidth data communication signals which through a converter use a TV receiver as a receive-only video terminal), two-way video and computer communications; inclusion of telephone functions on the cable, fire- and burglar-alarm systems, and many other interesting applications. Some of these applications require a two-way cable "street" (although in most cases the reverse traffic is of a low-bandwidth nature). The possibility of using fiber-optic cables in places of RF signals on coaxial cables also has many interesting and attractive ramifications, especially as one begins to think of the "cable" as more than just an entertainment-distribution medium.

7.9.2 Satellite TV

TV signals from private geostationary communication-relay satellites first became available in the early 1970s, and electronic experimenters immediately began constructing experimental home-sized TVRO (*TeleVision Receive Only*) systems to intercept these private communication signals. As the use of the relay satellites to replace terrestrial microwave and cable links grew, so did the number of home terminals. Although the legal status of intercepting and using these private signals was greatly in doubt (and still is to some extent), the sale of commercially produced home TVRO systems has become a major industry. Basically, the satellites are designed to relay network video programming to TV broadcast outlets and/or CATV system "headend" terminals. One relay satellite may carry several dozen such TV signals through its various transponders as well as relay stereo audio signals for a number of radio broadcasting networks.

The relay satellite operates by having a high-powered ground station transmit an *uplink* to it in the 5.9- to 6.4-GHz band, with the satellite-to-earth *downlink* operating in the 3.7- to 4.2-GHz region (this frequency region is also utilized for terrestrial microwave-relay uses). This 484-MHz-wide band is divided into 12 40-MHz-wide TV frequency "slots" and 1 4-MHz "guard" channel placed between satellite TV frequency slots 1 and 2. The guard channel is used for tracking, control, and signaling purposes. The 12 TV frequency slots can become 24 actual channels via the use of polarization diversity techniques (i.e., transmitting one downlink signal with one direction of electomagnetic polarization and another downlink channel with the opposite polarity direction).

In the RCA F1 satellite, for example, the odd-numbered channels are transmitted with vertical polarization, and the even-numbered channels are transmitted with horizontal polarization. The EIRP of the signal on each channel radiated from a typical relay satellite is in the vicinity of 25 to 40 dBW at the center of the downlink beam. When this is combined with a 4-GHz free-space loss of about -196 dB, the received power density at the earth's surface is only about 10^{-12} Watt per square meter, which obviously requires a large antenna and some very special low-noise receiving equipment. For example, for a 30-MHz noise bandwidth (typical of the bandwidth of a home TVRO receiver), the equivalent excess noise power of a "low-noise" receiver with about a 3-dB noise figure is about 1.3 $\times 10^{-13}$ Watt. Thus, with an antenna having an A_{eR} of about only 1 m² and using this receiver one would obtain less than a 10-dB predetection *S/N* ratio (an 11-dB predetection *S/N* ratio is considered to be the acceptable minimum). For this reason the A_{eR} of the antenna must be somewhat greater than 1 m² and/or the noise figure of the receiver less than 3 dB. Typically a 10-ft-diameter antenna

(A_{eR} of about 6 or 7 m²) and a receiver with about a 2.6-dB noise figure (N_{eq} ~ 240° K) are needed to obtain an adequate predetection S/N ratio in locations where it is possible to receive a signal which when transmitted from the satellite's antenna has an EIRP of 36 dBW or greater.

The modulation method used on the satellite TV downlinks is wide-band FM modulation of the microwave carrier by a signal consisting of the TV video signal plus FM-modulated audio subcarriers at about 6.2, 6.8, and/or 7.4 MHz (6.8 MHz is the most commonly used for monaural audio, but the availability of three audio channels permits stereo audio and/or multilingual audio for the TV programming). The entire FM signal covers approximately 36 MHz (at the -40dB-down points), but many home TVRO receivers use IF passbands of only 27 to 30 MHz to obtain improved predetection S/N ratios (but at the expense of some loss of picture detail due to the rejection of the higher-order FM widebands). Receivers for home TVRO systems usually consist of a microwave bandpass filter (3.7 to 4.2 GHz) followed by a low-noise two-stage gallium-arsenide FET (field effect transistor) amplifier and a mixer (usually also a GaAs FET) that converts the 3.7- to 4.2-GHz signal to a 70-MHz IF signal. The RF amplifier/mixer assembly and the voltage-tuned local oscillator are all usually located at the dish antenna, with the power supply and the LO tuning voltages being brought from inside the house to this assembly, and the IF signal being taken from this assembly to the IF amplifier located inside the house, where the video and audio detectors, the power supplies, the LO tuning control (channel selector), and so on are also all located. The 70-MHz FM signal out of the IF amplifier is usually detected with a PLL type of FM detector whose output is directed to a 4.2-MHz low-pass filter to recover the video signal and to 6.2-, 6.8-, and/or 7.4-MHz audio-subcarrier FM demodulators to recover the desired audio channels.

At this writing, experiments have been carried out with DBS (*Direct Broadcast Satellites*) at frequencies in the 800-MHz, 2.5-GHz, and 12-GHz bands. Most of these satellite channels had EIRPs of 50 to 55 dBW, thereby permitting the use of smaller receiving antennas than are presently being used for receiving TV signals from the relay-type satellites. The use of a 20- to 25-MHz RF bandwidth FM-modulated signal in the 12-GHz region has been proposed for a North American operational DBS system. The use of digitally encoded TV signals and the broadcast of *High Definition TeleVision* (HDTV) signals is also being considered. The latter (HDTV) would provide a TV signal with approximately twice the pixel resolution of the current NTSC standard TV picture.

A number of different proposals have been advanced concerning whether the DBS system would be advertiser and public supported or subscriber supported. In the latter case, it would be necessary for the viewer to rent or purchase in some manner the necessary decoding equipment and/or decoding signals or software needed to produce a viewable picture. Although a number of organizations seem to be interested in DBS-TV, the problem at the present time seems to be one of obtaining sufficient capital for the needed start-up and initial operational expenses of a DBS system, especially in view of the other options now available to the viewer (regular broadcast TV, CATV systems, the interception of TV-relay-satellite signals, and the rental or borrowing of videotape cartridges from private or public libraries).

7.10 DIGITAL TELEVISION SYSTEMS

If we look at what appears to be a "noise-free" NTSC TV video signal channel (4.2 MHz at a 48 dB S/N ratio), we find that the maximum channel capacity is given by

$$R_{I,\max} = 2 \text{ (BW) } \log_2 (\sqrt{1 + S/N})$$

$$\approx 67 \text{ Mbits/sec}$$

(7-5)

If we look at the $(525 - 28) = 497$ viewable lines in the NTSC frame and consider the 4:3 frame aspect ratio, we find that there is the possibility of having $(497)(4/3)(497) \approx 329{,}000$ square pixels (picture elements) in a standard 525-line video display. However, this would require transmitting about 662 pixels in a time period of $(0.82/f_h)$, which would give a maximum frequency of about 6.47 MHz (treating each *pair* of alternate bright and dark pixels along a line as one cycle at the highest frequency). This is the baseband video bandwidth which would be required for *square* pixels, and is very often approximately the bandwidth used on a medium-resolution computer graphics monitor. The 4.2-MHz bandwidth limitation of the NTSC signal effectively allows us only $(4.2/6.47)(662)$ or 430 pixels per visible line, which means that in the NTSC TV picture, we are working with *rectangular* pixels having a 1.54:1 length-to-height ratio. We will refer to the square-pixel case as a *monitor-quality* picture and the rectangular-pixel case as an *NTSC-quality* picture. It should be noted here that most consumer-type color TV receivers further limit the luminance-signal bandwidth to about 3 MHz, reducing the horizontal resolution to about 307 pixels per visible line (a length-to-height ratio of about 2.16:1). We will refer to the latter case as a *receiver-quality* picture.

If we assume that we can carry all the information about the intensity, hue, and color saturation of a pixel in 8 or 10 bits of digital data, and assume a frame rate of about 30 frames per second, the data bit rate associated with each quality level of picture would be as shown in Table 7-4. A fourth column indicates what bit rate would be required for a 1365-pixel wide by 1024-pixel high high-resolution color graphics monitor (HRCG monitor). Table 7-4 represents the actual data rates that would be required to send the video information using "brute-force" digitizing techniques (actually there would be some additional bits per second needed for synchronization and error-correcting functions). As can be seen, sending even a "receiver-quality" signal at 8 bits/pixel via a binary digital channel would require a minimum baseband bandwidth of 18.3 MHz (or about 6 times the receiver's 3-MHz video bandwidth), although a lower S/N channel could be used for this binary digital signal (about 20 to 22 dB) as opposed to what would be required for the analog video signal (about 48 dB).

TABLE 7-4 DIGITAL-DATA-BIT RATE FOR "BRUTE-FORCE" DIGITIZED VIDEO SIGNALS
OF DIFFERENT QUALITY LEVELS

	Data Bit Rate in Megabits/second			
Bits per Pixel	"Receiver Quality"	"NTSC Quality"	"Monitor Quality"	HRCG Monitor
8	36.6	51.3	79.0	335
10	45.8	64.1	98.7	419
8 bits per RGB color	—	—	237	1006

All the foregoing discussions assume that the entire scene on one video frame will be very detailed and that the entire scene will change completely from one frame of video to the next at a rate of 30 frames per second. In actual video displays the presentation of new data is nowhere this fast. For example, no changes may occur in a scene for several seconds, or only a small fraction of the total scene area may be changing, with the remainder of the scene remaining static. Also, large areas of a given scene may have approximately the same luminance and/or color hue and saturation. By designing different types of adaptive coding techniques, enormous savings can be made in the number of bits per second to be transmitted without sacrificing picture quality in most cases.

The various techniques used to do this are referred to as digital video bandwidth-compression methods. As an example of one of these, some type of a time-differential coding scheme could be adopted wherein instead of repetitively transmitting all 8 to 10 bits of data for each pixel every $\frac{1}{30}$ second, a scheme is devised whereby a much smaller number of bits is sent telling how each pixel has changed since the last time it was scanned. The only problem which might arise here is that if a very abrupt change occurs in the pixel, it might be several frames (or a fraction of a second) before the corresponding pixel at the receiving end has the correct characteristics.

To handle the large-homogeneous-area situation, there are also a number of techniques available. One of these is to apply some type of *differential-coding* technique to adjacent pixels to describe how they vary among each other (in effect this is the same thing that is done in an analog manner with the horizontal color detail by limiting the chrominance-signal bandwidth in the NTSC system). Another digital technique is *run-length coding* wherein if a number of adjacent pixels in the same line are all the same, the pixel data code is sent only once, accompanied by the information as to how many of the following pixels are the same.

All this special digital coding can, of course, become very complex and time consuming. Even with the advent of very fast special-purpose microprocessors, it is still necessary to make some compromise between the most efficient bandwidth-compression coding algorithms and getting the job done in "real time" with an acceptable equipment cost. This is an area of great interest at the present time, especially as the speed of digital processing equipment goes up and the price comes down.

Another area of digital video processing is the processing of a received *analog* TV signal via digital techniques to produce an "enhanced" signal which *appears* to have lower noise, higher resolution, and better quality than the unprocessed version. For example, a scene with about 500 visible scan lines can be converted to a scene with about 1000 scan lines by appropriate two-dimensional interpolation techniques. These are called antialiasing techniques because they appear to reduce the aliasing caused by transmitting a limited-bandwidth signal (which is, in effect, the same as sampling at a limited rate lower than the Nyquist criterion would dictate for the actual picture information detail which the scene contains). This type of digital processing TV receiver is now available on the home TV market.

In general, the whole area of digital video transmission, recording, and signal processing will probably be one of the more active areas of research and development in the next few years, having applications not only in TV transmission, but closely tied in with developments in the areas of *Computer-Generated Imagery* (CGI) and *Computer-Assisted Image Analysis*. These last areas are of interest in fields as varied as designing, medical diagnosis, missile guidance, pilot training, natural resources evaluation, and self-paced learning.

PROBLEMS

7-1. Explain why the present stereo FM system is compatible with a properly designed monaural FM receiver. What could happen in older monaural receivers which were designed before the development of the stereo FM system?

7-2. Show that by transmitting a 19-KHz pilot carrier it is possible to have the 38-KHz carrier injected at the receiver to be in phase with suppressed original 38-KHz carrier signal and that would not have been true if 76-KHz (2 × 38) had been selected for the pilot signal.

7-3. How could you design a receiver that would automatically switch from stereo mode of reception to monaural mode if the received FM signal became too weak or noisy?

7.4. Define the amplitudes of the various components in a stereo FM signal if standard AM modulation of the 38-KHz subcarrier were to be used in place of DSBSC modulation and a 19-KHz pilot carrier, assuming other ratios and relationships remain the same.

7.5. In the standard NTSC color TV signal, what are the sideband frequency ranges for the various components of the signal (i.e., the Y, Q, and I video signals and the sound channel)? What carriers are present or suppressed? How is the problem of detecting AM signals with unsymmetrical sidebands compensated for at the receiver (prove that a diode detector can be used to detect the composite video signal without introducing undue distortion).

7.6. What is meant by each of the terms *frame rate, field rate, raster, aspect ratio*?

7.7. Why is *interlaced* scanning desired in the NTSC TV system? How is it accomplished?

7.8. What frequencies are involved in the NTSC composite TV signal? Why are there definite ratios among these frequencies? How could they all be obtained from a single master oscillator at either 4.5 MHz or at 3.57954506 MHz (i.e., sketch a block diagram)?

7.9. What is an AGC system in a TV receiver and how does it differ from the AVC system in an AM or FM audio receiver? What is meant by a *keyed* or *gated* AGC system?

7.10. Explain how the synchronizing pulses on the composite TV signal are used to synchronize the horizontal and vertical sweeps at the receiver with those of the TV camera.

7-11. Explain how it was possible to add color information (color hue and saturation) to the black-and-white TV signal without utilizing any additional spectrum space. What compromises were made in picture quality to do this?

7-12. Why are the particular specified proportions of red, green, and blue signals used to make up the luminance (Y) signal rather than using equal portions?

7-13. Explain the makeup of the I and Q signals. Why have I and Q signals rather than (E_R and E_B) or (E_R and E_G) or (E_B and E_G) or ($E_Y - E_B$) and ($E_Y - E_R$)? Why do the I and Q signals have different bandwidths?

7-14. Why *can* the color subcarrier be eliminated? Why *do* you want to *not* transmit the color subcarrier? How do you compensate for the fact that it is not transmitted?

7-15. Why must the Y signal be delayed in the color TV receiver? What would happen if it were not delayed?

7-16. What "trick" is used in the color TV receiver to get the ($E_R - E_Y$) and the ($E_B - E_Y$) signals directly from the color demodulators rather than having to combine I and Q signals in a matrix?

7-17. How is the subcarrier injection oscillator in the receiver synchronized with the subcarrier oscillator at the studio?

7-18. Explain how you would incorporate adjustments for (a) picture brightness, (b) color saturation, and (c) color hue into a color TV receiver.

7-19. Explain the purpose of the high-frequency bias signal in an audio magnetic tape recorder. How does it reduce the nonlinear distortion inherent in the magnetic record/playback process? In what other ways can this nonlinear distortion be reduced? (Hint: Most video magnetic recording techniques do not use high-frequency biasing.)

7-20. "Compact" optical audio disks record a *digitized* version of each audio channel. Video optical disks record an *analog* version of the video signal. Both systems use an on/off (i.e., reflecting versus nonreflecting) physical recording mechanism. Explain the relationships among these three statements.

7-21. Using the optical-disk parameters discussed in Section 7.8.1 (44,056 16-bit samples per second per audio channel and a track velocity of 1.3 meters per second), find what the *length* of each on or off spot is in the along-track direction (the spot *width* is about 0.5 microns).

7-22. Assuming that the optical videodisk described in Section 7.8.2 records the "cycles" of the FM video signal as a square wave, what is the range of the "on" spot size lengths for the *white*, *black*, and *sync* levels for the constant-angular-velocity and the constant-linear-velocity modes?

7-23. Assuming a 4.2-MHz bandwidth for the TV video luminance signal and the FM signal frequencies given in Section 7.8.2 and the characteristics of the black-and-white TV signal given in Figure 7-18, estimate and sketch the approximate spectrum of the resulting FM signal.

8

PUBLIC COMMUNICATIONS SYSTEMS

8.1 THE PUBLIC TELEPHONE SYSTEM

8.1.1 History

The very earliest public telephone systems involved a number of telephone instruments connected to a single "party" line. A simplified schematic diagram of the telephone instrument in use at that time is shown in Figure 8-1. It consisted of a local audio-frequency circuit comprised of a dry-cell battery, the telephone "receiver" (actually a small electrodynamic loudspeaker), the "transmitter" (a carbon-granule, variable-resistance microphone), and the primary winding of an audio transformer (this was, and still is, usually called the telephone set's "induction coil"), all connected in series so that sound entering the microphone would cause the "d.c." current in the circuit to be modulated, producing sound in the telephone "receiver" and also inducing a voltage in the primary winding of the transformer. A switch contact on the "hook switch" actuated by hanging the receiver on the hook when the telephone was not in use served to disconnect (open) this primary audio loop, thereby preventing unnecessary drain on the dry-cell battery when the telephone was not in use. Another set of contacts on this same switch mechanism

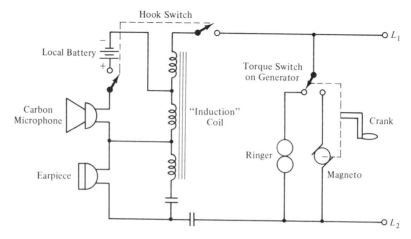

Figure 8-1 Early magneto telephone with local-battery power source.

disconnected the audio portion of the telephone from the line as long as the receiver was hanging on the hook.

With the receiver off the hook, the secondary of the "induction coil" transformer was connected to the telephone line and thereby to the secondaries of all the similar transformers in the other telephones on the line which had their receivers "off the hook." The voltage induced in the secondary of any telephone where a person was talking into the transmitter would appear across the secondary windings of all these other telephones, and this would in turn induce a voltage in the primary side of each of the transformers, which would cause the current in each telephone's primary loop to be modulated, thereby causing the speaker's voice to be heard in each telephone receiver which was "off the hook." As more and more people took their receivers off the hook to listen in to the conversation on the party line, the voltage induced in each of the primary windings became smaller and smaller, and the audio acoustical signal produced by each receiver became weaker and weaker. This was a perennial problem with party-line telephones with too many "eavesdropping" subscribers.

For one party on the line to signal another party that the first person wanted to initiate a conversation with the second, an a.c.-operated bell or "ringer" was provided, which when supplied with a low-frequency (20- to 40-Hz) a.c. voltage of about 40 to 150 volts would cause an armature to vibrate a bell clapper between two bells (incidently this is still the design of the telephone "bell" in the nonelectronic telephone instruments of today). In these early systems, each telephone instrument had a small a.c. generator (or *magneto*) to generate the voltage for ringing the bells of all the other telephones on the line. To designate which telephone was being called, each party line had an agreed-upon set of Morse Code–like signals (for example, one long ring followed by two short rings followed by two more longs might be the signal for a particular telephone on the party line). The magneto was operated by a hand crank that incorporated a torque switch that connected the output of the magneto to the line whenever torque was being applied to the crank.   Thus, it was fairly easy (with a little practice) to send the necessary code to "ring" the desired telephone with which you wanted to establish a conversation.   These telephone magnetos were much sought after by the experimenters and pranksters of that era, for they put out a sufficient voltage to deliver a good (but usually not lethal) electrical shock.   Along the same line, the partially worn-out dry-cell batteries (usually what were, and still are, referred to as "No. 6 cells") provided many boys of that era with their first chance to experiment with electrical circuits and equipment.

In addition to these side benefits, the party-line nature of these early telephone systems, especially in rural areas, provided not only "one-on-one" communication service, but also was the most efficient means for rapidly disseminating local news as well as a means of entertainment for the people on the "line."   Much of the time the line functioned in the "conference-call" mode of operation (at least until so many receivers were off the hook that the signals became too weak to be usable). This method of operation was not unlike that of a "CB" channel in today's world of electronics.

The limiting factor of the single party-line system was the number of telephones that could be attached to the line from a practical operational standpoint as well as the excess electrical loading of all the "ringers" upon the one magneto generating the ringing voltage.   The solution to this was to establish lines with only one to four users (businesses usually had "private" one-user lines while residential lines were usually two-party or four-party lines).   Each of these lines then went to a "central" interconnection site where a large number of lines from a given area each terminated at a pair of "phone" jacks on a switchboard panel.   In addition to being connected to the jacks, each line was also connected to a small light above the jacks.   Whenever any of the magnetos on a particular line was cranked, the light above the jacks would light up (in some cases a buzzer was also sounded). The switchboard operator (often referred to as "Central" by the telephone subscribers) would insert an "answering cord" into one of the jacks and answer the calling telephone.   The subscriber would then tell the operator which other telephone the subscriber wanted to call, and the operator would then insert a patch cord between the other jack and one of the jacks for the line going to the other telephone.   Then, depending upon the system, either the operator or the subscriber would crank their magneto to ring the signal for the desired telephone on that line. Once the telephone conversation was finished, one of the conversing parties would crank their magneto to "ring off," thereby notifying the operator that the patch cord could be removed.

As this type of centrally controlled telephone system was developed further, it was soon found that the local dry-cell battery at each telephone could be replaced by one large battery or power supply at the central switchboard by slightly modifying the design of the telephone instrument to permit the d.c. supply current to flow over the line along with the a.c. audio signal.   This was done by a simple modification in the design of the "induction coil" transformer circuit (see this part of the circuit in Figure 8–2) so that the d.c. currents would be balanced out in the primary and secondary circuits.   Since the d.c. current would only flow whenever the telephone was "off hook," the next step was to incorporate equipment into the

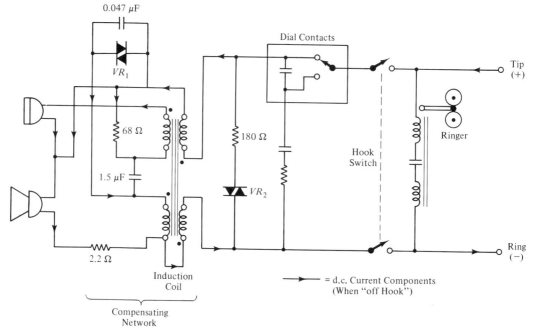

Figure 8-2 The modern rotary-dial telephone set.

switchboard to sense whenever there was d.c. current flow on a line. Thus, the need for the magneto at each telephone instrument for use in calling the operator was also eliminated. This was the ultimate stage of development for the manually switched telephone system.

Figure 8-3 shows a simplified schematic of a portion of the d.c. supply circuit at the central switching site. The 48-volt station battery bus supplies d.c. to the telephone line via the windings of the relay which senses the "off-hook" condition of the line (the "A" relay in Figure 8-3). This relay also serves another purpose: the inductance of its winding (about 5 Henries total) isolates the audio and ringing signals on each line from the common d.c. bus, thus preventing *crosstalk* among the various lines being supplied from the d.c. bus.

The pair of wires connecting a particular telephone instrument to its central office is referred to as the *local loop* and usually consists of #22AWG copper wire color coded so that the "tip" or positive wire is either green or black and the "ring" or negative wire is red or yellow (the terms *tip* and *ring* refer to the tip and ring on the phone plugs used on a manually operated switchboard).

The characteristic impedance of the local-loop line is usually either 600 or 900 Ohms, depending upon the operating company and the equipment it is using. Thus, the voice-frequency impedance presented to the line by an attached telephone set when it is "off-hook" on the far end of the line and by the local office equipment on the near end of the line should match this characteristic impedance to minimize signal reflections from the ends of the line.

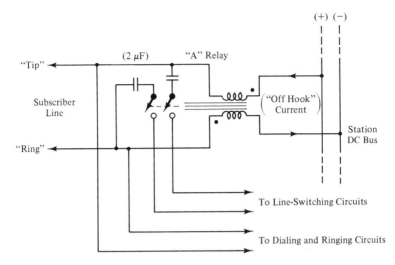

Figure 8-3 Typical central-office termination of a subscriber line.

The d.c. voltage on the line when the telephone is "on hook" and presenting a high d.c. impedance (10 megohms or more) to the line should be about equal to the central-office bus voltage (or about 48 volts), but when the telephone is taken "off hook" the d.c. voltage on the line at the terminals of the telephone drops to about 3 to 9 volts due to the d.c. voltage drop in the windings of the "A" relay and the d.c. resistance of the local-loop wires. The d.c. current drawn by the telephone set (part of which passes through the carbon microphone) will vary from about 20 mA up to about 70 mA, depending upon the length of the local loop (i.e., how far it is from the central office to the telephone set). The length of the local loop also affects the attenuation which the audio component suffers when passing along the line. To compensate for these variations in the d.c. operating current level and the audio attenuation, modern telephones incorporate nonlinear resistors (varistors, denoted as VR_1 and VR_2 in Figure 8-2) to adjust the d.c. current levels and audio signals within the telephone set itself. This tends to keep the acoustical audio level produced by the "receiver" reasonably constant, both for audio signals received from the line and for the audio *sidetone* signal produced in the receiver from whatever is spoken into the transmitter at the telephone itself.

The next major step in the telephone industry was the automation of the switching function. This was done by equipping each telephone instrument with a rotary dial that was manually rotated a fraction of a turn to a position corresponding to one of the decimal digits 1 through 9 and 0 (ten). When the dial was rotated clockwise to the desired stop position, a switch on the dial disconnected the audio portion of the telephone from the line and connected a series RC circuit in its place (the presence of the capacitor is for pulse-shaping purposes). When the dial is released and allowed to rotate counterclockwise back to its rest position, a pair of contacts connected in series with the resistor opens and closes the circuit a number of times corresponding to the stop position from which the dial begins

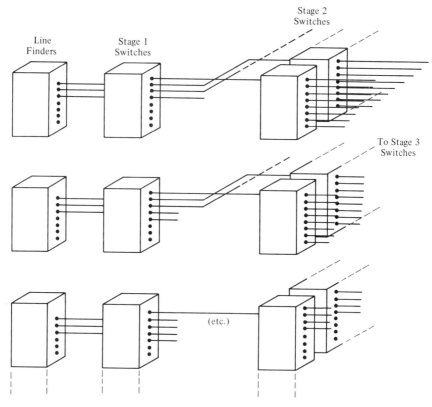

Figure 8-4 Step-by-step switching frame interconnections.

its counterclockwise motion, thus repetitively interrupting the d.c. current flow in the line. These interruptions in the current flow are sensed at the switching office and are used to control the switching of the caller's telephone line to the line to which the called telephone is attached.

The early automatic telephone switching exchanges were what are called Strowger[1] step-by-step switches. The method of operation of these switches is as follows (see Figure 8-4):

1. The first type of stepping switch encountered on an incoming line is the *line finder* relay. A group of local telephone lines comes into and shares a bank of this type of switch. The number of line finder switches versus the number of incoming telephone lines depends upon the average expected usage of each of the lines. There will always be more incoming lines than line-finder switches. Whenever a telephone attached to one of the lines coming into this group of line finders is taken off hook, one of the currently unused line-finder switches is assigned to search for and find the line with the off-hook telephone. It does this by moving

---

[1]Mr. Strowger, who invented the basic stepping-switch mechanism in Kansas City in 1889, was an undertaker by profession.

its main contact in a vertical direction until the off-hook line is sensed. It then rotates the contact in a horizontal direction across contacts attached to lines going to individual first stage stepping switches until an unused switch is found. It then connects the off-hook line to this unused stepping switch. Once this is done, the system sends a "dial tone" to the telephone which is attached to the off-hook line.

2. As the subscriber dials the first digit of the telephone number, the first-stage stepping switch connects the telephone's line to a selected (1 out of up to 100) lines going out from the switch. It does this in two steps—as the first digit is dialed the movable contact of the switch moves vertically upward to the selected row (out of 10 rows total, with each row representing the particular bank of *second-stage* stepping switches that corresponds to the first digit dialed) and then the contact is rotated along the contacts in that row until a line going to an unused stepping switch in the selected second-stage bank is sensed, at which point the incoming line (from the line-finder relay) is connected to the "input" of that unused second-stage stepping switch.

3. Step 2 is repeated for each digit of the telephone number to be dialed. If at any point along the way an unused stepping switch is not available, or if the called telephone line is in use (i.e., "busy"), a *busy signal* is sent back to the calling telephone. This busy signal is slightly different for the case when it is not possible to complete the call because of congestion in the switching network (i.e., no stepping switches available) and for the case where the called telephone line is in use. If a connection can be made to the called line, this connection is maintained through all the stepping switches involved until the conversation is ended, at which point each stepping switch and the line-finder relay all return to their "home" positions.

The last four digits of the telephone number dialed usually represent four final stages of switching located at one central location or *local exchange*. This local exchange may have up to about 10,000 possible telephone lines connected to it. A call originating in this same exchange will first pass through three initial outgoing switching stages that either route the call to the first of the four final stages or else route it onto *trunk* cables (or microwave or optical links) going to other local exchanges. Which occurs depends, of course, upon whether the first three digits dialed are the code for this local exchange or for some other local exchange. We will be discussing the trunk links between local exchanges more in a later section.

The next major step in telephone switching development was the invention of the *crossbar switch*. The heart of this type of switching system was the *crossbar frame*, which was essentially a matrix of vertical and horizontal rods, each of which could be rotated through a small angle by applying a current to an electromagnetic coil which attracted a steel armature attached to the rod. Along each rod was a series of contact fingers. Activating one of the electromagnets driving a horizontal rod would cause the contact fingers on that rod to move in the direction of the contact fingers on all the vertical rods in that row, but not far enough to close any of the contacts. However, when the coil driving one of the vertical rods was similarly energized and that rod rotated, its contact fingers would move toward all the horizontal-rod contacts in that particular column. At the point where the two rotated rods crossed, both the horizontal-rod finger and the vertical-rod finger would move sufficiently that contact would be established for that pair of contact fingers only. Actually, each contact finger carried several contacts, so that at the point where the two rotated rods crossed, a set of several electrically independent contact pairs closed simultaneously. Since there were similar sets of contacts at each point where a horizontal and a vertical rod crossed, by selecting one horizontal-rod coil and one vertical-rod coil for energizing, one of the many sets of contacts could be closed. This made it possible to connect one of a number of incoming telephone lines to one of a number of outgoing lines. By having a large number of such frames interconnected together, it was possible to route a call through a central switching office.

One of the major differences between the stepping-switch exchange and the crossbar exchange was that in the stepping-switch exchange, the call was routed as it was dialed, with the pulses generated by the rotating-dial switch on the telephone actually "pulsing" the stepping-switch mechanism and causing it to move vertically or to rotate horizontally to the desired position while in the crossbar exchange, the dial pulses were recorded as numbers in a "register" as they were dialed and arrived at the exchange. Once the entire number had been dialed, the logic system associated with the crossbar exchange would determine how the call was to be routed through the various crossbar frames in the exchange. This routing was determined by the telephone that was calling, the number of the telephone called, and which circuits and frames on the various possible routes through the exchange were already in use.

The very early crossbar exchanges used electromechanical-relay logic circuits for this function, with electronic wired-logic systems later replacing the electromechanical ones. The next step in the development of this type of "space" switch was to replace the wired logic with microprocessors implementing "software logic" by means of routing-algorithm programming. The actual crossbar-frame electromechanical mechanism was also replaced by a *crosspoint* matrix (note the slight change in terminology) of reed-relay switches, each of which was activated by two windings. When both the windings around a particular switch were pulsed with additive-polarity currents at the same time, the reed relay contacts would close. The contacts were then kept closed by the presence of a temporary "permanent" magnetic field produced by two pieces of magnetic material having a square hysteresis-loop characteristic. When the two coils that surrounded a particular reed-relay switch were similarly pulsed with subtractive-polarity currents, the magnetism on one of the two pieces of metal would be reversed and the net magnetic field in the contact area of the reed relay would be reduced to approximately zero, resulting in the opening of the set of contacts. One of the coils of all the relays in a given row in the matrix would be connected in series, and the other coil on each relay in that row would be connected in series with the similar coils of all the relays in the same column in which that relay was located. By pulsing one of the "vertical" wires and one of the "horizontal" wires going into the matrix with additive or subtractive current polarities, it was possible to either close or open the reed-relay contacts that were situated at the point where the two wires "crossed" in the matrix.

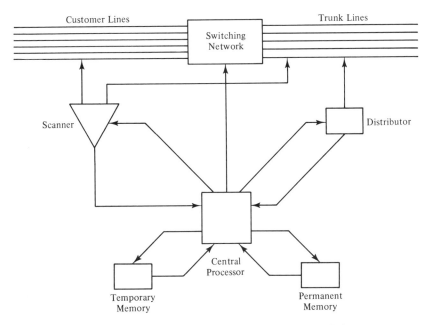

Figure 8-5   Organization of the A.T.&T.   ESS-1 switch.

This type of system was the basic design of the Western Electric ESS-1 electronic switching exchange of the mid-1960s which employed an 8×8 crosspoint matrix at each stage.   The organization of the customer-lines-to-trunks portion of the ESS-1 switch is shown in Figure 8-5.   The *scanner* unit scans the customer lines for an off-hook condition and notes when this occurs so that a dial tone can be presented to the customer.   As the customer dials the number of the called telephone, it is stored in a portion of the temporary memory of the switch.   This information is used in conjunction with information received from the *distributor* (which is scanning the trunk lines going out of the switch to determine which ones are not busy) in the routing-algorithm program in the computer's permanent memory to determine and set up a route through the switch to connect the calling subscriber with the appropriate outgoing trunk line (perhaps based upon the first three dialed digits).   The scanner also has the job of determining when the subscriber telephone is put back "on hook" so that the connection can be "torn down" and the portions of the route through the switch as well as the trunk line can be used by another subscriber.

One of the features which the implementation of electronic switching exchanges permitted was the elimination of the rotary pulser-type dial on the telephone set and its replacement by an electronic signaling scheme called *dual-tone multifrequency* (DTMF) dialing (often referred to as *Touch-Tone®* dialing).   In this scheme, each digit to be "dialed" is represented by simultaneously transmitting a pair of audio tones from the telephone to the switching equipment.   All 10 of the digits (1 through 9 and 0) on the standard rotary dial plus 6 additional "digits" (∗, #, A, B, C, and D) can be thus represented by simultaneously sending one

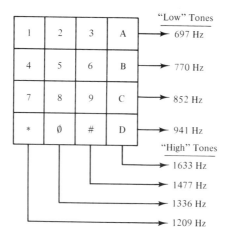

Table 8-1 The DTMF Tone Chart (the ABCD column is not implemented on most standard telephone sets).

tone from a *low group* of tones and one from a *high group* of tones, as indicated in Table 8–1.   These tones are usually generated by a digital programmable frequency divider that "divides" an input reference frequency from a 3.579. . .-MHz oscillator down to the desired tones (see the section on digital frequency synthesizers in Chapter 5).   At the switching-office end, the tones are detected by selective-filter or phase-locked-loop (PLL) DTMF decoders.

Standard Touch-Tone® telephones usually do not incorporate the A, B, C, and D digits and may or may not incorporate the * and # digits on their "dial." The inclusion of the DTMF tone oscillator into most of today's telephone sets not only permits its use for dialing, but also allows control of various other convenience telephone functions (such as call-forwarding, speed dialing, etc.) as well as providing a means that the caller may use to control various devices attached to a called telephone set (assuming that the called telephone is equipped with automatic answering equipment).

8.1.2 The Modern TDM-Switched Exchange

The stepping-switch exchange and the different types of crossbar or crosspoint exchanges discussed in the preceding section are normally classified as *space switches.* Essentially this means that the routing of the telephone signal is done by establishing an actual dedicated continuous analog channel for each signal routed through the exchange.   In the early systems, this channel was an actual physical two-wire circuit, but in later versions of the space-switched telephone system, a portion of the route taken by a particular channel might have consisted of a frequency-division-multi-plexed (FDM) channel on a superaudio or RF carrier, particularly if a portion of the route involved passage over trunk "circuits" between local switching offices. These FDM carrier circuits were mentioned briefly back in Section 4.8.   Whatever the exact mechanism for the switching and transmission, these space-switched systems were characterized by working with *continuous* analog signals.

In contrast, the next major step in telephone-switching development involved working with discrete-time (but still analog) signals.   The most important part of this technology area was the time-division-multiplexed common-bus switch or *time switch* as it is often called.   In this type of switch a wide-bandwidth common bus is time-division multiplexed in such a manner that a large number of analog TDM slots are available within each time frame.   If the TDM switch is designed for voice-bandwidth channels only, then since the Nyquist criterion requires that the sampling rate be twice the information bandwidth of the individual signals to be multiplexed onto the bus, we must have about 8,000 frames per second.   Since each slot can carry only one side of a two-way telephone conversation, we must have twice as many slots per frame as we have two-way telephone conversations to be switched if we are to have *nonblocking* quality of operation with a path for each one-way voice signal always available.

However, it is possible to take advantage of the statistical fact that in normal telephone conversations, each one-way signal is active, on the average, less than 50% of the time.   This permits us to assign the available time slots dynamically on an as-needed basis.   This results in being able to make better utilization of the available TDM slots, but may occasionally result in the nonavailability of a time slot to be assigned as soon as it is needed (i.e., as soon as someone begins a phrase of speech).   This usually results in the loss of the beginning syllable or two of the speech sequence and is referred to as *blocking*.   Since it usually does not occur too often and most of the time does seriously impair the conversations being carried on, a small amount of such blocking is usually tolerated in exchange for the higher utilization efficiency of the switching system.

This analog TDM method requires that the bus bandwidth be quite wide (in the Megahertz region) if the effective switching-matrix size is large enough to be usable.   In a typical small private branch exchange (PBX) system there will usually be 50 to 64 time slots per frame, giving a time-slot or pulse-repetition rate of 400,000 to 512,000 per second.   Since these are PAM analog pulses rather than binary PCM digital pulses, the accurate reproduction of the analog voice signal after demultiplexing the pulses and passing them through a low-pass filter requires that several orders of harmonics be carried with each pulse.   This is the reason for the high-bandwidth requirement on the switching bus.   The application of TDM switching techniques can also be applied to the switching of PCM binary digital signals, and this was the next direction that the electronic telephone system was to take (and still is taking).

However, before we get into a discussion of digital telephone electronics, it would be well to get a "bird's-eye" view of the structure of the typical public telephone network in general.   This is what we will do in the next section.

8.1.3 Telephone-Network Hierarchy

The nationwide telephone network in the past has been comprised primarily of one company operating in each geographic area with each such company interconnecting with other companies serving other areas according to a well-standardized hierarchy of interconnections so that the entire system operated very much as a single entity.   In recent years the advent of a multitude of competing companies, both large and small, in the same geographical area providing long-distance services via their own routes and interconnecting with the local telephone systems at many different points has caused the structure of the effective total telephone system to become much more complex.   For the purposes of this discussion, we will drop back to the era when things were somewhat less ambiguous and discuss the system that existed then.   The present-day system is just this system with many additional other-company routing possibilities superimposed upon it.

The Bell Telephone/AT&T System designated a number of different types

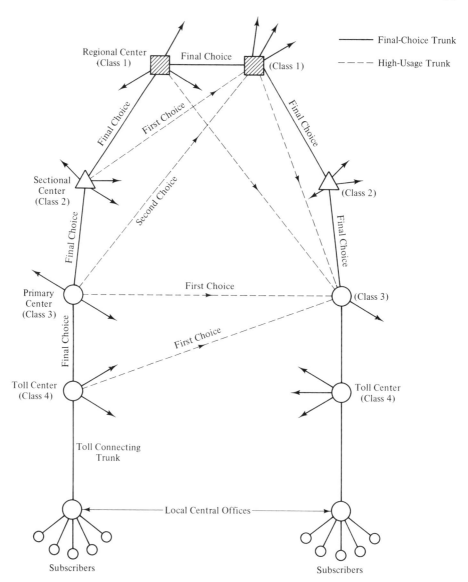

Figure 8-6 Telephone network switching hierarchy.

and classes of switching offices and centers. The local switching exchange (or Central Office) has already been discussed. In general, this is the one where the last three to five digits of the dialed telephone number are utilized to switch a call to the designated called telephone. The area covered by a Central Office may range from a few square miles in an urban area to over a hundred square miles in a rural area. A typical urban Central Office may have as many as 40,000 to 50,000 subscriber lines, with perhaps 5000 or so trunk lines going out to other switching offices.

The interconnection of these trunk lines (usually via underground cables, microwave relay systems, or optical fiber cables) among the Central Offices in a geographical area makes up what is known as the Exchange Area Network. A portion of this network is made up of high-usage direct-trunk connections between various pairs of Central Offices. These direct connections exist where there is a large amount of traffic between two particular offices. Otherwise the connection between two Central Offices is made through *Tandem Exchanges* (which are the "switchboards" for Central Offices much as the Central Office is the "switchboard" for individual telephone subscribers). Even where direct trunks are available between two Central Offices, there may be times when all these direct trunks lines are in use. At these times, the additional traffic is routed through other trunks to and from Tandem Exchange(s) to which the two Central Offices are also connected. Also a part of the Exchange Area Network are the Toll Centers, Primary Centers, and Sectional Centers. These are all interconnected with other similar exchanges at the same level and at the next-higher and the next-lower levels. Each such Exchange Area Network is connected to one or more Regional Centers as well as having interconnections at various levels with adjoining Exchange Area Networks.

Figure 8-6 shows an example of the types of interconnections that can occur. The switching or routing algorithm of the system is based upon making the connection from the calling to the called telephone via the shortest route possible at the lowest hierarchy level(s) possible, going up through the Sectional and Regional Centers only when no other route is available at a lower level or when all of the lower-levels routes are in use.

8.1.4 *Digital Techniques in the Voice Telephone Network*

The conversion of an audio signal into a PCM digital signal has already been mentioned in Chapter 4 (in connection with PCM modulation systems) and in Chapter 7 (in connection with digital audio recording systems). In the latter case, the object was to obtain a very high *S/N* ratio and a wide audio bandwidth to ensure sound reproduction with as high fidelity as possible. In the case of telephone systems, the main objective is to obtain only a "satisfactory" *S/N* ratio with an audio baseband bandwidth of only about 4 KHz or less. This can be done using a sampling rate of 8000 samples per second and about 8 bits per sample. The digitizing noise introduced by this small number of bits per sample can be reduced by companding the audio telephone signal level prior to sampling and digitizing it

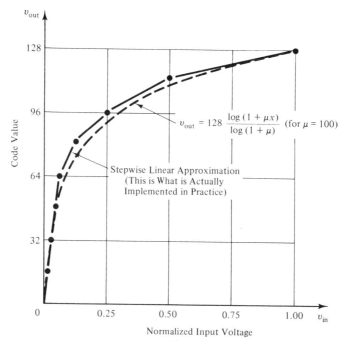

Figure 8-7 The μ-law companding curve.

(see Section 5.1.3).   The companding curve which is most commonly used in the United States is what is called the μ-*law companding curve* and is shown and explained in Figure 8-7.   At the present time the conversion from an analog signal to a PCM digital signal is usually done at the Central Office (if it is a digital switching office) or higher level, but in the not-too-distant future, it is planned that the digital signal will be brought directly to the subscriber's telephone set (see the discussion on the Integrated Services Digital Network in Chapter 11).

One of the primary reasons for converting the analog audio signal to PCM digital is the ease with which it may be switched.   The second reason is that there is almost no transmission noise introduced from the point in the system where the digitizing is done to the point where the PCM signal is converted back into an analog signal.

An example of the TDM digital switching exchange is AT&T's ESS-4 switch. It uses 8000 8-bit samples per second from a μ-law companded analog telephone signal as the digital signal to be switched.   The time frame of the ESS-4 switch has 128 8-bit "slots" and uses 120 of these for 120 input telephone signals (the other 8 × 8 bit positions are used for various other purposes).   The result is an 8.192-Mb/sec (8 × 128 × 8000/sec) PCM signal that can be sent via a wide-bandwith cable to time-multiplexed switches (called TMS switches) or time-slot-interchange switches (called TSI switches) where the entire group of 128 slots may be switched as a single entity or where switching may take place within the group of 128 slots, respectively.   Each new 8.192-Mb/sec PCM signal resulting from this combination of switching operations is then demultiplexed and converted back into 120 analog telephone signals with each analog signal going out on its appropriate line. In some cases the telephone signal is in digital form only while going through the switch at one geographical location and goes out on an analog trunk to the next switching office, but in many cases today the digital switching functions are spread out over many Central Office and/or higher-level switching exchanges so that the conversion of the PCM signal representing a particular analog input telephone signal back to an analog form may take place many miles or thousands of miles away from the point of initial digitalization.

For transmission purposes between exchanges, and in some cases even for transmission to or from a subscriber, PCM digital transmission techniques are now usually employed. One very common TDM/PCM transmission standard that is now used and that commonly can be made directly available to subscribers in most locations is the T1 digital "carrier" standard. This consists of 8000 193-bit frames per second (192 data bits and 1 sychronization bit). This can carry 24 telephone conversations using standard 8-bit PCM encoding schemes (or more using special data compaction or *concentrator* or *statistical multiplexing* equipment to do the digitizing and thereby accepting the possibility of the occurrence of some amount of blocking), or it can be used by the subscriber to send any type of 1.544 Mb/sec

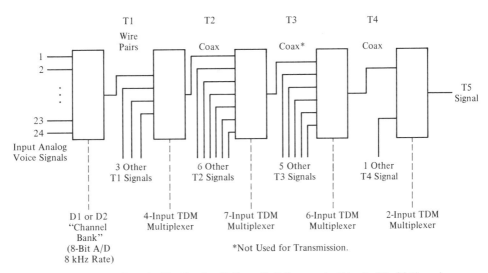

Signaling on T1 Done by "Stealing One Bit Every Sixth Frame or by Using 1 of the 24 Channels.

T1 Signals can also be 1.544-Mbps Data

T2 Signals can also be 6.312-Mbps Data or "Picturephone" Slow-Scan TV

T3 Signals can also be 44.736-Mbps Data or Commercial Quality TV

T4 Signals can also be 274.176-Mbps Data

T5 Signals can also be 560.160-Mbps Data

WT4 (Microwave) Consists of 58 T4 Signals (Total Bit Rate = 18,500 Mbps)

Figure 8-8   The AT&T svstem "T-carrier" organization.

PCM digital information. When the telephone company uses a T1 channel "in house" for transmitting independent telephone conversations, it either "steals" one bit from each of the 24 channels every sixth frame for control and signaling purposes, or else it uses one of the 24 digital channels for that purpose. Within the telephone system other digital transmission standards also exist.

For instance, 4 T1 channels can be TDM combined into a single T2 channel (at 6.312 Mbps), 7 T2 channels into one T3 (44.736-Mbps) channel, 6 T3 channels into one T4 (274.176-Mbps) channel, 2 T4 channels into 1 T5 (560.160-Mbps) channel, or 58 T4 channels into one 18,500-Mbps WT4 channel. The transmission medium normally used for each of these channels is twisted-wire-pair for T1, coax cable for T2 and T4, and microwave carrier for the WT4. The T3 is not usually considered a transmission standard and is used only for short distances between pieces of equipment within a switching plant. The T1 PCM digital signal can be generated in the standard manner from 24 analog telephone via what is called a "channel bank" (essentially a TDM A/D converter with companding). Figure 8-8 shows in diagram form the relationships among the various "T-carrier" transmission standards.

8.2 MOBILE COMMUNICATION SYSTEMS

By *mobile* communication systems we mean any type of receiver-transmitter combination that can be easily moved about, either by hand-carrying it or by moving it in some type of vehicle. The former small types are usually referred to as "hand-helds" or "handi-talkies" (or HTs for short), while the latter type usually infers some type of system either temporarily or permanently mounted in a vehicle and designed for use while the vehicle is in motion. Other types, such as "backpack" radios and "portable" communication stations that may be temporarily set up at some location (such as a disaster site or temporary construction or other activity site), might also be included within this definition, although we will usually be restricting our definition to the hand-held or in-motion vehicle-mounted types. The types of mobile systems that are normally considered are usually classed as land mobile (in land vehicles), aeronautical mobile (airplanes, balloons, blimps, etc.) and maritime mobile (boats or ships of any size). Hand-held transmitters are usually classed in with the land-mobile group. This is the group that we shall consider first.

8.2.1 Land-Mobile Radiotelephone Systems

For a long time, the largest groups of land-mobile users were the public safety forces (police and fire departments, etc.). As the equipment became more reliable and cheaper, many public service agencies, such as taxicab operations, utility companies, industrial firms, construction and repair companies, railroads and trucking companies, and other types of land transportation operations began utilizing land-mobile radio systems. Almost all these land-mobile systems operate in the VHF or UHF frequency region from 25 MHz up through and slightly above 1 GHz.

Although standard AM was used as the modulation method in some of the early systems, the standard for the present-day land-mobile systems is narrow-band (having a maximum FM modulation index M_F of about 1 to 2) FM modulation.   The main reason for using frequencies in the VHF and UHF regions for land-mobile units is the ability to utilize fairly small-sized antenna structures and still obtain good antenna gain and efficiency.   Because of the limitation on generating sufficient RF output power at the higher frequencies, most of the early mobile systems operated in the low-VHF region from about 25 to 60 MHz.   Automobiles and trucks carrying these types of systems could be easily identified by the long (three- to 8-foot) "whip" antennas attached to the vehicle.

Most of these early systems were single-frequency half-duplex systems (but referred to in mobile-radio terminology as *simplex* systems) such as is shown in Figure 8-9a.   The simplex system works fine as long as all the users of the single frequency are fairly close to each other (less than several miles apart for reliable communications between vehicles on level terrain and operating in the lower part

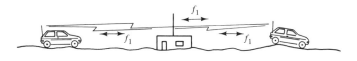

(a) Local Simplex System

(b) Dispatched Simplex System (Mobile "A" may not
be able to Hear "B" and Vice Versa)

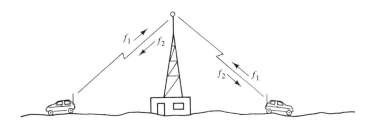

(c) Repeatered Mobile System

Figure 8-9   Different types of land-mobile radio systems.

of the VHF region). The range of this elementary type of simplex system becomes smaller as higher VHF frequencies or frequencies in the UHF region are used. This is due to the basic "line-of-sight" propagation limitations at these higher frequencies. One solution to this which is often employed when the mode of communication basically consists of messages between a central dispatcher and one or all of the mobile units is shown in Figure 8–9b.

The fixed-location dispatching station is usually characterized by having a high-gain antenna located on a high point of land and/or mounted on the top of a high tower so that it is essentially line-of-sight to any of the mobile units. Usually the central fixed station also has considerably higher transmitter power and better receiver sensitivity than do the mobile-unit transceivers. The operational protocol which is usually observed by the mobile units is that of not transmitting a message until they are either queried by the dispatcher or until they request and obtain permission to send a message by first transmitting a very short identification message (usually just their unit number). This protocol is necessary because many times a particular mobile unit may not be able to hear another mobile unit that is already transmitting to the dispatcher. One of the problems with this type of simplex system is that often the dispatcher must serve as the relay for any message that is to go from one mobile to another.

Some of the problems associated with the simplex system may be eliminated by going to a some form of a *repeatered* system such as is shown in Figure 8-9c. In this case, the mobile units all transmit to the centrally located repeater site via one *uplink* frequency which the repeater receives, processes, and retransmits on another frequency which is received by all the other mobile units. The repeater may be either dispatched or undispatched. In the former case the mode of operation is merely one mobile unit talking to all of the others, while in the latter case the central dispatcher can "grab" control of the repeater transmitter to talk to any or all of the mobile units. Even if the repeater itself is not wired for dispatching, one of the mobile units can function as the dispatcher for the system. A number of optional features may be included into this type of repeater unit.

One of the most popular ones is the use of subaudible (low in both frequency— around 100 to 150 Hz—and signal level compared to the voice signal level) to "open" the repeater and permit it to receive and retransmit an incoming signal. This feature prevents the repeater transmitter from starting up on incoming noise bursts or weak signals from other more distant systems operating on ("sharing") the same pair of frequencies. Another related feature is the use of transmitted subaudible tones or digital codes to control the squelch on the mobile receivers and permit the dispatcher to talk to only one or some subgroup of the mobile receivers at a time. In some cases this can even permit two or more groups of light-traffic users to share the same repeater system and pair of frequencies (a lockout is usually employed to prevent any transmissions from a mobile unit until the system is not in use by some other unit or by one of the dispatchers).

At the present time the frequency bands which are most commonly used for mobile systems of the types just described are 25 to 50 MHz, 150 to 162 MHz, 450 to 470 MHz, and in the vicinity of 800 MHz. Radio amateurs also conduct similar types of mobile operation in the frequency ranges around 28 MHz, 147 MHz, 220

MHz, 420 MHz, and upwards, and CB mobile units operate in the region around 27 MHz and 48 MHz (CB is one of the few mobile domestic services that uses either AM or SSB in place of FM). The military services also use frequencies in the VHF and UHF region for some of their mobile communication systems. The mobile radio telephones that are a part of the public telephone system are an important part of the land-mobile type of user and are destined to become even more important in the next few years. This type of service will be discussed in Section 8.2.3.

8.2.2 Marine and Aeronautical Mobile Systems

Maritime radio was historically one of the very first practical applications of radio, dating back to the second decade of this century. The early shipboard radios were spark-gap radiotelegraph transmitters operating in the LF and MF regions and intended for long-distance ship-to-ship and ship-to-shore communications. The ships that were large enough to afford a radio operator and equipment were also large enough to permit fairly extensive low-frequency antenna systems. As technology improved, the mode of operation became one of using CW radiotelegraph and later radiotelephone transmitters in the LF, MF, and HF regions for ship-to-shore and ship-to-ship communications over long distances. The maritime radiotelephone systems first used AM modulation, later switching to SSB type of modulation. As congestion in the major harbors and canals of the world became more acute and the need for radio communications over relatively short distances became apparent, the VHF region also became a part of the maritime–mobile radio technology. Maritime radiotelephone and data communications via geostationary communication-relay satellites are also becoming very important. This will be discussed further in Chapter 9.

Aeronautical radio first started out using the LF region below 500 KHz for a combination of radio communications between aircraft and the ground and for basic radionavigational purposes. The navigational functions were fulfilled either by means of using direction-finding equipment on the aircraft to determine the relative bearing of the transmitting station from the aircraft or else by the pilot following along one of the four legs of a low-frequency *Adcock* radio beam (the pilot, listening in his headphones would hear a continuous tone as long as he was flying in the center of the beam, but would hear the Morse code for the letter "A" if he got off to one side of the beam or the letter "N" if he got off to the other side). Voice transmissions were sometimes superimposed upon these continuously transmitted navigational signals (either the *nondirectional beacon* transmissions or on the signals of the four-legged *radio-range* transmitters). Airport control towers at the busier airports also had frequencies assigned to them for direct on-demand transmission to aircraft in the vicinity. The air-to-ground frequencies were relatively few in number initially since few aircraft had transmitters, and those that did had transmitters that could transmit on only a small number of separate frequencies.

Following World War II the aircraft navigation and communication systems began the transition to the VHF region, with the basic "backbone" of the navigational structure being based upon VHF *omnirange* stations that transmit in the VHF aeronautical navigational band from 108 to 118 MHz. These stations provide the pilot with information on his bearing relative to the range station rather than relative to his aircraft and thus permit a much more flexible navigational route structure. They are complemented by *distance measuring equipment* (DME) operating in the UHF region which gives the slant range from the station to the aircraft and by VHF/UHF *instrument landing systems* (ILS). The *localizer* portion of these ILS systems gives the pilot lateral steering information for "lining up" with the centerline of the runway and operates in the lower portion of the VHF navigation band while the *glide-slope* portion that helps the pilot to obtain a smooth vertical descent to within 100 to 300 feet above the runway surface operates in the UHF region. Although these navigational transmitters have the capability of having audio transmitted along with the navigational system modulation, in most cases this audio either consists of Morse code or voice station identification or continuous recorded weather and/or airport information data. The actual controller-pilot communications are usually conducted in the aeronautical voice band from 118 MHz to 136 MHz, usually using simplex (one-frequency half-duplex) channels with a 25-KHz channel spacing and employing AM modulation.

Other HF and VHF frequency bands are employed for air-traffic-control purposes on long-distance (transcontinental flights) and/or for communications of airline aircraft directly with company dispatchers. Some of these systems use SSB and some use NBFM modulation systems. Another type of system which is used for transoceanic flights is aircraft-ground links via geostationary communication-relay satellites. Other aeronautical electronic systems include satellite-based wide-coverage-area navigational systems, various types of *hyperbolic* large-area navigational systems using transmissions in the LF and VLF regions, operational and proposed aeronautical anticollision systems, air-traffic-control primary and secondary (beacon) radar systems, weather radar, ground-to-air digital data links, and automated touch-down systems. All these would be interesting topics for discussion, but that would require several textbooks; all we can do here is mention that they do exist.

8.2.3 Public Radiotelephone Systems

As radiotelephones became more prevalent in the public service and business areas, the public (especially the more affluent and influential members of the public) also desired to have this convenience at their disposal. This was provided by the public telephone companies setting up a radiotelephone service in the larger cities that could be interfaced with the public dial telephone network. Unfortunately this interface was at first not very direct. The services of an operator were required either for placing a call from the mobile unit or for the mobile unit to receive a call from somewhere in the telephone network. In addition, a person who wanted to call a mobile unit had to know that it *was* a mobile radiotelephone that was being called and in most cases had to first know the operator's number to be dialed and then had to be able to give the mobile operator the code number of the particular mobile unit to be called.

Although the system was somewhat automated over a period of years, it was still based upon the conventional mobile-radio scheme of having a powerful central transmitting site covering an entire metropolitan area with a relatively small number of available two-way radiotelephone channels. This led to much congestion and was very costly to the user. This problem is now in the process of being solved by a new approach to wide-coverage-area mobile radiotelephone systems called *cellular radio*. This approach appears to be the "coming" system of the future, not only for public radiotelephone systems, but also for certain wide-area private mobile-radio systems.

The basic concept of the cellular mobile system is to divide the service area to be covered up into *cells* of a few dozen to several hundred square miles each. Figure 8-10 shows such a service area divided into nine cells. Each cell in this example is assigned 60 distinct *voice-frequency pairs*. Each of these is actually two paired FM radio frequencies, one for the path from the mobile unit to the cell receiver and the other for the outlink from the cell transmitter to the mobile unit. For simplicity, we shall assign each pair only one designation (i.e., f_1 actually means a *pair* of carrier frequencies). It will be noted in Figure 8-10 that adjacent or nearby cells always are assigned different groups of frequency pairs, while cells which are located some distance from each other may use the same set of frequency pairs. This concept of *frequency reuse* is one of the features of the cellular concept which increases its ability to handle a large number of simultaneous conversations within a given service area. The size and shape of each of the cells will depend upon the geographical (primarily the topological) characteristics of the area as well as upon the transmitter power and the location and height of the antenna at each cell's transmitter/receiver site. These parameters are adjusted to obtain cell sizes and shapes that are consistent with the projected average density of mobile users which will be within that cell any given time. By keeping the cell site's transmitter power low and its antenna height low, the size of each cell can be made smaller, and thus the same set of frequency pairs can be reused more times within the geographical service area. It will be noted in Figure 8-10 that the boundaries between cells are shown as being fuzzy rather than as distinct sharp dividing lines. This is due to the fact that propagation conditions change from time to time and also because the antenna installations on most mobile units do not have perfectly omnidirectional radiation patterns in the horizontal plane.

A major concept of the cellular system is that mobile units should be allowed to move around freely within the service area without any action being taken on the part of the mobile radio user as the unit moves from one cell to another. This latter feature is implemented by means of an automated *hand-off* scheme that involves having receivers in adjacent cells listening to the frequency pairs involved in a given (initial operation) cell. The system uses these receivers to make continuous comparisons of the signal strength being received from a mobile unit by the cell receiver in the initial cell and the signals strengths received from that same mobile unit at the cell sites of all the cells adjacent to the initial cell. Whenever these comparisons indicate that more reliable communication with the mobile unit can be effected via the transmitter/receiver site of one of the adjacent cells, the mobile unit is *handed off* to the new cell. This hand-off operation involves first

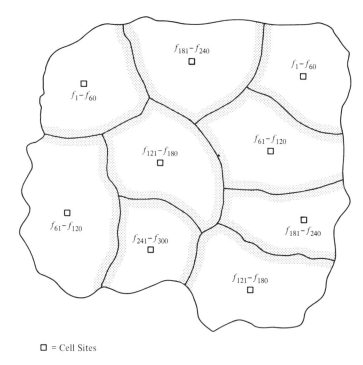

□ = Cell Sites

Figure 8-10   Six-cell cellular mobile system utilizing omnidirectional antennas.

determining if the new cell has an available unused frequency pair to assign and, if this is the case, sending instructions to the mobile unit to switch its operating frequencies to that new pair. If the system is operating properly, this hand-off operation should take place without any interruption in the service and without even being noticed by either party involved in the conversation with the mobile unit that is handed off.

The coordination of all this within a given service area is handled by what is termed a *Mobile Telecommunications Switching Office* (or MTSO as it is shown in Figure 8-11). The MTSO combines the features of a telephone switching exchange with additional computer equipment to process the data (about signal strengths, the status of all active and standby mobile units, available frequency pairs at the site, etc.) received from each cell site to determine which cell site and which frequency pair to use to communicate with a given mobile unit. It then sends the necessary control signals both to the cell site and to the mobile unit involved.

The cellular system operates in the 800-MHz region and therefore is very nearly optical line of sight (if intervening nonscreening objects are neglected). It utilizes 213 pairs of FM frequencies carrying voice conversations plus a number of control frequencies in each of two bands. Each pair of voice-channel frequencies occupies 60 KHz of spectral space. The few frequencies that are assigned for control purposes are shared among many mobile units in such a way as to permit up to about 100,000 individual mobile units to be controlled by a single group of

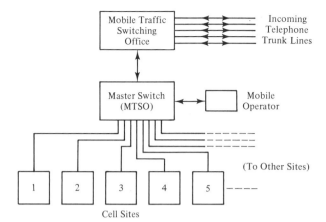

Figure 8-11   Cellular mobile switching organization.

these frequencies called a *Forward Control Channel*.   The frequencies in this channel are often termed the *setup* frequencies.   Those that go from the mobile units to the system are used to indicate to the system the status of each mobile unit (whether it is turned on, whether it is "off hook," etc.), while those from the system back to the mobile units are used to indicate whether a frequency pair can be assigned for use and if so what pair, and so on.   The initial signals that occur when power is applied to the mobile unit as well as signals which are sent occasionally after that are used by the system to identify the mobile unit and to determine in which cell it is currently located.

Figure 8-10 shows a system where all the frequency pairs assigned to a single cell are used for the entire cell.   This is usually characterized by having the cell site be equipped with an omnidirectional antenna and located near the center of the cell.   This gives the most flexibility in the use of the assigned frequency pairs since, even if all the users happen at some point in time to congregate near one spot in the cell, all the assigned frequency pairs are still available for use.   However, the use of an omnidirectional antenna at the cell site restricts the reuse of sets of frequency pairs within the systems service area.   By dividing each cell into 90-degree or 120-degree sectors and assigning one-fourth or one-third of the set of frequency pairs associated with the cell to each sector, it is possible to increase the number of times a frequency pair is reused in the total system and still maintain the FCC-specified 17-dB carrier-to-interference (CI) ratio.   This will thus increase the total number of mobile units that may be serviced simultaneously as shown in Figure 8-12.

The penalty that is paid for this increase in total system capacity is that the flexibility of the system is reduced (i.e., now only a fourth or a third of the users in a given cell may congregate at one point in the cell).   This says that the system will tolerate only smaller deviations from the statistical average distribution of users within the service area than will the system that employs omnidirectional cell-site antennas.   Since, whenever a channel is not available for immediate use, the grade of service deteriorates from an ideal of 99% availability to something less (perhaps as low as 90% in the worst case), the design of the system and the decisions as to

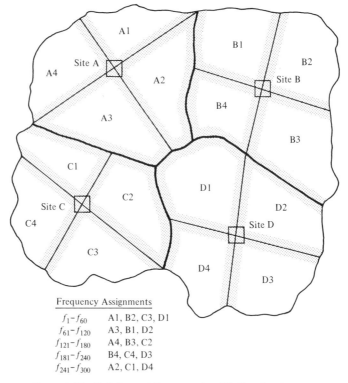

Frequency Assignments

f_1–f_{60}	A1, B2, C3, D1
f_{61}–f_{120}	A3, B1, D2
f_{121}–f_{180}	A4, B3, C2
f_{181}–f_{240}	B4, C4, D3
f_{241}–f_{300}	A2, C1, D4

Figure 8-12   Cellular mobile area using 90° directional antennas.

cell size, the use of omnidirectional versus directional antennas, and so on, need to include a careful study of the projected statistics of the user demands to be placed upn the system.   This allows projections to be made of the degree of availability (i.e., no *system* "busy" signals), the reliability (lack of conversations being interrupted as the mobile unit moves about), and the system frequency-utilization efficiency for different proposed system configurations.

It might be well in summarizing the concept of the cellular land-mobile radiotelephone system to list some of its major characteristics and design goals.   First, we list the design goals: the cellular mobile radiotelephone system should basically be just an extension of the public dialed telephone network so that the mobile user can be called from any telephone simply by the caller dialing the telephone number assigned to the mobile unit.   The only difference desirable here might be a different "busy" signal if the mobile unit is not turned on or is out of the service area.   The operation of the mobile unit itself should be essentially the same as that of a standard wired telephone and should provide most of the features of a standard telephone. The system must be designed for continuous growth of the system within the geographical service area without service disruptions or the replacement of user equipment.   The level of availability and reliability should be comparable with that obtainable from a standard wired telephone.

Second, we list those characteristics of the cellular system that are different from the characteristics of the mobile radiotelephone systems of the past: the past mobile systems depended upon coverage of the entire service area by means of a single high-powered base or repeater station with an antenna located at a high elevation. In contrast, the cellular system utilizes a number of cell sites, each of which has a low-powered transmitter and an antenna located at a much lower height (this, incidentally, makes cell sites much easier to locate and install than conventional radiotelephone base stations) and involves the reuse of the available set of frequencies and automatic switching of a single conversation from cell to cell as the mobile unit moves about the service area covered by the system. This latter concept makes the system easily expandable by gradually making the cells smaller and smaller (the cell sites less and less "powerful") until a limit (with the present number of allotted channels) of about 300 subscribers per square mile is reached.

PROBLEMS

8-1. What are the differences and similarities among
   a. A simplex mobile communication system?
   b. A duplex/base mobile commuication system?
   c. A duplex/repeater mobile communication system?
   d. A cellular mobile communication system?

8-2. What is a subaudio-tone-addressable communication system and how does it work?

8-3. Explain the differences and similarities among
   a. An analog sampled-voice TDM communication link (radio, wire, or cable) using PAM or PDM modulation.
   b. A digitized sampled-voice TDM communication link using PCM.
   c. A frequency-division-multiplexed analog voice communication system.
   Discuss such items as the spectral space required, noise immunity over long distances, equipment complexity, and so on.

8-4. How can the statistical properties of a normal telephone conversation be utilized to permit increasing the channel capacity in a TDM link carrying many telephone conversations simultaneously?

8-5. What is a *vocoder?* How does a vocoder operate? How does a vocoder differ from a *codec* (analog-to-digital/digital-to-analog conversion device used in digital telephone systems)? What might be the advantages and disadvantages of using one type of device over the other?

8-6. What does the *compander/expander* portion of a codec do? Why is it used (for what purpose)?

8-7. What is meant by *step-by-step* versus *stored-program* telephone switching?

8-8. Differentiate between *space-switching* and *time-switching* in an analog telephone switching office.

9

SPACE COMMUNICATIONS AND RADIO ASTRONOMY

9.1 SPACE COMMUNICATIONS

In general, the topic of space communications can be broken into two main areas: (1) the sending of information and data back from a space vehicle and the sending of control commands to it and (2) the use of a satellite as a relay station for sending information between two points on the earth. We will consider both these situations in this chapter, although the discussion of encoding the information sent back from a space vehicle (telemetry techniques) will be left to Chapter 12.

9.1.1 Space Vehicles and Orbits

When discussing a space communications system, it becomes necessary for the communications systems engineer to become at least somewhat acquainted with the basics of orbital mechanics and vehicle configurations. In general, we may divide vehicle trajectories into four basic types:

    1. The first encompasses the various low-altitude orbits. These are the orbits

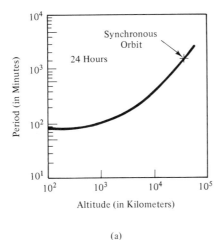

Figure 9-1(a)   Satellite period versus altitude above the earth (for a circular orbit).

(a)

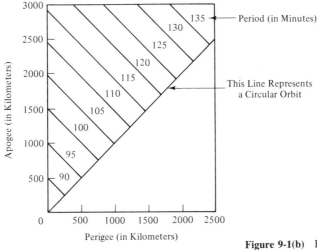

(b)

Figure 9-1(b)   Period versus altitudes for low-level elliptical orbits.

which are followed by satellites which are in constant motion with respect to the earth's surface.   Figure 9-1 shows the relationship between the orbital period of a satellite and its altitudes above the earth's surface.   Such orbits are widely employed with satellites designed for some type of survey of the earth's surface (weather surveillance, earth resources determination, military intelligence, etc.). Because of the constant motion of the satellite with respect to the surface of the earth, it may be necessary to have several ground terminals if the satellite is to be communicated with at all times, and each station must have an antenna system that can be rapidly slewed to track the satellite as long as it remains in view of that station. Conversely, some type of data storage may be employed within the satellite so that the data acquired during one or several orbits may be "dumped" at a rapid rate whenever the satellite is within view of a particular ground station.

2. The second type of trajectory is what is called the *stationary* or *synchronous* orbit. Here we have an orbit whose altitude above the earth's *surface* is approximately 22,000 statute miles (19,100 nautical miles). The stationary orbit is very popular for communication-relay satellites because the satellite is always in view of the ground stations the system is designed to work with, and elaborate ground station antenna-tracking systems are not needed. However, even though the orbit is called a "stationary" orbit, the satellite is never quite at rest with respect to the earth's surface. Even if we would put a satellite into synchronous orbit and bring it to a complete stop with reference to the earth's surface, we would soon find that it would begin to move again due to various disturbing forces (the nonuniformity of the earth's gravitational field; the gravitational fields of the sun, moon, other planets; etc.). This motion would be of two types, a periodic rocking motion (which causes the satellite to move in a figure-8 pattern centered on the earth's equator) and a slow drift along the equator to a spot of low gravitational potential over the Indian Ocean (the graveyard of synchronous satellites which have outlived their usefulness). However, because these forces disturbing the synchronous orbit are very weak, it is possible to largely offset them by including in the satellite system a station-keeping system which can be controlled from the ground. This may take the form of small valved gas jets, or it may consist of electric current coils which interact with the earth's magnetic field.

3. The third type of trajectory is the deep space probe. This trajectory may consist of an orbit where the perigee (low point) of the orbit is close in to the earth but where the apogee (high point) of the orbit may be quite distant from the earth. This produces a very elliptical orbit during which the satellite periodically comes near the earth and then disappears for quite some time into outer reaches of the solar system. Quite often a satellite of this type gathers information while away from the earth, stores it on magnetic tape or some other storage medium, and then "dumps" this information through its communication channel when near the earth. Similar types of orbits may also be obtained with the sun, moon, or some other solar system body as the body being orbited. In this case, and in the case of space probes which are not put into orbit, but which are put on trajectories that for all practical purposes takes them away from the earth forever, the communications system which sends the information back to the earth station must be designed to operate over extremely long distances.

4. The fourth basic type of orbit is a local orbit about the moon or one of the planets in the solar system. In this type of orbit, the satellite would be "visible" to an earth station only when the body about which it is orbiting is visible and when the satellite is on the side of that body which faces the earth.

No matter which type of satellite orbit is being discussed, the motion of any satellite about a much more massive body is described by Kepler's laws of orbital mechanics,[1] which stated briefly in words are:

I. The satellite motion is an ellipse with the large body at one focus.

II. The line between the center of the large body and the satellite cuts equal areas of the orbital plane in equal time periods.

III. The square of the period of revolution is proportional to the cube of the time-averaged distance from the center of the large body to the satellite.

These laws all assume that the mass of the large body is uniformly distributed (not quite true for the earth) and neglect the effects of other bodies or satellites.

In addition to the particular trajectory which the space vehicle is following, there is the problem of the orientation of the vehicle or its antenna with respect to the direction of the propagation path down to the earth. In the case of orbiting satellites, the satellite is usually designed to present one side constantly to the earth. This permits the use of relatively high-gain antennas on the satellite, although it does create extra problems of vehicle control and orientational stability because of the narrow beamwidth of the antenna(s). Quite often orientational stability is achieved in part by spinning the satellite like a top so that the gyroscopic action tends to stabilize the angular orientation of the satellite. In the case of surveillance satellites, some type of stabilization is also necessary for the operation of the surveillance devices on the satellite as well as for the operation of the communications system, and in some cases the orientation that is desirable for the surveillance devices is not the optimum orientation for the communication system antennas.

9.1.2 Ground-Station Antenna Tracking

In the case of most space communication systems, some form of ground-station antenna tracking is necessary. For satellites in synchronous orbit, the tracking requirements may be slight enough that any necessary adjustments in the antenna pointing direction at the ground station can be done by manual control. In some instances the ground station is built with the antenna not movable, and the adjustments to the system are made by keeping the satellite in the beam of the antenna rather than turning the antenna beam toward the position of the satellite.

For nonsynchronous satellites, space probes, and so on, more elaborate tracking systems are needed. One approach is to use computer-controlled antenna mounts which point the antenna in the direction where the satellite is *calculated* to be at a certain time in its orbit. This system, called programmed tracking, can be quite effective if the satellite orbital parameters are known accurately enough. A second type of antenna-tracking system known as monopulse tracking has been discussed in Chapter 6. Often this form of tracking system is used as the fine

---

[1]Johann Kepler (1571–1630), a German astronomer.

correction to a programmed-tracking system, especially in cases where the orbital parameters of the satellite are not known to a high degree of accuracy.

9.2 COMMUNICATION SATELLITES

9.2.1 Communication Satellite Transponders

In the case of signals which are sent one-way to or from a satellite, the normal forms of transmitters and receivers are used, the major difference between the space and the ground terminals being in the ruggedness, the size and weight, the transmitter power, the antenna gain, and the efficiency of the devices used aboard a space vehicle versus those used at the ground terminal. However, in the case of a communications-relay satellite, some special pieces of equipment have been developed which are called *communication satellite transponders*. The purpose of these devices is to receive a signal from the originating earth station, process that signal (usually without detecting it) at the satellite, and retransmit it on a different frequency to the receiving earth stations.

    Figure 9-2 shows the block diagram of a typical communications transponder. Here we are assuming that the uplink frequency is greater than the downlink frequency. A single oscillator with appropriate frequency multipliers is often used to supply the local oscillator signals for both the uplink and downlink conversion mixers, not just to save weight, reduce power consumption, and increase reliability, but also to reduce the effects of any drift in the oscillator frequency. This latter effect can be seen if we let the oscillator frequency be such that the relationship between the uplink and the downlink frequencies is given by $f_{DOWN} = f_{UP} + (M2 - M1) f_{LO}$ in Figure 9-2. This is the case if we let the relationship among the uplink, downlink, IF, and LO frequencies be given by either

$$f_{DOWN} = f_{M2} - f_{IF} = f_{M2} - (f_{M1} - f_{UP}) \tag{9-1a}$$

or

$$f_{DOWN} = f_{M2} + f_{IF} = f_{M2} + (f_{UP} - f_{M1}) \tag{9-1b}$$

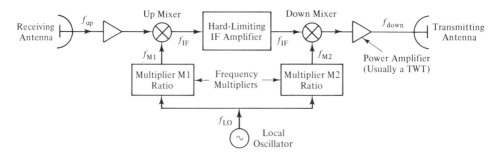

Figure 9-2   A communication satellite transponder.

In these cases a change in f_{LO} will change the output frequency only by the amount of (M2 − M1) times the change in f_{LO}. The other possible conversion schemes (Equations 9-2a and 9-2b) which produce an output frequency given by $(f_{M1} + f_{M2}) - f_{UP}$ will produce a frequency change in the output of (M1 + M2) times the frequency change in the local oscillator (which would be quite undesirable).

Another frequency factor which must sometimes be considered in choosing frequencies is the effect of Doppler shift due to the relative motion between the satellite and the ground station. If we let Δf_{UP} be the uplink Doppler shift and Δf_{DOWN} be the downlink shift, it can easily be shown that both of these have the same sign at the same time (positive for satellite moving toward the earth station and negative when it is moving away). If we look at Equations 9-1a and 9-1b, we see that Δf_{UP} is passed through without a change in sign (because f_{UP} has a + sign) and thus the uplink and downlink Doppler shift effects would be additive. However, if we use

$$f_{DOWN} = f_{M2} - f_{IF} = f_{M2} - (f_{UP} - f_{M1}) \tag{9-2a}$$

or

$$f_{DOWN} = f_{M2} + f_{IF} = f_{M2} + (f_{M1} - f_{UP}) \tag{9-2b}$$

we find that the sign of Δf_{UP} is reversed when it appears as part of the output frequency, and thus Δf_{UP} and Δf_{DOWN} would partially cancel (but then the effects of local oscillator frequency drift would be more severe since the factor (M1 + M2) Δf_{LO} would be present rather than $|(M1 - M2)| \Delta f_{LO}$).

If the uplink and downlink frequencies are already specified (by frequency allocations or other legal requirements), all the foregoing factors must be considered in selecting the IF frequency, the local oscillator frequency, and the multiplier ratios M1 and M2. For a stationary satellite, either Equation 9-1a or 9-1b may be used; the multiplier ratios M1 and M2 should be such that they can be formed from the products of multiplying the factors 2 and 3 together (representing frequency doubler or tripler stages), and the IF frequency should be such that image problems are minimized and that the stage can be conveniently constructed with the desired bandwidth by the use of conventional design.

Figure 9-2 indicates that a limiter amplifier is used in the IF part of the transponder. This is usually a "hard" or clipper-type limiter. The result of hard or clipper-type limiting is that there is some reduction of the noise on an RF signal passing through this type of limiter due to the removal of amplitude variations on the signal (i.e., the signal "captures" the limiter—see Chapter 4). This is essentially the same thing that happens in a limiter stage which precedes the FM discriminator in an FM receiver. If the signal-to-noise ratio of the signal into the limiter is already good (approximately 10 dB or better), it is much improved at the output. However, if the input signal-to-noise ratio is not already good, little or no improvement is provided by the limiter amplifier. Once a satellite is put into position, the RF level needed to be transmitted to provide good limiter action can be confirmed experimentally by transmitting a carrier signal at various levels and recording the signal received back from the satellite. Eventually a point will be found where the received signal power levels off and does not increase with increasing uplink transmitter power. This is the limiter saturation point in the satellite, and for good performance, the transmitter power level of a communications signal fed through the satellite should be somewhat higher than the carrier power which produced this saturation.

Figure 9-3 shows the saturation curve of a typical satellite as measured from the ground. The abscissa is the product of the ground station transmitting power and transmitting antenna gain (the *effective isotropic radiated power*, *EIRP*), and the ordinate is the received power from the satellite divided by the effective aperture area of the ground receiving antenna (the received power density, ρ_R). Another reason that a hard limiter in the satellite transponder is used is that there is usually time-variable attenuation on the uplink between the earth and the satellite which is due to the motion of the satellite, movement of the ground-station antenna, or atmospheric "scintillation" effects. This introduces amplitude modulation onto the signal received at the satellite and the hard limiter in the satellite effectively removes this AM modulation from the signal. Of course, it is obvious from this discussion that AM cannot be used as the means of modulating intelligence on the signal to and from the satellite, so some form of angle or pulse-time modulation must be used.

Returning to Figure 9-3, when no uplink signal is transmitted, the transponder IF and output stages amplify the noise which is received by the satellite's receiving antenna and that generated in its receiving RF amplifier and mixer stages. As the level of the uplink signal increases, this noise output drops and is replaced by the amplified received uplink signal, until the AM noise power level in ρ_R is primarily due to noise power introduced on the downlink signal. However, there *will* be some *phase noise* on the downlink signal transmitted by the satellite which does result from the AM noise power which goes into the transponder's hard limiter along with the receiver uplink signal. This phase noise will ultimately contribute to noise on the received signal *after* it is FM or PM detected at the receiving earth station.

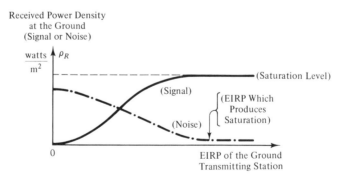

Figure 9-3 Characteristics of a hard limiting communications satellite.

9.2.2 Communication Satellite Design

At the present time almost all communication-relay satellites are being designed for geostationary synchronous orbits. Besides the obvious advantage of having the satellite always available to a given ground station and at the same approximate location in the sky as viewed from that station, there are other advantages to this type of orbit: (1) the satellite is almost always in the sunlight, which reduces the problems of temperature variations and their effects as well as meaning that there is almost always sunlight available for the production of the needed electric power from the satellite's solar cells (the geostationary satellite is shadowed by the earth for only short daily periods for a few days before and after the spring and fall equinoxes), (2) the satellite is well above the earth's surface, and thus is not affected by atmospheric drag, and (3) the satellite is above the Van Allen radiation belts, and thus is shielded from much of the variation in the earth's magnetic field as well as not being greatly affected by the intense particle radiation of the lower altitudes where the Van Allen belts exist.

The desirability of using a geostationary orbit leads to a problem of crowding more and more communication satellites together in the same orbit over the earth's equator. Placing satellites too close together in longitude in this orbit means that if the same frequencies are used by nearby satellites, the uplink stations on the earth must transmit very narrow beams so as to hit only the desired satellite. Otherwise, different uplink signal frequencies must be employed. Similarly, different downlink frequencies must be employed or else the satellite's antenna must form a narrow beam or beams that are directed to the point(s) on the earth where the receiving station(s) for that satellite are to be located. This also implies that the receiving stations for different satellites using the same downlink frequencies must be widely separated geographically. Also, it must be assured that the downlink transmissions from a satellite will not interfere with any terrestrial communication or navigational systems using that same frequency.

One way to get greater utilization of the available frequency spectrum under these conditions is by the use of polarization diversity: the use of different electromagnetic polarizations for different channels transmitted on the same or nearly the same carrier frequency(ies). For example, one of the downlink channels from a satellite might transmit its signal using right-hand circular polarization, and another channel from the same satellite on the same RF frequency might transmit its signal using left-hand circular polarization. Because of the coupling from the other polarization, there is bound to be some crosstalk effects appearing on each of these two RF signals after each of them is "picked out" of the receiving antenna dish by either an RHCP or a LHCP feed ("pickup"), respectively. However, the fact that the signals are angle modulated allows the crosstalk effect to be virtually eliminated by passing each signal through a hard-limiting amplifier at the IF frequency in the receiver prior to demodulating it. This technique is quite effective as long as the opposite-polarity RF (of IF) signal level is 10 dB or less than the level of the desired-polarity signal.

Referring back to Chapter 6, we remember that for given antenna aperture areas for the transmitting and receiving antennas, the received power level varies as a function of frequency squared. Thus, considering this factor alone, it would seem desirable to go to as high an operating frequency as possible for the RF links to and from the satellite. However, we find that as the frequency increases, the attenuation from the atmosphere (including rain, fog, etc.) also increases, as does also the noise temperature due to these sources. When all these different factors are considered, a frequency of approximately 3.7 GHz is about the optimum, with the less than 20-dB attenuation "window" extending from about 400 MHz up to about 11 GHz under adverse weather conditions (heavy clouds and rain) and low (approximately 15 degrees) elevation angles of the satellite as viewed from the ground site. With better weather conditions and higher elevation angles, this window extends upward somewhat above the 11-GHz limit.

All the factors mentioned in the preceding two paragraphs must be considered in assigning or licensing the satellite uplink and downlink frequencies, the ground-station siting locations, and the required transmitting antenna patterns of both the ground station and the satellite. Since this is a problem of international magnitude, it is handled by committees of the *World Administrative Radio Conference (WARC)* of the *International Telecommunications Union (ITU)*. Satellite frequency-band assignments that have been made thus far include the regions from about 3.7 to 4.3 GHz (domestic downlink), 5.7 to 6.2 GHz (domestic uplink), 7.2 to 7.8 GHz (military downlink), 7.9 to 8.5 GHz (military uplink), most of the region from 11.7 to 14.8 GHz (various up- and downlinks), 18 to 21 GHz (downlinks), 27 to 30 GHz (uplinks), 31 to 33 GHz (satellite-to-satellite "crosslinks"), 37 to 40 GHz (downlinks), 42 to 44 GHz (uplinks), 47 to 50 GHz (uplinks), 52 to 63 GHz (crosslinks), 72 to 76 GHz (uplinks), 82 to 86 GHz (downlinks), 92 to 96 GHz (uplinks), and 102 to 106 GHz (downlinks).

Communication satellite transponders fall into two general categories insofar as both uplink and downlink are concerned. On the uplink, the transponder receiving section and antenna may be designed to work with only one earth transmitting station, or it may be designed to work with many earth transmitters within a given geographical area. Similarly, the downlink from the transponder may beam a signal to only one earth station, or it may broadcast simultaneously to many earth receiving stations. Some examples of the uses of these are

1. *Transoceanic telephone, TV, and data relay* (say, from Paris to New York)—Both the uplink and downlink are designed to work with only one earth station each.

2. *North Atlantic air traffic control (ATC)*—In this system the link from the ATC controller to the pilot would be via a single ground station up to the satellite (at microwave frequency) with the satellite broadcasting to the entire North Atlantic area via VHF, UHF, or microwave. When the pilot talks to the controller the uplink is from the A/C anywhere in the North Atlantic (VHF, UHF, or $\mu\lambda$) to satellite and back down to a single ground station via microwave.

3. *TV broadcasting to a remote area*—The network originating station transmits uplink to the satellite (microwave) and the satellite broadcasts to an entire area at a power level sufficient for pickup by relatively small low-cost terminals (either attached to a single TV set or to a cable distribution system).

4. *Direct-broadcast-TV satellites (DBS)*—These satellites would have the capability of transmitting directly to individual home TV receivers which are equipped with a simple converter module and a relatively small indoor receiving antenna. Although the technology is already available for this type of service, the economic advantages of the direct reception of satellite-broadcast TV signals as opposed to the use of cable distribution, the reception of standard VHF and UHF TV broadcast signals, or the interception of satellite-relay TV signals by the use of large (10-foot) dishes and low-noise receivers (see Chapter 7) does not exist at this point in time, and therefore the DBS-TV idea seems to be currently "stalled out" for lack of financial interest and support.

5. *Remote-area telephone systems*—Each terminal (one terminal per village or small telephone exchange) transmits up on a microwave link either on a narrow channel using only a portion of the transponder bandwidth (FDM) or on a time-shared basis during its assigned time slot (but with a wider bandwidth) (TDM) with all transmissions being retransmitted by the satellite to all ground stations. Each ground station makes use of only those transmissions addressed to it.

Other examples could be given, but these will suffice at this point to illustrate some of the variety possible in the usage of communication satellites.

As mentioned previously, some type of angle modulation is used on satellite links. The actual modulating signal which produces the FM or PM modulation might be a standard TV video signal, or it might be a frequency or time-division-multiplexed composite of a large number of telephone conversations. Or the modulating signal may be digital (although not necessarily binary) in nature, producing FSK or PSK modulation (the digital equivalent of analog FM or PM). If a nonbinary digital signal is used, then the resulting modulated signal is multifrequency FSK or multiphase PSK. One of the advantages of using FSK or PSK is that it permits using the same channel for transmitting data, voice, or even digitized TV as the need arises. Also, some of the future satellite transponders may have the capability of reading the station to which a particular digital record (message or part of a message) is addressed and switching the transponder output to an antenna to beam the message in the direction of that station. Later satellites may have the capability of temporarily storing messages and assigning (switching) them to a particular ground station whenever that downlink becomes available (satellite "mail" being sorted at a "post office in the sky"). The advantages of using digital rather than analog signals and channels are so great that this is the direction that most satellite links will be taking in the future (with the possible exception of direct-broadcasting and mobile-radio (such as the ATC example)–type links).

9.3 *OTHER TYPES OF SATELLITES AND SPACE VEHICLES*

Although this is primarily a book about *communication* systems, it seems appropriate at this point to at least mention some of the other uses of satellite and space-probe vehicles. These fall into several general categories: navigational system satellites, surveillance satellites, research satellites, and deep-space probes.

9.3.1 *Navigational Satellites*

Most satellite-based navigational systems rely upon some form of triangulation method involving measuring the time delay (range) to three or more satellites from the point or vehicle which is to be geographically located. From these data a three-dimensional geographic "fix" can be determined *if* the positions of the satellites used can be accurately determined. The NAVSTAR GPS (Geographic Positioning System) is one such very sophisticated system which is presently being implemented. It will have a total of 18 operational and 3 "spare" satellites in six different low-altitude orbits (12-hour orbits inclined 55 degrees to the earth's equator and located 60 degrees apart from each other, with 3 equally spaced operational satellites in each orbit). The 21 satellites will be launched 7 per year in the years 1986, 1987, and 1988. A prototype system for test and evaluation purposes is now operational but does not provide continuous navigational coverage.

The operational system will make available a minimum of four satellites within view at any time at any point on the earth's surface (in most locations six to eight satellites will be simultaneously visible). Each satellite will transmit two *L*-band *spread-spectrum* (frequency-hopping) signals in bands centered about 1575.42 MHz (called L_1) and 1227.60 MHz (called L_2). The frequency-hopping algorithm will be in accordance with pseudorandom-noise (PRN) code sequences that will be known to the user and that will be executed according to a precise time schedule utilizing atomic clocks on board the satellites to obtain the necessary precision. (Note: This type of spread-spectrum frequency-hopping modulation actually amounts to nothing more than multilevel FSK modulation by a pseudorandom modulating signal.) The user will use the same PRN code sequence to control a conversion oscillator in the navigation receiver. By adjusting the timing of this local PRN sequence until the satellite signal is continuously converted to a constant IF frequency and then comparing this timing to an accurate time reference, the user can determine the propagation time (and hence the range) from the satellite to the receiver.

One problem that exists is that usually a locally available time reference of sufficient accuracy relative to the atomic clocks on the satellites is not directly available to the navigation receiver. However, by making observations on a minimum of four (rather than three) satellites, it is possible to obtain four equations in four unknowns (the three coordinates of the "fix" and the time reference). These equations may then be solved by a digital computer in the navigation receiver. To calculate the geographical position fix, calculations involving the satellite positions at the point in time at which the observations are made are calculated from the ephemeris data for each of the satellites (these ephemeris data are broadcast from each satellite on a periodic basis at a low data rate—50 bits per second—along with other telemetry and status data, and is accurate enough for use for about a 30-day period). By additionally measuring the Doppler shift on the received signal from each satellite, it is also possible to determine the velocity components of the vehicle bearing the NAVSTAR receiver, and by utilizing both of the L-band signals (which are 347.82 MHz apart), it is possible to correct for ionospheric propagation effects and thus obtain better accuracy.

The foregoing discussion describes the basics of the NAVSTAR GPS system operation. The details of the system are a little more complex. As already mentioned, two different L-band frequencies are utilized. Two different spread-spectrum modulations are utilized on these two frequencies. One mode of modulation uses a frequency-hopping (*chipping*) rate of 1.023×10^6 per second and a 1023-chip-long PRN code sequence lasting 1 millisecond. This is called the C/A (*coarse/acquisition*)–mode of modulation. The other modulation mode utilizes a chipping rate of 1.023×10^7 per second with a PRN that has a repeat period of seven days (making it very difficult to "crack" the code if you don't already know it). It is called the *precision* (*P*) mode of modulation. The L_2 frequency can be modulated by either one or the other of the two modulation types. The L_1 frequency is simultaneously modulated by both modulations. This is done by modulating the carrier-oscillator signal directly with one mode and modulating the carrier-oscillator signal phase-shifted by 90 degrees with the other mode. The carrier signal which is modulated by the C/A mode is made 3 dB greater than is the one modulated by the P mode. The 50-bps data modulation which transmits the satellite's ephemeris data is accomplished by phase modulating both of the L-band carrier-oscillator signals prior to applying the spread-spectrum FSK modulation to them. When the P modulation on the signal is utilized and ionospheric corrections are applied and six to eight satellites are used with statistical "curve-fitting" solutions to the equations, accuracies on the order of 5 meters (rms) are obtainable from the system. Simpler receiving systems utilizing only the C/A modulation may obtain accuracies on the order of 50 to a couple of hundred meters.

The NAVSTAR satellite ephemeris data are calculated from data obtained by various ground stations which track the NAVSTAR satellites by the use of optical laser and radar techniques. The ephemeris data thus obtained are periodically transmitted up to the satellite they concern for retransmission by that satellite. The NAVSTAR system thus requires a fairly intensive network of ground-support stations for its operation. It is one example of a satellite-based navigational system, and it appears to be one of the most accurate and useful ones proposed to date. It is also one illustration of the uses of nonsynchronous-orbit satellites (navigational systems satellites in geostationary orbits do not provide good coverage in the higher latitudes, although they are satisfactory for equatorial and midlatitude use).

9.3.2 Surveillance Satellites

Surveillance satellites may be placed in either geostationary or in lower orbits. The geostationary orbits give continuous coverage of the lower and middle latitudes and are especially useful for satellites which are used to track hurricanes, typhoons, and other types of tropical weather disturbances. Lower-altitude satellites (often in polar orbit) give good coverage of the higher-latitude and polar regions and are useful for periodic mapping of weather, earth resources, military activities, and so on. The major problem with these satellites is that they must be continuously tracked and that some provision must be made to store the surveillance data which they acquire until they are within sight of a ground-control station to which they may "dump" the data. The security aspects of controlling this data-dumping operation must be taken into account to prevent the loss of stored data due to either unintentional or intentional unauthorized access of the satellite.

The data obtained by surveillance satellites may involve visual and infrared imaging, electromagnetic spectrum surveillance, detection of upper-atmosphere chemical and physical constituents, astronomical and solar observation, and the like. These data (as well as voice signals and physiological and other data in the case of "manned" satellites or vehicles) must be transmitted from the satellite to the various ground stations involved in the particular system of which the satellite is a part. Most of the data will be encoded into some type of digital format for more reliable and error-free storage and transmission. Some of the data may be preprocessed on board the satellite prior to storage and transmission to best utilize the available storage and communication-channel capacities.

For satellites in nongeostationary orbits, the problem of satellite tracking and data acquisition implies the existence of a worldwide network of many tracking and data-acquisition sites. This "ground support" for a satellite system for any purpose (communications, navigation, research, or surveillance) may be a major factor in the design, cost, and operation of the system.

9.3.3 Space Probes

A space probe begins its travels as a satellite in a very eccentric high-apogee orbit. If the orbit is eccentric enough, the satellite (space probe) may come under the influence of other bodies (the moon, the sun, or other planets) which influence its path more than does the earth's gravitational field. At this point it essentially becomes free of the earth, at least for some extended period of time. As it moves farther away from the earth, the reception of data signals from it and the transmission of control signals to it become quite difficult and necessitate the use of large high-gain ground antennas (dishes on the order of 100-foot diameter or so) and very-high-power microwave transmitters. The probe's transmission of telemetry data must be at relatively low rates so that narrow predetection receiver bandwidths can be utilized to maximize the predetection S/N ratios. The use of high-gain antennas on the probes is desirable, but this implies that some type of active system must be provided to keep the narrow beam of such antennas always pointed in the direction of the earth. In many cases systems must also be provided for adjusting the attitudes of some of the data sensors on the spacecraft as well as for keeping its solar power panels properly oriented toward the sun. The problems associated with the propagation time delay between the transmission of a command or control signal from the earth and reception of the telemetry signal confirming the satellite's response to that command must also be taken into account.

All these difficulties notwithstanding, the past couple of decades have witnessed some remarkable achievements in the communication of control and data signals over very long distances. Future improvements in receiver and transmitter technology should make communications over even longer distances possible in the near future.

9.4 RADIO ASTRONOMY

Radio astronomy is of interest to the communications engineer because it is the study of extraterrestrial radio noise sources which may be strong enough to cause interference with some types of terrestrial communication systems as well as being a possible major factor in the design of systems employing satellites and other types of space vehicles either as communication-relay stations or as signal emitters.

Radio astronomy had its beginnings in the early 1930s when Carl B. Jansky of Bell Telephone conducted antenna noise studies in the region of 20 MHz. Later work in the late 1930s in the 150- to 160-MHz region was done by Grote Reber using a more directional parabolic-dish antenna. The development of microwave radar equipment in the early 1940s provided great technological impetus to the study of extraterrestrial radio sources at frequencies in the vicinity of 1 GHz to 10 GHz, where the background radio noise temperature of the sky (due to atmospheric/ionospheric attenuation and galactic radiation) is only about 10° Kelvin (see the antenna noise temperature discussions of Chapter 6).

Radio astronomy observations fall into two major categories: sky radio-noise-temperature mapping (scanning) and long-time observations (hours at a time or days at a time for some objects observed with high-latitude telescopes). The first type of observations are usually made with telescopes that have antennas that project a narrow beam which is adjustable only in elevation angle along a north-south line through the telescope site. This allows adjustment of the beam elevation relative to the earth's equatorial plane (adjustment of the *declination angle* which goes from −90 degrees (south) to +90 degrees (north)). The earth's rotation then scans this beam in a conical sweep once every 23 hours, 56 minutes, and 4 seconds (the length of the *sidereal* day in terms of solar time) varying the *right-ascension angle* (RA) which is based upon the sidereal day and given in hours/minutes/seconds of 24-hour *sidereal time* (where "sidereal midnight" is defined as the direction of the sun from the center of the earth at the time of the spring equinox). The declination angle range which is possible from a given radio telescope site with a particular type of antenna is limited by the local horizon toward the equator and by the heavy atmospheric attenuation which occurs at most sites as the horizon is approached. For instance, the tiltable-plane-reflector antenna at the Ohio State University radio telescope near Delaware, Ohio (latitude 40 degrees north), can reliably cover a −36- to +63-degree declination-angle range. This particular observatory has catalogued over 20,000 individual extraterrestrial radio sources. Since the typical beamwidth of a large radio telescope is somewhat less than 1 degree, it may take well over a year of continuous observation to make one complete scan of the available sky at a particular location.

The second major type of telescope is the large parabolic dish antenna with a very narrow "pencil" beam. It may be mounted either on an azimuth/elevation type of mount or on an *equatorial* type of mount (the latter has one axis of rotation parallel to the earth's axis of rotation and the adjustments in the antenna pointing angle are made in terms of declination angle and RA angle). The equatorial mount makes tracking a particular "radio star" easier because only the RA adjustment needs to be changed to compensate for the earth's rotation, and it changes at a constant rate. Tracking with an azimuth/elevation mount usually calls for some type of computer-controlled tracking system.

The receivers used with radio astronomy telescope systems must have very low noise figures. Since the background sky temperature can be as low as 10° Kelvin, it would be desirable to have the receiver's excess noise temperature to be in this range also. This would correspond to a standard noise figure of about 0.15 dB (or $F_o = 1.0345$). In the 1960s liquid helium– and liquid hydrogen–cooled maser amplifiers were utilized to obtain these low noise figures. Liquid nitrogen–cooled parametric amplifiers have also been used (with noise temperatures of approximately 90° to 100° K and F_o of about 1.17 to 1.29 dB). Room temperature–operated gallium-arsenide field effect transistor amplifiers with noise temperatures of 35° to 50° K are now available, and these greatly simplify the operation of the radio telescopes.

Some of the extraterrestrial radio sources apparently have very high radio noise temperatures, but since the solid angle they subtend as seen from the earth is much less than the solid-angle beamwidth of the radio telescope antennas, the *observed* radio noise temperature is proportionally less. In an attempt to get narrower *effective* antenna beamwidths to estimate better the temperature and extent of a radio source as well as to study the spatial variations in the radio noise temperature of more extended sources, radio interferometer techniques are often employed. These techniques involve the use of two or more large radio telescope antennas located some distance (thousands of feet to thousands of miles) apart and then comparing (in amplitude and phase) the outputs of these antennas. This can be done in either real time via cables or radio links or in nonreal time by recording each signal (along with an accurate time reference) and bringing the recordings together in a laboratory for later correlation analysis.

There are many interesting aspects of radio astronomy, both of astronomical and of communication-technological interest which could be discussed here if space permitted. But since this is primarily a book on communication systems, the reader is referred to the many fine texts and periodicals available about the subject of astronomy in general and radio astronomy in particular.

PROBLEMS

9-1. Prepare a table listing the various types of possible space vehicles, the types of orbits they have, their uses, and other characteristics. Discuss the problems associated with communicating with each type from the ground.

9-2. For a *synchronous* satellite, why is station-keeping equipment necessary? What types of devices or equipment might be used to maintain a satellite's station and altitude?

9-3. What is the "Indian Ocean graveyard" for satellites? Why does it occur?

9-4. How would you employ a *monopulse* antenna tracking system with a nonsynchronous space vehicle? (Hint: See Chapter 6.) What other types of antenna tracking systems could be used?

9-5. Why is a *hard-limiting* amplifier usually used in the transponder of a communication-relay type of satellite? What does this imply regarding the type(s) of modulation which may be used on the communication signals which the satellite is used to relay? Explain what is meant by the *noise quieting* effect of a hard limiter.

9-6. Discuss some of the factors involved in selecting the uplink and downlink frequencies for a communications-relay satellite link.

9-7. What is an equatorial antenna mount? With what types of space communication systems might you want to use this type of antenna mount?
(See also the problems at the end of chapter 6.)

10

OPTICAL COMMUNICATION LINKS

Optical communication links may be of several different types, but the major categorization is *guided propagation* versus *unguided propagation*. The first category encompasses mostly the various types of *fiber optic* links while the latter category may refer to simple infrared links intended for only a few meters or tens of meters, or it may refer to long-distance links using high-power solid-state or gaseous lasers and complex lens systems.

10.1 FIBER-OPTICAL LINKS

The concept of using IR or visual-light transparent fibers to guide light propagation over fairly long distances dates back to the late 1960s (shortly after the era in which much of the early laser work was done). Work was begun at that time on the development of very transparent optical materials. Most of the common "transparent" materials (window glass, etc.) are so lossy or scatter the light so much that light can pass through a thickness of less than one to several meters without being almost totally absorbed. This high absorption is due primarily to the presence of impurities within the glass which serve to absorb or scatter the light. Early attempts at producing pure glasses resulted in materials with losses still on the order of 1000 dB/km, but progress was made rapidly, and by 1970 glasses were produced with losses on the order of 20 dB/km. By 1975 this was down to about 2 dB/km, and at this writing fiber optic material with losses considerably less than 1 dB/km are available. The loss factor cannot be made much lower than about 1 dB/Km at visual-light wavelengths (where λ < 800 nm or 0.8 μm) because of the Rayleigh scattering due to the amorphous character of the glass (crystalline structured materials are not desirable because lattice defects present within the crystalline structure serve as scattering "centers" for the propagating visual or IR signal). The Rayleigh scattering phenomenon loss decreases as the wavelength increases. At the present time the most desirable range of wavelengths for propagation along optical fibers is in the near-infrared region from about 800-nm to about 1500-nm wavelength (this corresponds to frequencies between about 200 and 375 THz (TerraHertz)). Two of the most popular wavelengths for use are at about 1300 nm and 1500 nm, where the attenuation is primarily due to the Rayleigh scattering limit and is about 0.35 and 0.18 dB/Km, respectively.

10.1.1 Construction and Operation of the Fiber-Optical Path

Figure 10-1 is a representation of some of the various propagation modes which may possibly occur within a circular *multimodal* optical fiber. The fiber illustrated is what is termed a "stepped index" fiber wherein the glass which makes up the central *core* of the fiber has a larger index of optical refraction than does the glass that makes up the *cladding* that surrounds the core. As long as the *angle of incidence* γ_i is larger than some limiting *critical angle of incidence* γ_{crit}, wherein sin $\gamma_{\text{crit}} = n_2/n_1$, the energy in the incident ray will be totally reflected at a *reflection angle* γ_r, which is equal to γ_i (see Figure 10-2). If γ_i *is* smaller than γ_{crit}, then the ray will be transmitted into the second medium (the cladding in this case) at some *refractive angle* (see ray 3 in Figure 10-1). In the case of the optical fiber, it will travel to and through the outer wall of the cladding and be absorbed in the *jacket* of optically absorptive material which surrounds the cladding. In most multimode optical fibers n_1 is about 1 to 2% greater than is n_2 and thus γ_{crit} is about 78.6 to about 81.9 degrees, whereas in single-mode fibers n_1 and n_2 may differ by only 0.1 to 0.2 percent and γ_{crit} is about 84 to 87 degrees.

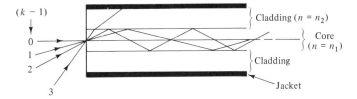

Figure 10-1   Propagation modes in an optical fiber.

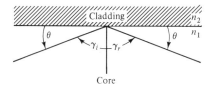

Figure 10-2 Total internal reflection ($\gamma_r = \gamma_i$ as long as $\gamma_i > \gamma_{crit}$).

Propagation of light energy will not occur for all possible values of the angle θ. Due to the boundary conditions imposed upon the electromagnetic wave propagation equation by the diameter of the core relative to the wavelength of the light, only certain angles of θ will occur. These are given roughly by

$$\sin \theta_k = (k - 1) \left(\frac{\sqrt{2}\,\lambda}{\pi\,d} \right) \tag{10-1}$$

Thus, if d is large compared to λ, $\Delta\theta_k = (\theta_k - \theta_{k-1})$ will be small, and many modes of propagation will be able to occur for values of θ from 0 degrees (where $k = 1$) up to the point where total internal reflection no longer occurs (at θ about 8 to 12 degrees for multimode fibers or 3 to 6 degrees for single-mode fibers). However, if d is made small enough so that $\Delta\theta_k$ is larger than this latter limit (for example, if (λ/d) is made greater than 0.2322 so that $\Delta\theta_1$ is greater than 6 degrees), the modes which occur for k greater than unity cannot propagate, and only the $k = 1$ mode (where $\theta = 0$ degrees) can propagate. In this case we have what is known as a *single-mode* fiber. For operation at about 1300-nm wavelength (in air), this would require that d be less than $(1300/0.2322) = 6000$ nanometers to obtain single-mode operation.

The 6000 to 12000 nm diameter of a single-mode fiber is equal to about 0.000236 to 0.000472 inch. This is a very-small-diameter glass fiber when one considers that an average human hair is about 0.004 inch in diameter. Multimode stepped-index fibers, on the other hand, are about ten times as large in diameter (on the order of 50,000 to over 100,000 nm), and thus $\Delta\theta_k$ is about 1/10th of what it is for the single-mode fiber, leading to on the order of about ten possible modes of propagation. Both the single-mode and the multimode fibers have their respective advantages and disadvantages as will be discussed in the following paragraphs.

The multimode fiber has the advantage of being able to accept light coming into the end of the fiber within a much wider *cone of acceptance* than can the single-mode fiber (see Figure 10-3). The many modes of this type of fiber "light

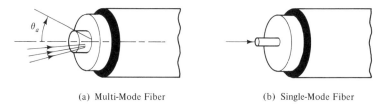

(a) Multi-Mode Fiber   (b) Single-Mode Fiber

Figure 10-3 Acceptance of light into the optical fiber waveguide.

pipe" allow the half-angle of the acceptance cone, θ_a, to be given by the relation $\sin \theta_a = \sqrt{n_1^2 - n_2^2}$. For the case of $n_1 = 1.48$ and $n_2 = 1.46$ for the multimode fiber, this would be about 14 degrees, while for the single-mode fiber, the only radiation that will propagate is that which enters at approximately zero degrees (i.e., highly collimated radiation).

However, there is a problem with the multimode fiber accepting and propagating light at so many different value of the angle θ: the light travels along longer paths for larger values of θ. Thus, if a pulse of light enters the fiber and certain portions of the pulse energy go into different modes of propagation, these different portions of the pulse energy will arrive at the far end of the fiber at different times, giving rise to what is known as *multimode pulse dispersion* (see Figure 10-4). This places a limit upon the pulse repetition rate (PRR) that can be used for a given length of fiber and thus limits the binary data PCM transmission rate which can be achieved. Thus, the multimode fiber has the advantages of having a larger cross-sectional area for light input (about 100 to 400 times that of the single-mode fiber) and a fairly large cone-of-acceptance half-angle (about 10 to 20 degrees), but it has the disadvantage of large pulse-dispersion effects due to the multimode propagation.

The single-mode fiber does not have the multimode pulse-dispersion problem (thereby permitting longer line lengths and/or higher data rates), but its small diameter intensifies the mechanical problems of splicing, connecting, and so on and makes it more difficult to couple radiation into its small cross-sectional area. In addition, its single mode of propagation requires that the light source "feeding" the fiber be highly collimated, which usually means that a laser source must be used (although recently *edge-emitting* light-emitting diodes have found some application in this regard).

Still another type of optical fiber is what is known as the *graded-index* type of fiber. In this type of fiber the core is large as in the multimode fiber, but it has an index of refraction which decreases with the radial distance from the center of the core. This leads to a multiplicity of curved paths as is shown in Figure 10-5. Light entering at different angles as shown by the rays 0 through 3 will propagate in curved paths (except for the axial-path ray 0) due to the gradual bending of the rays by the radially changing index of refraction. Although the higher-numbered ray paths in Figure 10-5 travel longer paths than do the lower-numbered ones, their velocity of propagation increases as they pass through regions of the core with lower values of n. If the change of n with radial distance in the core is carefully engineered, it is possible to make all the rays (there are an infinity

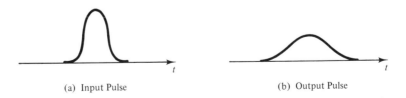

(a) Input Pulse   (b) Output Pulse

Figure 10-4   Dispersion of the energy in a transmitted pulse.

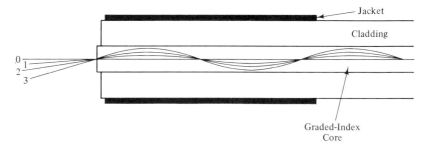

Figure 10-5   The graded-index fiber.

of them since the graded-index fiber does not have a propagation modal structure) have nearly the same *average axially directed* velocity component.   This greatly reduces the pulse dispersion that is present due to different rays traveling different distances.

Although the graded-index fiber greatly reduces the *modal* pulse dispersion due to multipath propagation as compared to the stepped-index multimode fiber, and although the single-mode fiber eliminates dispersion due to this mechanism completely, there is another pulse dispersion mechanism which must be considered: that due to the change of propagation velocity at different optical frequencies. This is referred to as *chromatic* dispersion and must be considered since none of the light sources currently in use are truly monochromatic (single-frequency) in nature.   These effects will be discussed at the end of the next section.

10.1.2 Light Sources for Fiber-Optical Communication Links

The two most commonly used sources for fiber-optical communication links are the light-emitting diode (LED) and the injection laser.   In the near-infrared regions of current interest (at about 1300- and 1500-nm wavelength), the LED is commonly of Aluminum/Gallium–Arsenide (AlGaAs) or Indium/Gallium–Arsenide/Phosphide (InGaAsP) construction and produces radiation from a "well" (in the more conventional design) etched perpendicularly into the diode junction area (see Figure 10-6).   The efficiency of the LED is rather poor (about 0.2 to 2 mW of light output for about 150 to 1500 mW of electrical power input).   For a given junction temperature, the optical output is approximately a linear function of the diode current.   This makes the LED easy to operate (i.e., its operation is stable).   The LED can be PCM on/off modulated at rates up to 10 to 100 Mbps and is reasonably low in cost (a few dollars each).   These latter two advantages are balanced by the disadvantages of not having the optical power well collimated and of having its spectrum spread out over a relatively wide range of wavelengths (in comparison to the spectrum of the solid-state injection laser).   These characteristics cause problems of coupling the power into the fiber and cause multimodal and chromatic pulse dispersion.   The latter limitations, coupled with the ease of use and the low cost of the LED, make it most useful for use with multimode or graded-index

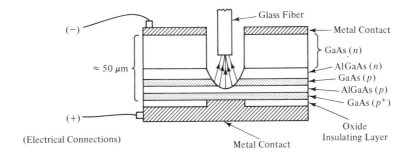

Figure 10-6   The AlGaAs light-emitting diode optical source.

fibers for applications over relatively short distances (less than 1 or 2 Km) at moderate data rates (less than about 10 to 20 Kbps).

The injection laser is essentially another form of the Al/GaAs/P diode with the radiation taken out of the edge of the junction region. When highly polished and made closely parallel to each other, the sides of the semiconductor "chip" form an optical cavity within the junction region, permitting the diode to operate in the *stimulated emission of radiation* (laser) mode(s). This has several net effects: a higher conversion efficiency is obtained (about ten times better than the LED), the radiation is more monochromatic in nature (it is still not single frequency, because several different *oscillation* modes are usually possible and the laser may operate in several of them at once and/or jump between modes), and the output is highly collimated. All these make the injection laser suitable for use with single-mode fibers. Since it can put out on the order of 10 mW of highly collimated near-monochromatic radiation and can be pulsed at more than 10 Gbps data rates, it is possible to construct long-distance high-data-rate single-mode fiber optical data links up to many Kilometers in length without intermediate repeater stations. The main problems with using the injection laser are its higher cost (about $1500.00 each), its lower level of reliability and shorter lifetime, and the difficulty in obtaining a stable operating (biasing) point as compared to the LED. However, these latter problems are being intensively studied at the present time, and it is likely that in the near future, they will be greatly mitigated.

From the preceding it is apparent that two basic types of systems are emerging at the present time: a low-cost, short-distance, moderate-data-rate system based upon using LEDs and the multimode or graded-index types of fibers and a long-distance, high-data-rate, more costly system based upon the use of the injection laser and single-mode fibers. A somewhat intermediate type of system may be developed based upon the use of the edge-emitting diode (somewhat similar in construction to the laser but less monochromatic and efficient but easier to operate). Calculations, experiments, and field experience with the shorter-distance LED-based systems have shown that this type of system has a range (in kilometers) of about

$$[14 - 5 (\log_{10} R_D)] \text{ km} \tag{10-2}$$

(where R_D is the data rate megabits per second with an upper limit of about 100 Mbps due to multimodal pulse dispersion) when a 10-dB loss margin is allowed for.[1] For the longer-range laser-based systems the comparable equation is

$$[78 - 21 (\log_{10} R_D)] \text{ km} \qquad (10\text{-}3)$$

where in this case R_D is limited to about 1 Gbps due to chromatic pulse dispersion caused by the multimode and mode-hopping oscillation of the laser. At 100 Mbps, a 30-Km repeaterless link is easily obtainable (again assuming a 10-dB loss margin) at the present time. This makes this type of system currently feasible for telephone company trunk circuits because it eliminates the need for the repeaters which are required every couple of kilometers on a metallic circuit. As research continues to produce improvements in the solid-state laser, the cost-effectiveness of this type of fiber optic link will become even greater due to improved reliability and longer repeaterless link distances and/or higher data rates per fiber.

10.1.3 Optical Detectors

Two basic types of optical detectors are presently being used to convert the optical signal coming out of the optical fiber at the receiving end into an equivalent electrical signal. These are the *PIN diode* and the *avalanche diode*. Figure 10-7 shows a cross section of the PIN diode. It differs from the conventional p–n junction diode in that it has a relatively thick, lightly doped (almost intrinsic) n layer between the heavily doped p and n regions. This i region effectively increases the width of the junction depletion region, and therefore when the diode is reverse biased, the E field due to the reverse bias is spread out uniformly over a longer distance. If this distance is made larger than the mean absorption length of the optical radiation entering the diode from the left, then almost all the carrier pairs which the impinging photons cause to be created will be created within the region where the E field exists and can immediately contribute to the flow of reverse current through the diode. If a conventional p–n junction were to be used, some of the optical radiation would have gone on into the n region and produced carrier pairs there which would have had to first diffuse back to the junction region before they could have contributed to the reverse current flow. The inclusion of the i region therefore speeds up the response time of the photodiode, thus increasing its pulse sensitivity and decreasing the dispersion which it introduces upon the electrical output pulse. Typical GaAs PIN photodiodes operating in the near-IR region can effectively absorb and utilize (convert photons into a hole-electron pairs) about 70% of the incident optical radiation and can handle pulse rates up to 10^{12} pulses per second (or a PCM data rate of 10^{12} bps).

The ultimate sensitivity of an optical detector for a PCM system is determined by the quantum noise caused by the fact that at low optical signal power levels and high pulse rates, the number of photons per pulse becomes quite small. Since the photons are arriving at ("assigned" to pulses at) a random rate, determining with

---

[1] From data presented by Stewart Personik at various conferences.

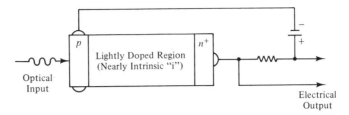

Figure 10-7   The PIN photodetector.

a high order of probability whether an "on" or an "off" PCM pulse has arrived requires a certain average number of photons per pulse. This number has been set at 21 photons per pulse period on the average. At a 100-Mbps data rate and 1300-nm optical radiation wavelength, this corresponds to minimum received power level of −65 dBm.[2] This is the minimum signal level which an ideal detector could detect.

The large amount of photon energy in visible and near-IR radiation compared to that at RF and microwave frequencies has another effect upon the detection process in the case of the PIN photodiode detector: each photon generates only one pair of carriers (quantum of electronic charge), and as a result, the amount of current produced by a given radiation power level is quite small (on the order of one nanoampere for one nanowatt of incident optical radiation). This small output current requires that the PIN photodiode be followed by a large amount of electronic amplification. It is the excess noise of this amplifier which limits the sensitivity of the PIN-diode/amplifier optical detector system. This causes the sensitivity of this type of detection system to be only −39 dBm at the data rate and radiation wavelengths given (or about 26 dB worse than the "ideal" detector).

A more sensitive detector can be built using another type of photodiode which produces a larger charge transfer per photon than does the PIN photodiode. This more sensitive detector is the GaAs avalanche photodiode which is shown in Figure 10-8. It uses the *avalanche* effect wherein some of the carriers which are created by an impinging photon of the optical radiation are generated in a region where there is a large electric field and are thus accelerated quite rapidly to a high enough velocity that when they collide with the structure of the crystal lattice they produce other "secondary" carrier pairs which also contribute to the reverse (output) current of the diode. The net multiplication is on the order of about ten times the charge transfer which is caused directly by the photon itself (as in the PIN photodiode). Thus, we find that the output current of the avalanche photodiode is about ten times that of the PIN diode, and thus it does not require quite as much electronic amplification as does the PIN diode, resulting in less amplifer noise being introduced. For the same data rate and optical wavelength as previously, the sensitivity of the avalanche diode/amplifier detection system is about −50 dBm.

---

[2] This minimum required power is calculated by

$$P_{\text{min, ideal detector}} = (21 \text{ (photons/pulse)}) \, (hc/\lambda \text{ (Joules/photon)}) \, (R_D \text{ (pulses/sec)})$$

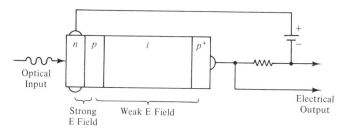

Figure 10-8   The avalanche photodiode detector.

Taking a quick look at how overall system performance is affected by the choice of detector as well as the choice of the optical radiation source and the type of fiber, we can come up with the following order-of-magnitude results. If we were to use an avalanche photodiode conservatively estimated to have a sensitivity of −45 dBm as the detector and an injection laser estimated to have a minimum output power of about 0 dBm as the source, we would then have at the most only about 45 dB of total link loss to budget among the various loss mechanisms of the link such as the loss in the optical fiber, connector and splicing losses, and a *loss margin* (i.e., "safety factor"). Again, we have assumed a 100-Mbps data rate and 1300-nm wavelength. This 45 dB of optical budget is much less than the 70 to 90 dB of budget which we might have to "play with" on a typical metallic circuit. However, the lower total budget figure for this laser/avalanche−diode optical case is somewhat compensated for by the fact that the single-mode fiber which we probably would use with this type of source and detector has a much lower dB loss per kilometer than does the metallic circuit (a fraction of a decibel per Kilometer for the optical case compared with several decibels per Kilometer for the metallic circuit).

10.1.4 Other Components of the Optical Fiber Link

In addition to the optical source which converts an electrical signal into an optical signal at the transmitting end, the photodetector which converts it back into an electrical signal at the receiving end, and the optical fiber which conveys the optical signal from transmitter to receiver, the optical communication link may also involve a number of other types of components depending upon the use to which it is put. Two of these will be almost certain to appear: fiber-optical cable connectors and splicers. Both these perform essentially the same function: to connect one piece of optical fiber to another piece (optical sources and detectors are usually manu-factured with a short piece of fiber attached). The difference between them is that connectors are usually used in situations where it may be desirable to disconnect the connection easily whereas splicers or splices are used where the connection is to be more or less permanent in nature.

In addition to these devices, some links may involve fiber-optical switches, fiber-optical directional couplers, and fiber-optical duplexors and multiplexors as well as other special-purpose optical devices. In the future we may also find that optical amplifiers (devices that directly amplify an optical signal without first converting it back to the electrical form for electronic amplification and the creation of a new optical signal from a new source) will also be part of the long-distance types of optical links.

10.1.4.1 Fiber connectors, splicers, and splices.

Because of their small size, the joining of optical fibers, either temporarily or permanently, requires extreme mechanical precision to have a low-loss junction.   This is especially true in the case of the single-mode fiber which is smaller than a human hair in diameter.   Theoretically, if the ends of two optical fibers are well polished perpendicular to the axis of the fiber and if the two polished ends are exactly aligned in both displacement and angle, the optical radiation will be transmitted across the junction with very little (much less than 1 dB) loss.

One device which is designed to do a good job of aligning the fibers to be temporarily or permanently joined is the biconical type of connector which is shown in Figure 10-9.   It relies on accurately machined conical plugs, one of which fits over each end of the fiber cable sections to be joined and is securely attached to it by clamping or epoxy adhesive.   The fiber itself passes through a very small hole which is exactly (to within machining or molding tolerances) along the axis of the cone.   When the two plugs are then inserted into an accurately machined or molded biconical receptacle, the axes of the two plugs are closely aligned, and thus the fibers which lie along these axes are also closely aligned.   If the ends of the fibers are polished after the cables are installed in the plugs, there should be very little loss in transmitting light from one of the fibers to the other.   This type of device is especially useful for connecting single multimode or graded-index (i.e., the larger-diameter type) fibers.   Standard quick–connect/disconnect connectors using this principle have been designed, thus allowing optical equipment from various manufacturers to be easily and quickly interconnected.

For connecting more than one fiber at a time (many fiber-optic cables have a large number of fibers in them), some variation of the groove-block type of connector or splicer such as is shown in Figure 10-10 may be used.   This technique

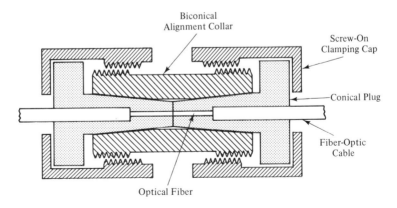

Figure 10-9   The biconical fiber-optical connector.

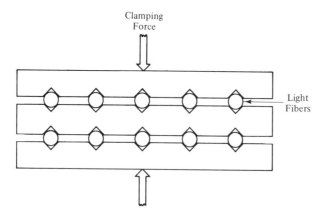

Light
Fibers

Figure 10-10 The groove-block connector or splicer.

for mechanically aligning a group of fibers is quite often applied to cables containing multimode or graded-index fibers. One common standard that is used has 12 fibers per plane and up to 12 planes (a maximum of up to 144 fibers in one groove block). If used for a splice, the fiber ends from the two cables are polished and then butted together within a single block and then clamped into place. For a connector, the individual fibers from one cable are inserted into a block with their ends protruding and after the cable is thus attached to the block the protruding fiber ends are trimmed and polished flush with the block surface which will mate with the similar surface of the block attached to the end of the other cable. An accurately machined connector assembly and indexing pins keeps the two groove blocks in accurate alignment.

Another type of connecting or splicing technique is to use some type of focusing system between the two fibers as is shown in Figure 10-11. The simplest of these techniques (shown in Figure 10-11a) is simply to round the ends of the two fibers to form in effect two semispherical lenses. This is done by carefully melting the ends of the two fibers in an electric arc. As can be seen from the drawing, the "lenses" help to focus and collect the light that might otherwise be lost by a displacement or angular misalignment of the fibers.

Figure 10-11b shows the use of a spherical lens between the ends of the two fibers. As long as the light coming out of the incoming-light fiber hits the spherical lens, the lens will concentrate it toward one point. If this focus point lies on or in the fiber which is to receive the light, most of the light will be coupled into that fiber.

In Figure 10-11c is shown another connector which is designed to be used for connecting single-mode fibers. Each side of the connector has a lens which either expands the very-small-diameter (about 8 microns) single-mode light beam into a beam of light which is about 200 times larger in diameter (about 1500 microns) or else focuses this larger beam back down to the diameter of the single-mode fiber. By increasing the diameter of the beam temporarily, the mechanical accuracies involved with aligning the two connector halves (at plane II in the drawing) need not be nearly as stringent as if the two fibers themselves had to be thus aligned. However, the positioning of each optical fiber with respect to the centerline of its

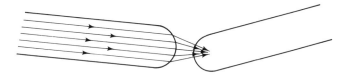

(a) Rounded-Fiber End Lens for Focusing

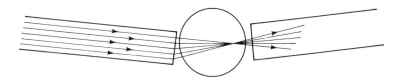

(b) Spherical Lens Between Fibers

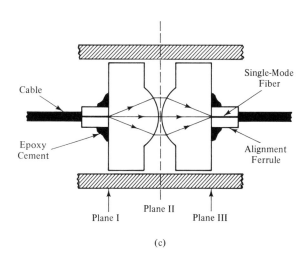

(c)

Figure 10-11 Lens-type connectors.

lens is very critical as each connector half is assembled onto its Fiber-Optic cable. The manufacturer of this type of connector provides a special piece of equipment to assist the technician in performing the connector-mounting operation.

Another technique for aligning fibers accurately involves the use of three smooth straight rods which when placed together parallel to each other and enclosed within a piece of heat-shrunk tubing, clamp the two fiber ends into close alignment. This low-cost, quick-join technique is usable for the larger-diameter fibers. Still another approach to permanently connecting two fibers is to melt thermally (weld) the two ends together via the use of an electric arc while the two ends are held in

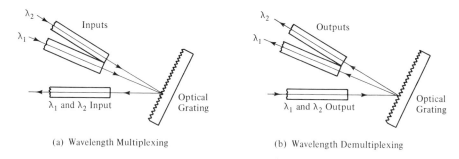

(a) Wavelength Multiplexing   (b) Wavelength Demultiplexing

Figure 10-12

precise mechanical alignment by the welding jig. This method is quite often used on long-distance single-mode optical fiber links. While being a time-consuming and expensive operation, the splice thus produced is very low in loss and is thus worth the effort on these types of links.

10.1.4.2 Wavelength multiplexors. Figure 10-12a shows the basic principle of using an optical grating[3] to separate a beam containing two (λ_1 and λ_2) wavelengths of radiation into two spatially separate beams. Since the process is reversible, the grating can also be used to combine two separate beams at different wavelengths into one common beam as is shown in Figure 10-12b. Or the same structure can be used as one type of optical duplexor as is shown in Figure 10-13 to enable one fiber to carry information in both directions simultaneously. Figures 10-12 and 10-13 show only a single collimated ray of radiation passing through the system. However, this arrangement would not work in practice, and it is necessary to include some type of a lens system to first collimate and then focus the radiation going to and from the grating. Figure 10-14 shows one very mechanically rugged type of structure which has been developed for this purpose.

Another type of duplexing structure is shown in Figure 10-15. In this case the fiber coming from the transmitter has a flat face upon which is formed an

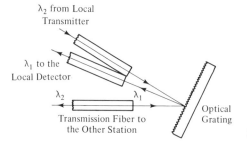

Figure 10-13

[3] The *grating* used here functions similarly to the way in which an optical prism splits a visible beam of light up into its spectral components. Optical gratings, however, function as well (in fact usually better) in the IR region as they do in the visible-light region whereas it is somewhat difficult to construct an effective low-loss prism at IR wavelengths.

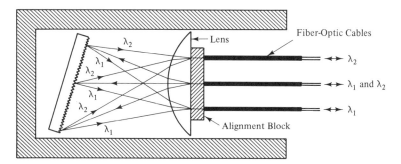

Figure 10-14

optical interference (wavelength-selective) filter which allows the transmitter-wave-length (λ_2) radiation to be transmitted through, but which reflects radiation at other wavelengths (specifically the wavelength λ_1 of the incoming radiation from the other end of the link).   The transmission fiber has a rounded end to help it collect the radiation coming out of the end of the fiber coming from the transmitter and the fiber going to the optical receiver also has a rounded end to help it collect the radiation from the other end of the link which comes out of the transmission fiber and either hits the receiver fiber directly or is reflected off the interference filter on the face of the fiber coming from the transmitter.   Since some of the power from the local transmitter (source) will invariably get into the fiber going to the local receiver, it is necessary that the local optical detector either be insensitive to the local source's wavelength of radiation or else an optical rejection filter of some type must be incorporated in the design of the local receiver.

Duplexing and two-wavelength multiplexing schemes using multimode or graded-index fibers commonly use radiation at the two near-IR wavelengths of approximately 800 and 1300 nm.   For multiplexing more wavelengths of optical radiation using the grating method it has been found[4] that it is possible to combine as many as ten different signals via wavelength multiplexing, using the IR region from about 1140 to 1500 nm, with the spacing between the channels being about 36 nm in wavelength and the "bandwidth" being about 20 nm in wavelength.

10.1.4.3 Fiber-optical switches.   These devices are basically highly precise connectorlike assemblies in which a sliding carriage can be moved to align one fiber with one of two or more other fibers which are mounted on the fixed surface across which the carriage moves (see Figure 10-16a).   This type of switch is primarily used with multimode or graded-index fibers in installations where it may be desired to change the destination of the optical signal or to be able to switch the signal quickly from a failed source or detector to a functional one.   An improvement upon the design can be obtained by using a pair of expansion/focusing

    [4] See *Wavelength Division Multiplexing in Fibre-Optical Systems* by B. Hillerich, O. Krumpholz, M. Rode, and E. Weidel in the July 1985 issue of *Telecommunication*, page 73.

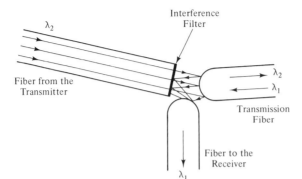

Figure 10-15 Duplexor using a wavelength interference filter.

lenses in the same manner as was shown in Figure 10-11c to lessen the mechanical-alignment accuracy requirements.

10.1.4.4 Optical directional couplers. Figure 10-17 shows a "directional coupler" made by fusing the claddings of two multimode fibers which have first been heated and stretched so as to reduce the diameter of the fibers' cores (thus causing some of the optical radiation power contained in the higher modes in the core of one of the fibers to leave the core and enter the cladding of that fiber, from which it is then partially transferred to the cladding of the other fiber through the fused region and thence into the core of the other fiber at the point where the core expands back to its original diameter). If $n_1 \neq n_2$, and if an electric field is applied across the coupler, the amount of the coupling can be made wavelength dependent, and incoming radiation at two different wavelengths on one of the fibers can be made to separate, with the radiation at one wavelength being almost completely coupled into the other fiber and radiation at the other wavelength

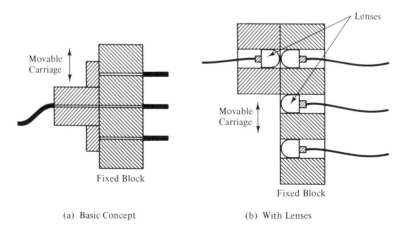

(a) Basic Concept   (b) With Lenses

Figure 10-16 Fiber-optical switch.

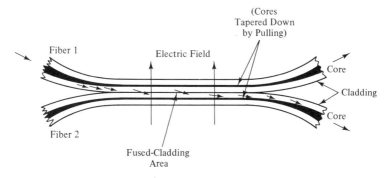

Figure 10-17   Fiber-optical directional coupler.

continuing to propagate along the original fiber.   Such an arrangement can be utilized as a wavelength multiplexor or a duplexor.

10.1.5 Fiber-Optical Link Calculations

Link calculations for fiber-optical links are very similar to those for ordinary wire or cable links.   Assuming that enough optical power will be available to the photodetector at the receiving end of the link, the maximum data rate which may be carried on a link will be dependent upon the maximum rate at which the optical source can be switched on or off, the response time of the photodetector, and/or the pulse dispersion which occurs on the optical fiber transmission line.   On short links, the first two of these factors usually predominate, while on longer-distance links, the pulse dispersion is usually the limiting factor.   In some cases two or even all three of these factors may be effective, and in that case their effects must be added by adding the rise or fall time of the pulse at the source, the response time of the detector, and the dispersion time of the pulse (i.e., the amount of time by which it is lengthened due to dispersion).   The maximum data rate is then the inverse of this sum if the assumption is made that there is enough power at the detector.   This assumption can be checked by using the relationship that the power level required at the detector must be at least[5]

$$P_{\text{dBm, min}} = P_{\text{dBm, 100/1300}} + 10 \log_{10}\left[130 \left(\frac{R_D}{\lambda_{\text{nm}}} \right) \right] \qquad (10\text{-}4)$$

where R_D is the data rate in Megabits per second, λ_{nm} is the wavelength of the radiation in nanometers, and $P_{\text{dBm, 100/1300}}$ is the minimum power for 1300-nm radiation and a data rate of 100 Mbps for whichever type of optical radiation detector is used (-65 dBm for the ideal detector, about -45 dBm for the avalanche diode, and about -39 dBm for the PIN diode).

---

[5] Equation 10-4 incorporates an assumed 10 dB loss margin.

Once this "target" figure is obtained, the power budget of the link may be calculated to determine if this amount of power will be available to the receiving-end photodetector. This is done in the conventional manner by stating the power output of the source in dBm and then subtracting off the various decibel losses (the optical fiber loss, the loss at each anticipated connector or splice and a *loss margin* figure (safety factor) to account for unforeseen losses or degradation such as a decrease in the source output, a decrease in the sensitivity of the detector, extra splices needed to repair cable breaks which might occur, etc.). If the resultant expected power at the receiving end is not enough to satisfy the minimum power requirement of Equation 10-4, then the required minimum power level must be lowered by decreasing the data rate R_D or using a more sensitive detector, or else the received power level must be increased by increasing the source power or decreasing the link losses.

Applying what we have learned about the requirements of the system and the characteristics of the various sources, detectors, and fibers, we can now discuss the major factors that have to be taken into account for various types of links. In most short-distance links, it will be usually found that there will be enough received power that the data rate is limited by the three factors discussed at the beginning of the preceding paragraph, but on long-distance links, the received power level and the available detector sensitivity, plus possible pulse dispersion on the optical fiber line, will be the factors which limit the available data rate. On short-distance links, where a very high rate is desired, the graded-index fiber is often used since it has good dispersion characteristics and is still very easy to work with insofar as splicing and connecting are concerned. The single-mode fiber used with the injection laser and the avalanche-diode photodetector is usually reserved for very long distances where the high cost of the laser and problems of working with the small-diameter single-mode fiber are worthwhile because they eliminate the need for repeaters on links up to several tens of Kilometers in length. The larger fibers (the multimode, stepped-index fibers and the graded-index fibers) are usually used where fairly short distances are to be covered, where connecting and disconnecting of the fibers may be required and where wide bandwidths (high data rates) are needed.

10.1.6 *Modulation of the Optical Carrier*

In all the preceding discussion it has been assumed that NRZ (nonreturn to-zero) PCM modulation of the optical carrier would be employed to transmit a stream of binary digital data. In some cases it may be desirable to use some other type of on/off digital coding such as the Manchester Code or a four-out-of-five code and so on. The use of these types of codes would mean that for the same pulse rate the actual digital data rate on the link would be less than for the more efficient NRZ code.

Because of the nonlinearities involved on the optical fiber link, especially in the sources, any type of analog amplitude modulation is not very suitable. However, the various types of analog pulse-time modulation such as pulse-frequency, pulse-duration, pulse-length, or pulse-position modulation may be used quite satisfactorily to transmit analog information signals over an optical link without first digitizing them. Because of the wide bandwidths available, standard TV signals can be easily transmitted in this manner, using FM (PFM) modulation similar to the type used to transmit TV signals over a satellite link (see Chapter 7).

10.1.7 Fiber-Optical Link Applications

The use of a fiber-optical link for long-distance repeaterless "trunking" of digital data, digitized voice, and so on has already been mentioned as one of the main areas of application for the laser/single-mode–fiber systems. The use of optical fibers for shorter distances where very wide bandwidths are required has also been mentioned. One such area as in local area (data) networks or LANs such as will be described in detail in Chapter 11. In this case, since the power output of the more economical LED sources is low in comparison to the power output of most of the electronic devices which might be used on a metallic LAN, and since the sensitivity of the fiber-optical photodetectors is low compared with electronic detectors, the possibilities of having one source "talk" to a large group of listeners as is done on an electronic bus-type LAN is very limited in the case of fiber optics since each listener requires a substantial amount of input power, and there is not all that much power available from the optical source (especially from LEDs). Thus, we usually find that the optical-fiber LAN takes on the form of a token-ring architecture (where each station on the network "talks" only to the adjacent stations on the ring) rather than a bus architecture where each station has to talk to all of the other stations on the network simultaneously. The more complex connecting arrangements and operational protocols required by the token-ring architecture are somewhat compensated for by the high data rates which are possible with the fiber-optical links.

Other places where fiber-optical links excel in performance are in those environments where there is a lot of electromagnetic interference (EMI), such as in a power plant, substation, factory, or large aircraft. The fiber link can pass data through these types of environments with little chance that it will be damaged by an electrical switching spike or other types of electrical impulse noise. Also, since no detectable optical energy escapes from a fiber cable, and since it is very easy to detect at the receiving end when any usable amount of optical energy has been lost or removed from the link, the security of the data passing over a fiber-optical link is much greater than it is for data passing over a metallic circuit.

At the present time fiber-optical communication links are a relatively new phenomenon in the communications area, but they have already taken a solid place in some applications. As more experience is acquired in working with this technology and as the cost of the components gets lower, many other applications for the use of fiber-optical paths will surely be found, both in replacing the use of present metallic-circuit applications and probably also in entirely new applications which are not possible to implement with metallic circuits.

10.2 Nonguided Optical Links

The use of optical radiation (IR and visible) in the atmosphere and in free-space above the atmosphere is, at present, not a very important application of optical communications. However, there are a few interesting applications and some interesting future possibilities. Therefore, we will very briefly discuss these as a conclusion to this chapter.

One of the applications of very-short-range IR line-of-sight transmission has been in the transmission of digital data between data devices in the same room where it has been desired to have at least one of the devices be semiportable in nature. An example of this was the use of IR between the keyboard and the main cabinet of a well-known model of personal computer. Other uses of the same principle have been used to go between two stationary data devices where it would not have been convenient to run cables across the floor or the ceiling. In the case of some types of remote radio and TV broadcasting from inside of large buildings, the concept has been utilized to get the signal from the remote "roving camera" up to the TV mixing console. Another rather valuable application of the short-range IR link has been in the case where it was desired to transmit data signals between two buildings in an urban area where it would have been very difficult to run cables either above ground or below ground. In these cases IR transmitter/receiver pairs are mounted on the top of each of the two buildings and "aimed" at one-another (or in some cases where there were windows facing each other in the two buildings, each transmitter/receiver pair was just mounted inside of a room looking out of a window at the T/R pair in a window of the other building). Other examples could be given where a short-range IR link would be the best solution to a given communication-link problem.

The other major category of nonguided optical communications involves the use of an IR or visible-light high-power gaseous laser as the optical source for a long-distance link between two elevated locations on the earth, or between the earth and a space vehicle, or between two space vehicles. The ground-to-ground application is worth the effort only when the amount of data to be transmitted is very high and where there is no other possible way of communicating those data via a conventional radio or microwave link. The space communication applications for this type of link are not very prevalent at the present time, but as the radio and microwave spectrum becomes more crowded and as the optical sources and aiming systems are better developed, this will undoubtedly become a major application area. The use of laser beams for satellite-to-satellite relay presents little in the way of problems, and for some areas of the world where the weather is dry and where the cloud cover is rather sparse both in duration and in thickness, the use of optical or IR for uplink and downlink communications is also a definite possibility (especially the use of IR links, since IR radiation is less affected by the moisture in the air).

The technical area of nonguided optical communications will undoubtedly continue to grow in importance in the future, but not at the same rate as is now envisioned by most people for the guided (fiber-optical) branch of optical communications.

PROBLEMS

10-1. The wavelength of optical radiation is variably given in units of microns, nanometers, and Angstroms. Set up a table which shows how to convert among wavelengths given in these units and which also shows how to convert the wavelength in a medium of given refractive index to the equivalent frequency in TerraHertz. Refer to the CIE diagram to determine the free-space wavelength of red, yellow, green, (see Figure 7-20) and blue light, and find the frequency of each.

10-2. A 400-micron-diameter optical fiber is to be used with an infrared signal of 1300 nanometers wavelength in free space. How many modes are possible in this fiber at this wavelength? What diameter fiber is required to have single-mode propagation at this wavelength? Let $n_1 = 1.015$ times n_2 and $n_2 = 1.455$.

10-3. In Problem 10-2, what is the cone-of-acceptance half-angle, and how many of the possible propagating modes will be excited?

10-4. Discuss the problems associated with coupling optical energy into a single-mode fiber as opposed to coupling it into a multimode fiber. What factors affect this coupling and what must be done to maximize it?

10-5. Assuming an "ideal" detector limited only by the quantum-statistical noise, what is the maximum length of link possible with radiation having a free-space wavelength of 1300 nm, a 1.544×10^6 pulse rate, $+5$ dBm of power coupled into the fiber from the source laser, 0.8 dB/km of fiber loss, 12 dB of "loss margin" to account for connector losses, and so on? Recalculate the problem to find the maximum link length if a "practical" detector with a sensitivity of -40 dBm is used in place of the "ideal" detector.

10-6. Discuss the use of a typical single-mode fiber-optical link 8 statute miles long carrying a 10-Mbps data signal versus the use of a coaxial-cable RF link for the same purpose. Assume a 1-watt RF power input into the transmitting end of the coax and a 7-dB standard noise figure for the detector electronics at the receiving end of each section of the coaxial link. A 23-dB predetection signal-to-noise ratio must be present in both cases to assure an acceptably low bit error rate. Assume a coaxial cable loss of 30 dB/km and refer to problem 10-5 for the other FO system parameters.

11

DATA COMMUNICATIONS

In present-day usage, the terminology data communications is generally understood to mean the use of binary-encoded digital code streams to allow the communication of information from one digital data device to another. A digital data device could be simply a computer, a computer terminal (a typewriterlike keyboard with either a cathode-ray-tube (crt) display or a character printer), a high-speed printer, a point-of-sales terminal, a data-logging device, a graphics CRT display or plotter, or any other type of device whose input or output is a pulse-code-modulated (PCM) digital data stream. In many cases a number of these devices (for example, a computer, some disk and/or tape drives, one or more CRT terminals, and a printer or two) may be directly connected (or *interfaced*) to form a complete local computing system. For the purposes of discussing digital data communication systems, we would view such a directly interfaced local computing system as a single *digital data device* or *station* on a *digital data link* or a *digital data network*.

11.1 ORIGINS OF DIGITAL DATA

In this chapter we shall consider digital data to consist of a series of one or more *m*-bit binary data strings, where each such string might represent a particular text *character* (number, alphabetical character, punctuation mark, etc.), a computer data *word* (usually 8, 16, 24, 32, 36, 64, or 72 bits long), a command signal (such as line-feed, carriage-return, sound "beep," etc.), the contents of an *n*-bit-long binary data buffer (without regards as to what the binary data in that buffer represents), or a *control string* (of bits) needed for control of the data link or data network over which the communication is to be conducted.

Historically, the first type of digital-data communications system was also the first electrical communication system ever invented—the Morse telegraph system. It used a variable-length binary (on/off) coding system as will be discussed in the following paragraph. It was followed by the Edison stock "ticker" system and a little later by teleprinter systems using several different types of binary coding schemes. Most of these systems were fairly simple, were largely based upon electromechanical logic and signal-processing systems, and were adequate for data communications until several years following the commercial appearance of electronic computing systems. However, as computers became faster and capable of handling more data and as their use expanded into areas other than simple "number crunching," it became obvious that large volumes of digital data would have to be electrically transported from one place to another. This is when drastic changes in the digital-data communications area really began to occur.

11.2 INFORMATION CODES

11.2.1 Types of Information Codes

The earliest telegraph code used was the *Morse Code*. This was later modified slightly to produce the *International Morse Code* (see Figure 11-1). This code consists of an unique combination of dots and dashes for each character. The

A	· −	O	− − −	1	· − − − −
B	− · · ·	P	· − − ·	2	· · − − −
C	− · − ·	Q	− − · −	3	· · · − −
D	− · ·	R	· − ·	4	· · · · −
E	·	S	· · ·	5	· · · · ·
F	· · − ·	T	−	6	− · · · ·
G	− − ·	U	· · −	7	− − · · ·
H	· · · ·	V	· · · −	8	− − − · ·
I	· ·	W	· − −	9	− − − − ·
J	· − − −	X	− · · −	0	− − − − −
K	− · −	Y	− · − −	.	· − · − · −
L	· − · ·	Z	− − · ·	,	− − · · − −
M	− −	(Pause)	− · · · −	?	· · − − · ·
N	− ·				

Figure 11-1 The International Morse Code.

length of a dash is specified at three times the length of a dot (the basic unit of length) with the interdot/dash interval being one unit long and the intercharacter interval being three units (one dash length) long. However, since this code was, and in many cases still is, generated ("sent") manually by a human operator, the actual spacing depends greatly upon the sending characteristics or "fist" of the operator doing the sending.

One of the distinguishing characteristics of this code is the nonuniformity of the length of the coding for different characters of the alphabet (1 unit long for the letter "E" to 19 units long for the zero digit). This was done to increase the efficiency of the code—the most commonly used letters require less transmission time than do the less used letters, numerals, punctuation marks, and so on. Also, the blank space between words is just a pause (specified to be 9 units long) in the sending rather than an unique character as it is in the other codes that we will be discussing. Even with these somewhat irregular characteristics, it is possible to construct electronic equipment to generate Morse code from a keyboard and convert received Morse code into characters presented on a CRT or typed onto paper. Thus, it may be considered to be a valid binary coding scheme.

With the advent of the electromechanical teleprinter, the five-unit (5-bit) teleprinter code (often called the *Baudot Code*) was developed. This is available in several different versions for different applications (communications service, brokerage-house communications, weather reporting, etc.) as is shown in Figure 11-2. The Baudot Code is basically an *asynchronous* system code. The asynchronous method is also called the start-stop method, wherein when no data are being sent, the binary output is held at the *mark* level (one of the two binary levels—the other level is called the *space* level), which keeps the receiving machine's mechanism "locked out." Whenever a character is to be sent, the transmitting equipment changes the binary output to the space level for one pulse-unit length. This pulse is called the *start pulse*, and its purpose is to tell the receiving equipment that a character is arriving. Following the start pulse there are five binary data pulses representing the character or command function being sent. All six of these pulses are of the same length, but they are followed by a longer (approximately 1.41 units long) *stop pulse* at the mark level which tells the decoding equipment to stop decoding the binary stream and to print the received character or perform the specified function. The stop pulse was made longer than the other six pulses to allow the original mechanical receiving mechanisms to "catch up" with the sending mechanism in the event that the motor in the receiving machine was running a little bit slower than the motor in the sending machine. The Baudot Code is used at several different standard speeds which are specified in terms of average-length (five-character) words per minute. Figure 4-15 shows the 60-word-per-minute teleprinter code stream and Table 11-1 (page 342) lists the length of each of the first six pulses (the data-pulse length) and the length of the stop pulse for several standard teleprinter speeds.

Another code very often used in the start-stop mode is the *American Standard Code for Information Interchange* (or ASCII, often pronounced *as' key*). It is also extensively used for data storage in some computer applications and/or for storing digital characters on paper or magnetic tape, magnetic disks, and so on, and also

Code Signals							Lowercase	Uppercase for Different Teleprinter Systems			
	Denotes "Space" Polarity							CCITT Standard International Telegraph Alphabet 2	United States Teletype Commercial Keyboard	AT&T Fractions Keyboard	Weather Keyboard
Start	1	2	3	4	5	Stop					
	X	X				X	A	–	–	–	↑
	X			X	X	X	B	?	?	$\frac{5}{8}$	⊕
		X	X	X		X	C	:	:	$\frac{1}{8}$	○
	X			X		X	D	Who are you?	5	5	↗
	X					X	E	3	3	3	3
	X		X	X		X	F	Note 1	!	$\frac{1}{4}$	→
		X		X	X	X	G	Note 1	&	&	↘
		X		X	X	X	H	Note 1	'		↓
		X	X			X	I	8	8	8	8
	X	X		X		X	J	Bell	Bell	'	↙
	X	X	X	X		X	K	(	(	$\frac{1}{2}$	←
		X		X	X	X	L	)	)	$\frac{3}{4}$	↖
			X	X	X	X	M	.	.	.	.
			X	X		X	N	,	,	$\frac{7}{8}$	⊕
				X	X	X	O	9	9	9	9
		X	X		X	X	P	0	0	0	0
	X	X	X			X	Q	1	1	1	!
		X		X		X	R	4	4	4	4
	X		X			X	S	'	'	Bell	Bell
				X	X	X	T	5	5	5	5
	X	X	X			X	U	7	7	7	7
		X	X	X	X	X	V	=	;	$\frac{3}{8}$	⊕
	X	X		X	X	X	W	2	2	2	2
	X		X	X	X	X	X	/	/	/	/
	X	X		X	X	X	Y	6	6	6	6
	X			X	X	X	Z	+	"	"	+
						X	Blank				–
	X	X	X	X	X	X	Letters Shift				↓
	X	X		X	X	X	Figures Shift				↑
			X			X	Space				■
				X		X	Carriage Return				<
		X				X	Line Feed				≡

Figure 11-2 The 5-unit Baudot code.

for *synchronous* data transmission. When it is used in the asynchronous (start-stop) mode for data communication, it consists of a start pulse, seven data pulses, a parity-check pulse, and a stop pulse that may be either one or two units long. Odd parity is usually employed for the parity-check bit (or an odd number of the eight pulses must be space pulses). The start and stop pulses do not enter into the parity-checking scheme. Figure 11-3 tabulates the ASCII code and Table 11-2 lists the code, explaining the standard control-character assignments and giving the way an operator can generate these control characters on many of the standard computer terminal keyboards which are equipped with a *control-key* option.

Another data code in common use is the *EBCDIC Code* (Extended Binary-Coded-Decimal Information Code) shown in Figure 11-4. This is very commonly used in communicating between computer mainframes and interfaced peripheral devices. Other codes which are also used for character/numeral (or *alphanumeric*) representation are the *Hollerith Code*, which is used on computer punch cards, and the various codes used to represent alphanumeric data in internal storage in computer registers and memories. Quite often it is necessary to convert from one coding scheme to another as different pieces of equipment or subsystems must communicate with each other.

Two other data codes which are in use for special data communication purposes are the *6-bit transcode* (*SBT Code*) shown in Figure 11-5 and often used in interfacing with printers and other peripheral devices, and the 7-bit *ARQ Teleprinter Code* shown in Figure 11-6, wherein the representation for each valid character always has four 0 bits and three 1 bits (it is used on teleprinter systems requiring a low undetected-error probability). Figure 11-7 shows the Hollerith Code used with "computer punch cards." Row 12 is at the top edge of the card, followed by row 11 and then rows 0 through 9 with row 9 being at the bottom of the card. This gives a 12-bit code which has 4096 possible characters. Only 63 of them are defined as standard in the version shown in Figure 11-7.

At this point it might be well to discuss what is known as *encoding efficiency*. Since with an m-bit binary number it is possible to represent 2^m different possibilities, if an m-bit binary data code is used to represent less than 2^m possible items (characters, functions, or control operations) to be sent, the n-bit code will not be fully (100% efficiently) utilized. Thus, from Chapter 1, if there are $\log_2 p$ bits of

TABLE 11-1   BAUDOT AND ASCII ASYNCHRONOUS SPEED PARAMETERS

Baudot teleprinter speed (words per minute)	"60"	"75"	"100"
Data pulse length (milliseconds)	22	17.57	13.47
STOP pulse length (milliseconds)	>31	>25	>19.18
Baud rate	≈45	56.9	74.2
ASCII link speed (words per minute)	100	300 1200	9600
Pulse length (milliseconds)	9.091	3.333 0.8333	0.1041
Baud rate for NRZ (per second)	110	300 1200	9600
Bits per character (asynchronous)	11	10 10	10

Bit 4	Bit 3	Bit 2	Bit 1	Bit 7	0	0	0	0	1	1	1	1
				Bit 6	0	0	1	1	0	0	1	1
				Bit 5	0	1	0	1	0	1	0	1
				Column / Row	0	1	2	3	4	5	6	7
0	0	0	0	0	NUL	DLE	SP	0	@	P		p
0	0	0	1	1	SOH	DC1	!	1	A	Q	a	q
0	0	1	0	2	STX	DC2	"	2	B	R	b	r
0	0	1	1	3	ETX	DC3	#	3	C	S	c	s
0	1	0	0	4	EOT	DC4	$	4	D	T	d	t
0	1	0	1	5	ENQ	NAK	%	5	E	U	e	u
0	1	1	0	6	ACK	SYN	&	6	F	V	f	v
0	1	1	1	7	BEL	ETB	'	7	G	W	g	w
1	0	0	0	8	BS	CAN	(	8	H	X	h	x
1	0	0	1	9	HT	EM	)	9	I	Y	i	y
1	0	1	0	A	LF	SUB	*	:	J	Z	j	z
1	0	1	1	B	VT	ESC	+	;	K	[	k	{
1	1	0	0	C	FF	FS	,	<	L	\	l	¦
1	1	0	1	D	CR	GS	−	=	M	]	m	}
1	1	1	0	E	SO	RS	.	>	N	^	n	~
1	1	1	1	F	SI	US	/	?	O	−	o	DEL

Figure 11-3 The Seven-Bit ASCII Code.

information for p possible items to be sent using an m-bit data coding scheme (where p must be $<2^m$ of course), we can define the encoding efficiency as

$$\text{encoding efficiency} = \left(\frac{\log_2 p}{m}\right)(100)\% \qquad (11\text{-}1)$$

Later on in this chapter we will be discussing other data communication efficiencies (the *transmission efficiency*, the *protocol efficiency*, the *channel efficiency*, etc.).

11.2.2 Error-Detecting Codes

In the telegrapher's Morse Code and the teleprinter's Baudot Code, there is no error detecting or correcting built into the code itself, and the sender has to depend upon whether or not the receiving operator or user *thinks* a correct message has been received. This is not a very reliable method in general, although for plain

TABLE 11-2 AMERICAN STANDARD CODE FOR INFORMATION INTERCHANGE (ASCII) CODE AND TYPICAL (IBM PC®) KEYBOARD CONFIGURATION FOR THE CONTROL CHARACTERS

	ASCII Data Code and Typical Keyboard Configuration		
Octal/Hex Value		Character or Function	Typical Keyboard*
000 00	NUL	nul or blank (no op)	@ [or @P
001 01	SOH	start-of-header	@ A
002 02	STX	start-of-text	@ B
003 03	ETX	end-of-text	@ C
004 04	EOT	end-of-transmission	@ D
005 05	ENQ	enquiry	@ E
006 06	ACK	positive acknowledgement	@ F
007 07	BEL	bell (or "beep")	@ G
010 08	BS	backspace	@ H
011 09	HT	horizontal tab character	@ I
012 0A	LF	line feed	@ J
013 0B	VT	vertical tab	@ K
014 0C	FF	form feed (new page)	@ L
015 0D	CR	carriage return (to left side)	@ M
016 0E	SO	shift out	@ N
017 0F	SI	shift in	@ O
020 10	DLE	data-line ESCAPE	@ P
021 11	DC1	device-control-function no. 1 or XON	@ Q
022 12	DC2	device-control-function no. 2	@ R
023 13	DC3	device-control-function no. 3 or XOFF	@ S
024 14	DC4	device-control-function no. 4 or STOP	@ T
025 15	NAK	negative acknowledgement	@ U
026 16	SYN	synchronization character	@ V
027 17	ETB	end-of-text-block	@ W
030 18	CAN	cancel	@ X
031 19	EM	end-of-medium	@ Y
032 1A	SUB	substitute	@ Z
033 1B	ESC	ESCAPE	ESC, @ \, @K
034 1C	FS	file separator character	@ =, @ L
035 1D	GS	group separator character	@ {, @ M
036 1E	RS	record separator character	@ −, @ N
037 1F	US	unit separator character	@ ', @ O
040 20	SP	space	space bar
041 21	!		$1
042 22	"		$'
043 23	#		$3
044 24	$		$4
045 25	%		$5
046 26	&		$7
047 27	'	closing single quote	'
050 28	(		$9
051 29	)		$0
052 2A	*		$8
053 2B	+		$ =

TABLE 11-2 Continued

		ASCII Data Code and Typical Keyboard Configuration		
Octal/Hex Value		Character or Function		Typical Keyboard*
054 2C	,	comma		.
055 2D	-	hyphen		-
056 2E	.	period		.
057 2F	/	forward slash		/
060 30	0			0
061 31	1			1
062 32	2			2
063 33	3			3
064 34	4			4
065 35	5			5
066 36	6			6
067 37	7			7
070 38	8			8
071 39	9			9
072 3A	:			$;
073 3B	;			;
074 3C	<	less than		$,
075 3D	=	equal to		=
076 3E	>	greater than		$.
077 3F	?			$/
100 40	@			$2
101 41	A			$A
102 42	B			$B
103 43	C			$C
104 44	D			$D
105 45	E			$E
106 46	F			$F
107 47	G			$G
110 48	H			$H
111 49	I			$I
112 4A	J			$J
113 4B	K			$K
114 4C	L			$L
115 4D	M			$M
116 4E	N			$N
117 4F	O			$O
120 50	P			$P
121 51	Q			$Q
122 52	R			$R
123 53	S			$S
124 54	T			$T
125 55	U			$U
126 56	V			$V
127 57	W			$W
130 58	X			$X
131 59	Y			$Y

TABLE 11-2 Continued

ASCII Data Code and Typical Keyboard Configuration			
Octal/Hex Value		Character or Function	Typical Keyboard*
132 5A	Z		$Z
133 5B	[		[
134 5C	\	backslash (correction)	\
135 5D	]		]
136 5E	^		$6
137 5F	_	underline	$_
140 60	‘	opening single quote	‘
141 61	a		A
142 62	b		B
143 63	c		C
144 64	d		D
145 65	e		E
146 66	f		F
147 67	g		G
150 68	h		H
151 69	i		I
152 6A	j		J
153 6B	k		K
154 6C	l		L
155 6D	m		M
156 6E	n		N
157 6F	o		O
160 70	p		P
161 71	q		Q
162 72	r		R
163 73	s		S
164 74	t		T
165 75	u		U
166 76	v		V
167 77	w		W
170 78	x		X
171 79	y		Y
172 7A	z		Z
173 7B	{	opening brace	$[
174 7C	|	vertical line	$\
175 7D	}	closing brace	$]
176 7E	~	tilde	$‘
177 7F	DEL	delete	DEL

*$ = shift key down

@ = control key down

The EBCDIC Data Code.

Bit Position 8 7 6 5 \ 4 3 2 1	0 0 0 0	0 0 0 1	0 0 1 0	0 0 1 1	0 1 0 0	0 1 0 1	0 1 1 0	0 1 1 1	1 0 0 0	1 0 0 1	1 0 1 0	1 0 1 1	1 1 0 0	1 1 0 1	1 1 1 0	1 1 1 1
0 0 0 0	NUL	SOH	STX	ETX	PF	HT	LC	DEL			SMM	VT	FF	CR	SO	SI
0 0 0 1	DLE	DC_1	DC_2	DC_3	RES	NL	BS	IL	CAN	EM	CC		IFS	IGS	IRS	IUS
0 0 1 0	DS	SOS	FS		BYP	LF	EOB	PRE			SM			ENQ	ACK	BEL
0 0 1 1			SYN		PN	RS	UC	EOT					DC_4	NAK		SUB
0 1 0 0	SP										¢	.	<	(	+	¦
0 1 0 1	&										!	$	*	)	;	¬
0 1 1 0	−	/										,	%	−	>	?
0 1 1 1												:	#	@	,	=	"
1 0 0 0		a	b	c	d	e	f	g	h	i						
1 0 0 1		j	k	l	m	n	o	p	q	r						
1 0 1 0			s	t	u	v	w	x	y	z						
1 0 1 1																
1 1 0 0		A	B	C	D	E	F	G	H	I						
1 1 0 1		J	K	L	M	N	O	P	Q	R						
1 1 1 0			S	T	U	V	W	X	Y	Z						
1 1 1 1	0	1	2	3	4	5	6	7	8	9						■

Figure 11-4   The EBCDIC data code.

@	Ø Ø	P	2 Ø	□	4 Ø	Ø	6 Ø
A	Ø 1	Q	2 1	!	4 1	1	6 1
B	Ø 2	R	2 2	"	4 2	2	6 2
C	Ø 3	S	2 3	#	4 3	3	6 3
D	Ø 4	T	2 4	$	4 4	4	6 4
E	Ø 5	U	2 5	%	4 5	5	6 5
F	Ø 6	V	2 6	&	4 6	6	6 6
G	Ø 7	W	2 7	'	4 7	7	6 7
H	1 Ø	X	3 Ø	(	5 Ø	8	7 Ø
I	1 1	Y	3 1	)	5 1	9	7 1
J	1 2	Z	3 2	*	5 2	:	7 2
K	1 3	[	3 3	+	5 3	;	7 3
L	1 4	\	3 4	,	5 4	<	7 4
M	1 5	]	3 5	−	5 5	=	7 5
N	1 6	↑	3 6	.	5 6	>	7 6
O	1 7	←	3 7	/	5 7	?	7 7

Figure 11-5   The SBT data code.

Letters Shift	Figures Shift	1	2	3	4	5	6	7
A	–			X	X		X	
B	?			X	X			X
C	:	X			X	X		
D				X	X	X		
E	3		X	X	X			
F				X			X	X
G		X	X					X
H		X		X			X	
I	8	X	X	X				
J	Bell		X				X	X
K	(				X		X	X
L	)	X	X				X	
M	.	X		X				X
N	,	X		X		X		
O	9	X				X	X	
P	0	X			X		X	
Q	1				X	X		X
R	4	X	X			X		
S	'		X		X		X	
T	5	X				X		X
U	7		X	X			X	
V	=	X			X			X
W	2		X			X		X
X	/			X		X	X	
Y	6			X		X		X
Z	+		X	X				X
Carriage Return		X					X	X
Line Feed		X		X	X			
Letters Shift					X	X	X	
Figures Shift			X			X	X	
Space		X	X		X			
Unperforated Tape						X	X	X
Control Signal			X	X		X		
Idle Signal β			X		X	X		
Idle Signal α			X		X			X

Figure 11-6 The 7-Bit ARQ teleprinter code.

text in a *natural* language (English, French, Spanish, etc.), it is not too much of a problem because of the *redundancy* of natural languages (i.e., letters can be left out or words misspelled and, in general, the text is still "readable"). However, when data such as addresses, telephone numbers, catalog numbers, and computer information are being communicated, a transmission error can occur and a character

@	8, 4	P	11, 7	Blank	(No Punch)	Ø	Ø
A	12, 1	Q	11, 8	- - - - - - - - - - -		1	1
B	12, 2	R	11, 9	"	Ø, 8, 5	2	2
C	12, 3	S	Ø, 2	#	Ø, 8, 6	3	3
D	12, 4	T	Ø, 3	$	11, 8, 3	4	4
E	12, 5	U	Ø, 4	%	Ø, 8, 7	5	5
F	12, 6	V	Ø, 5	&	11, 8, 7	6	6
G	12, 7	W	Ø, 6	'	8, 6	7	7
H	12, 8	X	Ø, 7	(	Ø, 8, 4	8	8
I	12, 9	Y	Ø, 8	)	12, 8, 4	9	9
J	11, 1	Z	Ø, 9	*	11, 8, 4	↑	11, Ø
K	11, 2	[	11, 8, 5	+	12	;	Ø, 8, 2
L	11, 3	\	8, 7	,	Ø, 8, 3	<	12, 8, 6
M	11, 4	]	12, 8, 5	−	11	=	8, 3
N	11, 5	↑	8, 5	.	12, 8, 3	>	11, 8, 6
O	11, 6	←	12, 8, 7	/	Ø, 1	?	12, Ø

Figure 11-7   The Hollenrith data code.

can be altered with little or no chance of being detected. Therefore, most modern data communication systems (and also digital data storage systems) use some form of elementary error detecting and/or correcting codes.

One of the simplest forms of error-detecting is to include an extra parity-check bit for each character. For example, in the ASCII code (7 data bits), an extra eighth bit can be sent, the value of this eighth bit being such that there is always an odd number of 1 bits in the total of eight (for *odd parity*) or an even number of 1 bits in the total of eight (for *even parity*). The adding of a parity bit to each character is called *vertical redundancy checking* (VRC). In addition, a second type of parity checking referred to as *longitudinal redundancy checking* (LRC) (sometimes also called *serial* or *run-length* parity) may be employed. In this method an extra 8-bit character is sent at the end of each group (usually a CRT line or some predetermined size block) of data characters. Each of the bits in this extra check character is determined so that the total number of 1 bits occupying the same bit position in the block is an odd number (for odd parity) or an even number (for even parity).

Another error-detection technique is that of generating a *checksum* number when a list of data numbers is being sent—the checksum is generated from the actual numerical data according to some numerical formula or algorithm at both the transmitting and the receiving ends of the data link. The number generated at the transmitting end is sent to the receiving end following the stream of actual numerical data. The receiving end compares this checksum number received from the transmitting end with the checksum number which it generated from the actual received data to see if the two checksums agree. If they do not agree, a transmission error has occurred either in transmitting the actual data bits or the checksum bits.

Another group of error-detection techniques exists which somewhat resembles the checksum method except that these techniques are based upon data *bit streams* (just sequences of binary data that have no particular meaning to the error-checking technique), whereas the checksum technique is usually based upon a sequence of bit groups which represent a sequence of decimal digits. In general any technique in this group of methods will produce an *r*-bit binary stream referred to as the *BCC* (*block check character*) or *FCS* (*frame check sequence*). We will use the term BCC in most of the following discussion. This sequence of *r* check bits is formed from a sequence of *k* data bits according to some binary algorithm. The total *codeword* which is transmitted consists of the *k* data bits followed by the *r* BCC bits, giving a total of $n = k + r$ bits in the codeword. One of the most commonly used methods for generating the BCC is via the use of the *polynomial code* or *cyclical redundancy code* (*CRC*) method.

The CRC method is based upon a technique of binary *modulo-2* long division of one binary number (the *k*-bit data stream with *r* zeros (or a specified *r*-bit sequence)[1] appended to it) by another binary number of $(r + 1)$ bits called the *generator function* to produce a quotient (in which we are not going to be the least bit interested) and an *r-bit remainder* which we use as the BCC to be sent following the *k* data bits. At this point we will introduce a form of notation that we will be using: to represent a string of *s* bits algebraically, we will construct an (s − 1)-order polynomial in the variable *x* where the coefficients of the polynomial have values of either 0 or 1 according to the bit values in the binary number. For example, we could represent the 8-bit binary number (10011011) as

$$(10011011)_2 <=> 1x^7 + 0x^6 + 0x^5 + 1x^4 + 1x^3 + 0x^2 + 1x^1 + 1x^0$$

$$<=> x^7 + x^4 + x^3 + x + 1 = R(x)$$

where the double-headed arrow means "is represented by." Using this technique, we can then represent the *k*-bit message by a $(k − 1)$ maximum-order polynomial $M(x)$; the $(r + 1)$-bit generator function by an *r*-order polynomial $G(x)$ (the *r*-order and the 0-order terms in $G(x)$ are always taken to have unity coefficients) and the *r*-bit BCC by the $(r − 1)$ maximum-order polynomial $C(x)$.[2] The transmitted *codeword* sequence consisting of $M(x)$ followed by $C(x)$ will be represented by the $(n − 1)$ maximum-order polynomial $T(x)$.

It should be noted that in handling these binary sequences serially, the bit representing the highest order in the polynomial is always sent first.

Before showing an example of an actual BCC generator using this technique, we need first to define modulo-2 binary arithmetic. Basically, it involves binary addition or subtraction without any "carrying" or "borrowing" (i.e., each column is "handled by itself"). It is best shown by a series of examples:

---

[1] The HDLC/LAP-B data-link protocol used in the X.25 packet network prescribes this to be 16 1 bits, with the 1's complement of the remainder being transmitted instead of the remainder itself (see Appendix G).

[2] Be careful here: we will use *M, G, C,* and *T* for the *message*, the *generator*, the *BCC*, and the *codeword* sequences, respectively. Other authors use [*M, G,* −, *T*] or [*û, ĝ, b̂, v̂*] or [*M, P, R, T*] or [*G, P, C, F*] or [etc. etc. etc.], respectively!

Examples:

$$
\begin{array}{r}
10110 \\
+11010 \\
\hline
01100
\end{array}
\qquad
\begin{array}{r}
10110 \\
-11010 \\
\hline
01100
\end{array}
\qquad
\text{(i.e., modulo-2 addition is the same}
$$
as modulo-2 subtraction!)

$$
\begin{array}{r}
110101 \\
101110 \\
+110111 \\
\hline
101100
\end{array}
$$
(i.e., an odd number of 1's in a column gives a 1 in the sum, an even number gives a 0 in the sum)

The behavior of modulo-2 addition is just the *exclusive-OR* (or ExOR) operation of binary logic-gate circuitry. This makes the electronic implementation of the modulo-2 long-division operations we are about to discuss quite simple to implement via conventional electronic logic circuitry.

To keep the first example simple, we will start with a fairly short (twelfth order) generator polynomial:

$$G(x) = x^{12} + x^{11} + x^3 + x^2 + 1$$

This 13-bit sequence is referred to as the standard *CRC-12 generator polynomial*. It is usually used with sequences of 6-bit transcode (SBT)-coded characters. *As an example* we will suppose a 12-bit BCC is generated for every four sequential characters (a sequence of $k = 24$ bits). Thus, letting $M(x) = (110101111101000101011011)_2$, appending 12 0's to this, and dividing by $(1100000001101)_2$, we have the process shown in the following example, which gives $(010010110010)_2$ as the BCC function represented by $C(x)$. If we append this check sequence, C, to the message sequence, M, to form the total codeword, T, and transmit T to the receiving end of the data link *without any errors* and then divide the received T by G, we should get a zero remainder,[3] indicating that no errors did occur in the transmission. (See the example on pages 352 & 353.)

As mentioned previously, the modulo-2 addition operation can be implemented via the use of an ExOR logic circuit. The CRC-12 logic operation can be implemented by the use of the circuit shown in Figure 11-8a. Other common generator functions are shown in parts (b), (c), and (d) of Figure 11-8. The basic operation of any of these CRC shift-register generators is as follows:

1. Begin with the switch in position A and 0's in all stages of the shift register (or else the register loaded with a known binary number as in the case of the HDLC/LAP-B error checking).

---

[3] It is very simple to prove that if no transmission errors occur, the remainder at the receiving end is zero: when $(2^r M)$ is divided by G at the transmitting end, we get $(2^r M)/G = Q + (C/G)$. The codeword T is then $(2^r M) + C$. Dividing this again by G at the receiving end gives $[Q + (C/G)] + (C/G) = Q + \left(\dfrac{C + C}{G}\right)$ where the numerator of the second term is the remainder left in the register (which is zero since the sum of any modulo-2 number added to itself is zero).

Example:

at the transmitting end:

```
                              10011010110111011001010
1100000001101    11010111110100010101101 1000000000000
                 1100000001101
   G             001011110111001   M        12 zeros
                 1100000001101
                 1111101101000
                 1100000001101
                  01110110010110
                  1100000001101
                   01011001101111
                   1100000001101
                    1110011000100
                    1100000001101
                     01001100100111
                     1100000001101
                      1011001010100
                      1100000001101
                       1110010110010
                       1100000001101
                        01001011111100
                        1100000001101
                         1010111100010
                         1100000001101
                          1101111011110
                          1100000001101
                           001111010011000
                           1100000001101
                            01101001010100
                            1100000001101
                             0010010110010
                             0000000000000
Contents of the CRC register at  }
the end of the message stream, M }  = C ——→ 010010110010
```

2. Put a data bit on the input data line and wait until the propagation of the logic feedback signal occurs (usually a small fraction of the bit period).

3. Shift the register and wait for the remainder of the bit period.

4. If the bit in step 2 was *not* the last data bit in M, go back through steps 2 and 3, or if the data bit in step 2 *was* the last data bit in M, throw the switch to position B to obtain the first bit in the BCC.

5. Shift the register at the same rate as the bit rate of the M data stream until the data in the leftmost register stage has been shifted into the rightmost stage and put onto the output line.

Example: (continued)

and at the receiving end:

```
                                    10011010101101110111001010
                 _____
1100000001101    11010111110100010101101101010010110010
‿‿‿‿‿            1100000001101
      G           001011110111001   M          C
                  1100000001101
                  1111101101000
                  1100000001101
                   01110110010110
                   1100000001101
                    01011001101111
                    1100000001101
                     1110011000100
                     1100000001101
                      01001100100111
                      1100000001101
                       1011001010100
                       1100000001101
                        1110010110011
                        1100000001101
                         01001011111000
                         1100000001101
                          1010111101011
                          1100000001101
                           1101111001100
                           1100000001101
                            001111000001110
                            1100000001101
                             1100000001101
                             1100000001101
                             0000000000000
                             0000000000000
```

Zero remainder shows that no } = C ⟶ 000000000000
transmission errors have occurred }

The operational states of a CRC register can be mapped for any input data stream by means of a table such as that shown in Figure 11-9. In this case we are showing the CRC-12 register states for the same 24-bit input data stream used in the modulo-2 binary long-division example. The use of the expanded form of the table shown in Figure 11-9a is as follows:

1. Begin with all zeros (or the preset code) in the top row of digits representing the shift-register states.
2. In the *data bit* column, enter the data sequence of M from top to bottom.

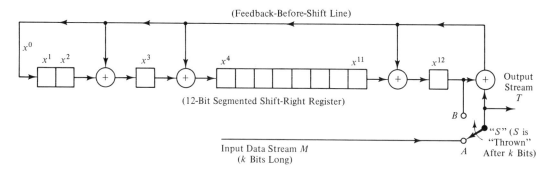

(a) CRC Register for Generating the CRC-12 BCC,
Where $G(x) = x^{12} + x^{11} + x^3 + x^2 + 1$

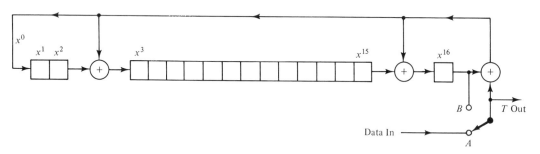

(b) CRC Register for CRC-16, Where $G(x) = x^{16} + x^{15} + x^2 + 1$

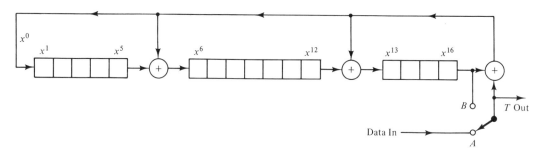

(c) CRC Register for CRC-CCITT, Where $G(x) = x^{16} + x^{12} + x^5 + 1$

Figure 11-8 Cyclical redundancy code generation by means of segnemted shift registers.

3. At the right-hand side, do the ExOR operation involving the state of the rightmost register stage and the input data bit and put the result in the *feedback bit* column at the right and also in the leftmost column of the *next* row down.

4. Use the feedback bit and the bit to the upper-left of each ExOR symbol in an ExOR operation and place the result in the position to the lower right of the ExOR symbol.

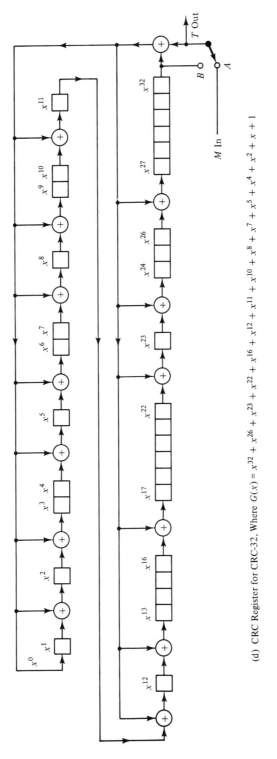

(d) CRC Register for CRC-32, Where $G(x) = x^{32} + x^{26} + x^{23} + x^{22} + x^{16} + x^{12} + x^{11} + x^{10} + x^8 + x^7 + x^5 + x^4 + x^2 + x + 1$

Figure 11-8 Continued

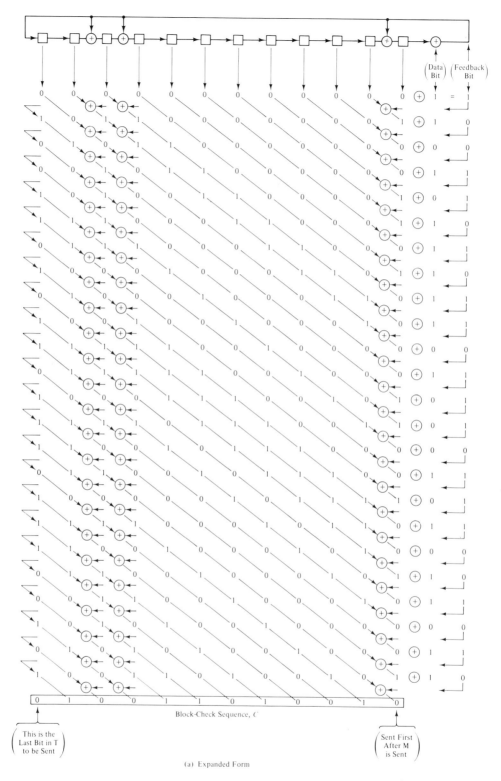

(a) Expanded Form

Figure 11-9 Operational table for the CRC-12 register with the message sequence of the example. Time runs from top to bottom.

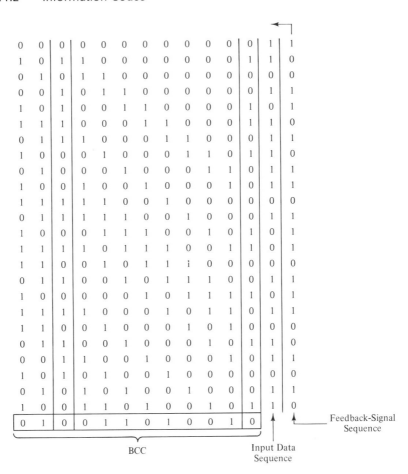

Feedback-Signal Sequence

Input Data Sequence

BCC

(b) Condensed Form

Figure 11-9   Continued

5. Shift the remaining bits not involved in ExOR operations downward to the right to fill in the remainder of the row.
6. Keep repeating steps 3 through 5 until all the input data bits have been utilized. The last row of the table is the CRC check sequence (reading from right to left).

Figure 11-9b is a more condensed version of this same table and is the form usually used in practice (some people prefer to displace the input data column downward by one row). The vertical dividing lines are placed at positions corresponding to the ExOR gate locations between the shift-register stages.

CRC generator circuits of the type just described are used at the transmitting end to generate the BCC sequence which is sent following the sequence of data bits, and also at the receiving end to verify that no bits in T have been mutilated during transmission. In the receive-end application, the switch is left in position A and *all* the bits in T (the BCC bits as well as the data bits) are "clocked through" the register. After thus shifting the register n times, its contents should be zero (or equal to some specified binary value – $(0001\ 1101\ 0000\ 1111)_2$ in the case of the HDLC/LAP-B protocol shown in Appendix G). If it is not so, then one or more bit errors have occurred in T and the data-link protocol (see Section 11.4.3) can begin to take corrective action.

It should be noted at this point that even though the final state of the receiver's CRC register might indicate that no bit errors have occurred, there is a very small finite probability that some combination of bit reversals occurring in T will not be detected. This probability increases as the number of data bits, k, increases relative to the number of check bits, r. A major area of theoretical study has been in the area of designing generating polynomials, $G(x)$, to optimize the detection of the most probable types of bit error combinations in a bit stream of reasonable length without using an excessively long BCC sequence.

It should also be noted in closing this section that the use of the most commonly accepted CRC register circuits (such as those shown in Figure 11-8) has become so widespread that most types of these register circuits are now available as single-package integrated circuits at very reasonable cost.

11.2.3 Error-Correcting Codes

Prior to considering error-correcting coding schemes themselves, let us look at some basic relationships among the number of possible k-bit *messages* and n-bit *codewords* that are possible with given values for k and n. If we have k bits in the message sequence, then there are 2^k possible valid messages which could be sent (they are all "valid" at this level of consideration even though many of them might be "binary nonsense" at some higher level of consideration).

After adding r bits of error-detecting or error-correcting information to create an n-bit codeword, where $n = k + r$, we find that there are 2^n possible codewords, or $2^n/2^k = 2^{(n-k)} = 2^r$ possible codewords for every possible valid message. Ideally, if we design our error-detecting or error-correcting code properly, there will be only one possible codeword associated with each possible message. Thus, out of the set of 2^n *possible* codewords, there is a subset of 2^k *valid* codewords. All the other $(2^n - 2^k)$ codewords are "invalid," and if one of them arrives at the receiving end of a data link, we may presume that one or more bit errors has occurred during the transmission of the codeword.

At this point we will define an important coding concept: the *distance* of a code. This is defined as the number of bits which must be changed to convert a given *valid* codeword into *another valid codeword*. For example, if $n = k + 1$, then there will be one invalid codeword for each valid codeword, and *on the average,* changing one bit in a valid codeword would produce an invalid codeword, but changing two bits would produce a *wrong* valid codeword.

For example, the valid codewords might contain an odd number of 1 bits (i.e., odd parity checking employed) and invalid codewords an even number of 1 bits.   An error of one bit would be caught by the parity-checking system, but two bits altered in a valid codeword would produce another valid (but wrong) codeword. Thus, we would say that the *average code distance* in the case where $n = k + 1$ is equal to 2 (two changes required to get from one valid codeword to another). For the case of using this extra bit for odd or even parity checking, the distance is *constant* at 2.   For other schemes where $n = k + 1$, the distance will depend upon *which* valid codeword is being considered (it might be 1 for some and 3 for others, for example).   For a given error detection and/or correction coding scheme, we will be interested in the *minimum distance* that occurs for *any* valid codeword in the set of all valid codewords.   The minimum distance of a code is commonly referred to as its *Hamming distance*.

One of the goals of designing an effective error-correcting and/or detecting code is to maximize the Hamming distance for the number of data and check bits to be used.   A code with a Hamming distance of 2 can be used as an error-detecting code if it can be assumed that no more than one bit error can occur in the sequence *T*, but it is unreliable if this assumption cannot be assured.   Similarly, if a Hamming distance of 3 exists, then there will be one valid and two or more one-bit-in-error invalid codewords associated with each possible message.   This would allow us to do some elementary error *correcting* if we can assume that only one bit in the sequence *T* is in error.

One type of error-correcting codes that does have a Hamming distance of 3 is the class of codes called *Hamming codes*.   They must have $r \geqslant 3$ and be related to *k* by

$$k_{\max} = (2^r - r - 1) \tag{11-2}$$

Thus, for $r = 3$, $k_{\max} = 4$; for $r = 4$, $k_{\max} = 11$; for $r = 5$, $k_{\max} = 26$; and so on.   The position of the data bits and the check bits in *T* is very important. Expressing *T* in polynomial form, we find that the check bits must occur at x^{n-1}, x^{n-2}, x^{n-4}, x^{n-8}, x^{n-16}, ...   Thus, for $r = 4$ and $k = 11$, bits 1, 2, 4, and 8 (counting from the left) would be check bits, and the remainder would be data bits.

This type of coding is often used in encoding a character set where something better than parity-checking (which is only error *detecting*) is desired.   For example, one might add 4 Hamming-code error-correcting bits to the 7 bits of the standard ASCII code (putting the check bits into positions 1, 2, 4, and 8 and the original ASCII data bits into positions 3, 5, 6, 7, 9, 10, and 11), making an 11-bit code sequence for each character in the ASCII set.   This would allow error correcting for one-bit-per-character errors to take place at the receiving end (or, alternatively, it could be used to allow the *detection* of one *or two* bits per character in error).

Another technique for generating error-detecting and/or -correcting code streams which operates on a continuous basis, producing a highly redundant code *stream* (as opposed to producing a BCC which is appended onto the end of an original data message to produce a code *word*), is known as *convolutional coding*.   Figure 11-10 shows the concept of a *parallel* convolutional encoder which will produce $n = 6$ outputs for every $k = 4$ input bits (i.e., $i = 1, 2, \ldots, k; j = 1, 2, \ldots, n$). It is called *convolutional* because the *n* outputs not only depend upon the *k* current

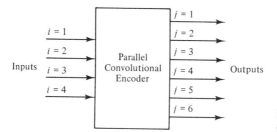

Figure 11-10 The parallel convolutional encoder circuit.

(i.e., $p = 0$) binary inputs, but also upon the past m states (i.e., $p = 1, 2, \ldots, m$) of the k input data lines ($p = m$ represents the oldest input data group affecting the current output while $p = 1$ is the most recent past group and $p = 0$ is the group currently on the k input lines).

Figure 11-11a shows a "(2, 1, 3)" convolutional encoder where $n = 2$, $k = 1$, and $m = 3$. Since this encoder utilizes three preceding $\mathbf{u}^{(1)}$ input states, a three-stage serial-in/parallel-out shift register must be utilized. Two parallel output bits on lines $\mathbf{v}^{(1)}$ and $\mathbf{v}^{(2)}$ are produced. Since parallel data lines are not very common, what is often done is to provide a shift-out register (the dashed lines on the drawing) which is loaded in parallel and "latched-in" as soon as $\mathbf{v}^{(1)}$ and $\mathbf{v}^{(2)}$ are available after the input data bit appears on $\mathbf{u}_1$ and the main shift register is shifted. These output data are then clocked out serially as the sequence ($\mathbf{v}^{(1)}$, $\mathbf{v}^{(2)}$). The bit rate on the serial output line obviously must be twice that on the input data line $\mathbf{u}^{(1)}$.

One way to describe the logic and operation of a convolutional encoder is via a directed state diagram. Figure 11-12 shows the state diagram for the encoder which we have been discussing. The numbers in the state circles represent what is stored in the shift register (most recent on the right and the oldest on the left), and the information along the transition line is the current input and output prior to the next shift of the register. Each output, $\mathbf{v}^{(1)}$ and $\mathbf{v}^{(2)}$, is given in terms of the convolution of $\mathbf{u}^{(1)}$ with a generating function

$$\mathbf{v}^{(1)} = \mathbf{u}^{(1)} * \mathbf{g}_1^{(1)} \tag{11-3a}$$

and

$$\mathbf{v}^{(2)} = \mathbf{u}^{(1)} * \mathbf{g}_1^{(2)} \tag{11-3b}$$

Since $\mathbf{v}^{(j)}$ can be represented as a time sequence where $v_p^{(j)}$ is the value of $\mathbf{v}^{(j)}$ at time increment p, we can express these convolutional relationships as (with $p_{\max} = m = 3$)

$$v_p^{(1)} = u_p^{(1)} g_{1,0}^{(1)} + u_{p-1}^{(1)} g_{1,1}^{(1)} + u_{p-2}^{(1)} g_{1,2}^{(1)} + u_{p-m}^{(1)} g_{1,m}^{(1)} \tag{11-4a}$$

$$v_p^{(2)} = u_p^{(1)} g_{1,0}^{(2)} + u_{p-1}^{(1)} g_{1,1}^{(2)} + u_{p-2}^{(1)} g_{1,2}^{(2)} + u_{p-m}^{(1)} g_{1,m}^{(2)} \tag{11-4b}$$

When we have more than one input as shown in Figure 11-11b, things work the same, but get to be a little more messy. In this case, since $m = 1$, the foregoing

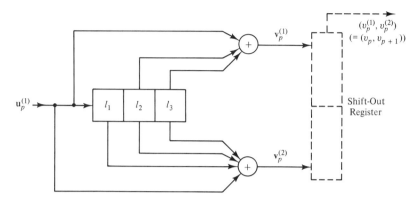

(a) A (2, 1, 3) Convolutional Encoder

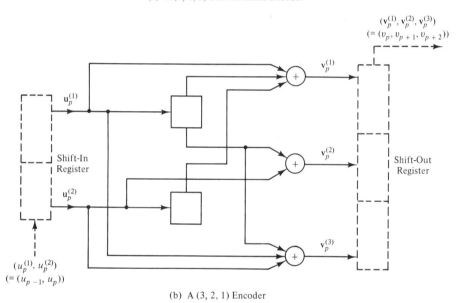

(b) A (3, 2, 1) Encoder

Figure 11-11   Convolutional encoders.

equations become, using $g_{i,p}^{(j)}$ elements

$$v_p^{(1)} = [u_p^{(1)}g_{1,0}^{(1)} + u_{p-1}^{(1)}g_{1,1}^{(1)}] + [u_p^{(2)}g_{2,0}^{(1)} + u_{p-1}^{(2)}g_{2,1}^{(1)}] \qquad (11\text{-}5a)$$

$$v_p^{(2)} = [u_p^{(1)}g_{1,0}^{(2)} + u_{p-1}^{(1)}g_{1,1}^{(2)}] + [u_p^{(2)}g_{2,0}^{(2)} + u_{p-1}^{(2)}g_{2,1}^{(2)}] \qquad (11\text{-}5b)$$

$$v_p^{(3)} = [u_p^{(1)}g_{1,0}^{(3)} + u_{p-1}^{(1)}g_{1,1}^{(3)}] + [u_p^{(2)}g_{2,0}^{(3)} + u_{p-1}^{(2)}g_{2,1}^{(3)}] \qquad (11\text{-}5c)$$

The formal mathematics of convolutional encoders involves working with the fore-going relationships in matrix form.   The design of convolutional generating-function matrices and the receiving-end error-correcting algorithms for specific types

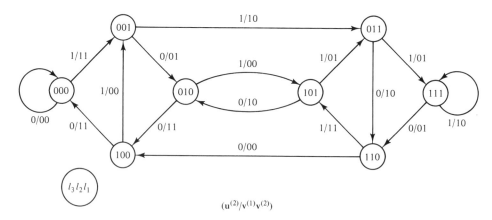

Figure 11-12 State diagram for the convolutional encoder of Figure 11-11a.

of data and predicted error characteristics and statistics becomes a very involved and highly mathematical subject.

In the first part of Section 11.2.2 we mentioned that a 16-bit BCC could do a good job of providing reliable error *detection* for data sequences ranging from a few dozen to several hundred bits, depending upon the link's BER (bit error rate). Thus, the "overhead" (or extra data bits which must be sent) is fairly low for error *detection*. Error *correction* (often called *forward error correction,* or *FEC*) is quite another matter: we often find that the overhead in terms of added bits may be as high as 30 to 100%. For example, in the "compact disk" optical audio digital recording system discussed in Chapter 7, 112 error-correcting bits are added to every 336 EFM-coded data bits (a 33.3% overhead). Because of this large overhead factor, we usually find that error *detection* is employed along with "go back and send it again" type protocols for fixing data errors except in those instances where it is not possible to, or it would be inconvenient to, back up and resend the data again; or else where the excess channel capacity needed to support forward error correction is readily available in the system under consideration.

11.2.4 Parallel versus Serial Data Transmission

Once a code has been established, one of the next factors to consider is whether the bits are to be sent along a single circuit one at a time, or whether all the bits for one character are to be sent simultaneously along as many circuits as there are bits in the character, or whether some combination of these two basic methods is to be employed. If the simultaneous method (called the *parallel method*) is employed, the bit rate on any one circuit is much slower, and therefore a narrower bandwidth is required for each circuit than would be needed for the single-circuit method (called the *serial method* of data transmission) for the same data-transfer rate. Or, conversely, much more data could be transferred by the parallel method if the bandwidth on each parallel circuit is the same as for the one serial circuit.

Parallel techniques are usually used for short-distance, high-data-rate links such as from a computer mainframe to high-speed peripheral devices such as magnetic tape and disk drives, communication-interface equipment, other nearby computers, instrumentation in a laboratory or factory, and so on, whereas serial data transmission is used over most long-distance links and for short-distance communications with slow-speed devices such as computer terminals, card and paper-tape readers, and punches and character printers.

Sometimes a combined parallel/serial technique is used, especially over longer-distance links where relatively high data-transfer rates are needed. These combined techniques quite often employ some form of multiphase quadrature-shift keying (QSK) or multiphase-amplitude (QAM) modulation methods on the link itself as is discussed in Section 11.2.5 and in Chapter 4. If, for example, 8-bit ASCII is used as the coding technique and if four-phase QSK is used as the modulation method, 2 bits can be sent simultaneously, and it takes only four pulses to transmit each ASCII character. If 16-phase/amplitude QAM is used, then only two pulses are required to transmit the 8 bits of information about the character. If 32-state QAM is used, then 5 bits are transmitted per pulse. When this is the case, the usual procedure is to combine a group of characters together (5 ASCII characters or 40 bits, for example) in what is called a *frame* of data in such a way as the total number of bits in the frame is some integer multiple of the number of bits represented by each transmission pulse (the 40 bits of our example could be transmitted by 8 successive 32-level pulses). The word *frame* is also used, of course, to mean a group of bits which are sent as a synchronously transmitted bit stream.

11.2.5 Generating the Baseband Signal

Whether the binary data are to be transmitted in parallel or serially, we still have the situation of converting a string of data bits intended for a particular circuit into an electrical time-function signal on that circuit. We call the resulting electrical signal the *baseband data signal*. This signal can then be directly transmitted over a pair of wires to the receiving site, or it can be used to modulate some form of carrier signal which is then sent over an audio, RF, or optical link to the receiving site. The former technique is called a *baseband link* or sometimes a *local-loop data link* and is usually now used only for relatively short distances (less than several miles), although in the past such types of links were used for telegraph and teleprinter communications over thousands of miles.

In Figure 11-13 are shown several common means by which the bit stream 111011001011 (for example) can be converted into an electrical baseband signal. Figure 11-13a shows the *bipolar* (or $+/-$) *nonreturn-to-zero* (NRZ) implementation of the bit stream. Note, first, that with the NRZ method, there is only one signal transition per bit (at the most, for the case of alternating 0 and 1 bits), whereas in the other four methods shown, there is a maximum of two transitions per bit. This is an important consideration because it affects the channel bandwidth requirements. Second, if there is not, on the short-term average, an equal number of 0's and 1's present in the bit stream, a positive or negative average d.c. level (called *signal bias*) will begin to be present on the transmission line, and this may cause problems on the line or at the receiving end. Also, the NRZ code is not

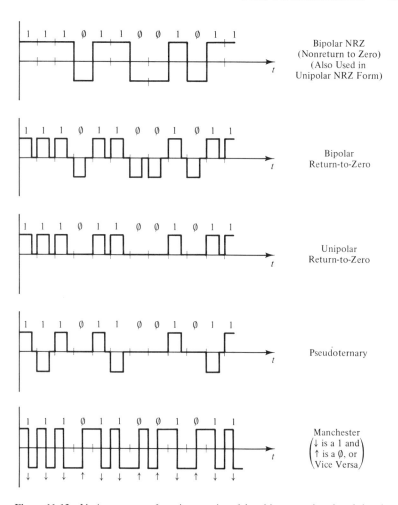

Figure 11-13   Various means of putting a series of data bits onto a baseband signal.

"self-clocking." This means that when there are long streams of 0's or 1's, some form of independent synchronism may be needed at the receiving end to decode a synchronous bit stream. The codes shown in (d) and (e) were designed to reduce or eliminate these types of problems.

At this point is is probably good to clear up some common problems associated with the various terms used to describe data-transfer rates. Probably the most basic measure of data rate is the bits-per-second (bps) rate, although for teleprinter communication links, the words-per-minute figure discussed earlier is often used. Transferring from one of these measures to the other may be a problem if we are talking about an asynchronous system where the stop-pulse length is not the same as the data-pulse length. Also, we have to make clear whether we are even including the start and stop pulses in the "bits" figure. If we are discussing the transmission system, then we *do* want to include them when calculating the *transmission bit rate*.

Once the actual rate of *transmission* bits per second has been established, then we can talk about the *Baud* rate, which may or may not be equal to the bit rate. Since the Baud rate is defined as the modulation rate (the number of electrical-signal state changes or "transitions" per second), we see that the Baud rate is equal to the bit rate only in certain cases. The most common of these is the NRZ type of signal encoding. The other types of signal encoding used in Figure 11-13 all have a Baud rate equal to twice the bit rate. On the other hand, when we use the data signal to QPSK or QAM modulate a carrier, we may have two, three, or more bits represented by each signal state, and consequently the Baud rate would be only one-half, one-third, and so on of the data bit rate.

Next to be considered is the relationship between the Baud rate and the channel bandwidth and S/N ratio. For a baseband signal, this is somewhat dependent upon how much distortion of the transmitted pulse train can be tolerated by the receiving-end (or baseband digital-repeater) equipment. If only the fundamental component of the electrical signal is passed through the circuit, then since there are two transitions per cycle at the fundamental frequency, it is possible theoretically to have the baseband bandwidth of the transmission channel be one-half of the Baud rate. Usually something more than this is needed, and it is often the case to pass at least the third harmonic component so that the bandwidth is at least $1\frac{1}{2}$ times the Baud rate. The actual bandwidth to be chosen will depend upon factors such as the design of the receiving equipment, the pulse phase dispersion on the channel, and the channel signal-to-noise ratio.

The problem of signal bias has already been mentioned. In connection with this, it is important to realize that as an electrical signal pulse propagates along a transmission line, it is distorted due to the limited frequency passband of the line and the phase distortion (dispersion) which occurs because the different frequency components present do not all travel down the line with the same velocity. As a result, the nice square pulse which starts down the line at the transmitting end arrives at the receiving end of the line with rounded corners and sloping sides (see Figure 11-14b). Since the receiving-end equipment (or the digital-repeater equipment) makes a decision on which polarity of pulse is present by comparing the incoming signal with some set reference level, any bias which is present on the line will cause apparent lengthening or shortening of the received pulse (see Figure 11-14c). For this reason, signal-generation methods which reduce or eliminate signal bias are preferred for longer-distance baseband circuits at higher speeds. The pseudoternary code and the Manchester Code shown in Figures 11-13(d) and (e) are two such codes. The Manchester Code is especially popular in what are called "baseband local-area networks" (or LANs) and is sometimes referred to as the *diphase* code. In the *Ethernet* LAN standard, the Manchester Code with voltage levels nominally at 0.0 and -2.05 volts with the upward voltage transition representing a binary 1 is used on a coaxial cable to achieve data rates up to 10 Mbps. It should be noted that in using either of these codes, a trade-off is being made between eliminating or reducing the line-bias problem and having a higher Baud rate (and thus needing a wider bandwidth).

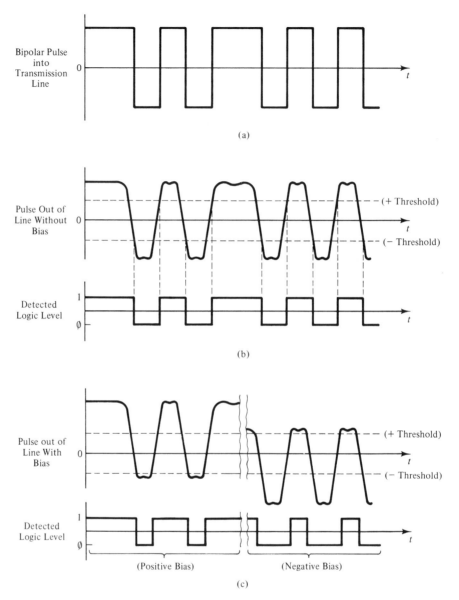

Figure 11-14   Effects of signal bias upon detected logic level.

11.2.6 *Transmitting the PCM Signal*

As has just been mentioned, the baseband PCM data signal may be put directly onto a coax cable or a pair of wires (or a group of wires for parallel or serial/parallel transmission) for transmission over relatively short distances. If longer distances are to be covered, this method is not too practical becuase it involves "conditioning" the baseband circuit by the installation of series loading coils and other techniques to minimize the phase dispersion effects.   For longer than a couple of miles, it is usually necessary to install some sort of digital-repeater stations that interpret the incoming distorted pulse train and regenerate a nice "fresh" train of square well-timed pulses.   However, in today's technology the usual practice is not to extend baseband links to the point where these techniques are required, but rather to use the baseband signal to modulate a carrier signal in some manner (see the discussion of PCM techniques in Chapter 4).

A very common method of data communications at the slower speeds is via the use of frequency-shift keying of an audio tone carrier (AFSK).   In the teleprinter field, one of the older standards was to use the audio tones of 2975 for the SPACE signal level and 2125 for the MARK level (giving a difference or SHIFT of 850 Hz), although later techniques began to utilize other closer-spaced pairs of frequencies, with shifts of less than 100 Hz now being common.   These same FSK techniques can also be applied to shifting an RF carrier (in fact, an AFSK signal applied to the audio input of an SSB voice transmitter will produce an RF/FSK signal output where the RF/FSK signal has the same shift as has the AFSK signal).

11.2.6.1 Modem standards.

Over the years, a number of FSK, PSK, DPSK, QPSK, and QAM standards have been adopted for various data communication purposes.   One of the standards for asynchronous and low-speed synchronous data communications over standard voice-grade telephone lines was the Bell System's 103/113 family of modems for data rates up to 300 bits per second.   The modems in this family were full duplex in design, permitting simultaneous two-way data transfer utilizing two pairs of AFSK frequencies.   One pair, called the *originating-end* transmission pair, is at 1270 for MARK and 1070 for SPACE, while the *answering-end* transmission pair is at 2225 for MARK and 2025 for SPACE.   Obviously, one end has to be designated the *originating end* and the other end the *answering end*.

In a simple link between two stations of the same type, this may be an entirely arbitrary decision.   In other cases, there may be a definite reason for designating one end as a particular type.   For example, in a dial-up time-share computer system, when the computer "answers" the telephone call, it does so by putting the MARK tone for the *answering-end transmit* pair on the line.   This choice was made because any tone in the frequency range from 2010 to 2240 Hz will disable any echo-suppression equipment that might be active on the telephone link.[4] Since the "answering" computer is the first station to place a carrier signal on the link,

---

[4] Telephone echo-suppression or echo-cancellation equipment is used on many intermediate-distance and almost all long-distance voice circuits to disable the reverse-direction circuit when a conversation signal is being carried in the forward direction in order to prevent a voice echo being sent back from a mismatch in the the receiving end's local-loop hybrid-coil circuit.   Since we want simultaneously to carry on a full-duplex two-directions-at-a-time data communication, these echo-suppression/cancellation facilities must be disabled.   The telephone company has provided a mechanism for the subscriber to do this by placing a tone of approximately 2100 Hz on the line.

the job of disabling the echo suppressors falls upon it, and hence it is assigned the higher pair of FSK frequencies.   The Bell 103 models and their equivalents made by other manufacturers have a switch which allows the selection of either the *originate* or the *answer* mode of operation, while the 113A and the 113D models and their equivalents are originate-end-only modems, and the 113B and 113C and their equivalents are answer-end-only types.   The bandwidth of the AFSK signal in either direction is set at about 400 Hz (from 100 Hz below the SPACE signal frequency to 100 Hz above the MARK signal frequency).

As we move up in speed, we find the Bell System 202 family of modems and their equivalents.   These are also intended for voice-grade line operation but in the half-duplex mode at speeds up to 1200 bps on a dial-up line or up to 1800 bps on a conditioned leased line.   They utilize AFSK with a 1200-Hz MARK signal and a 2200-Hz SPACE signal (note that the SPACE signal must occur often enough to maintain echo suppression).   In addition, there is a single-frequency reverse channel at 387 Hz which can be used to allow the receiving end to send back an RTS (request-to-send) carrier to enable data transmission at the transmitting end. In some cases this may be on/off or PSK modulated at a low data rate to send back other types of acknowledgment signals.

As we move into international modem standards, we find a number of CCITT standards in common use, among which are

V.21: A full-duplex FSK modem standard using the frequency pairs of 1180/ 980 and 1850/1650 for 300-bps communications.

V.22: A full-duplex two- or four-phase DPSK modem standard with one carrier at 1200 Hz and the other at 2400 Hz.   In the PSK mode, two bits at a time (i.e., a *dibit*) are transmitted at up to 1200 bps.

V.23: A two-channel modem standard with pairs at 450/390 and either 2100/ 1300 or 1700/1300.   Usually used in a modified half-duplex mode with the low-frequency pair used to return acknowledgment signals from the "receiving" end to the main transmitting end.   The data rate on the main channel is either 600 (with the 1700/1300 pair) or 1200 (with the 2100/1300 pair) baud.

V.26: A single-channel four-phase DPSK modem standard with the carrier at 1300 Hz used for half-duplex or simplex operation on one voice channel at 2400 bps.

V.27: A 4800-bps DPSK modem standard transmitting three bits simultaneously (i.e., a *tribit*) using phases spaced 45 degrees apart and a 1800-Hz frequency.

The highest-rate modems suitable for voice-grade operation are those which transmit four bits at a time at a Baud rate of 2400 transitions per second (a bit rate of 4 × 2400 or 9600 bits per second) using QAM keying.   Figure 11-15 shows two of the common QAM schemes used: the CCITT V.29 specification and the Bell System's model 209-A modem specification.

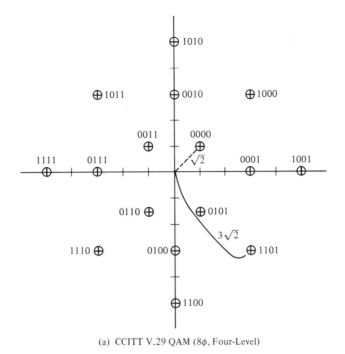

(a) CCITT V.29 QAM (8φ, Four-Level)

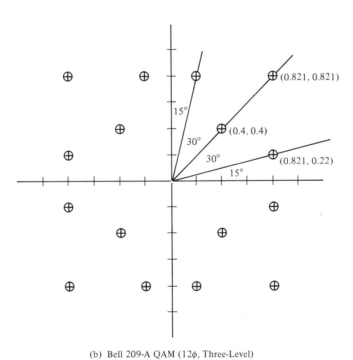

(b) Bell 209-A QAM (12φ, Three-Level)

Figure 11-15 Two typical QAM (QSK) transmitting constellation patterns.

11.3 SINGLE-LINK POINT-TO-POINT DATA SYSTEMS

11.3.1 Data-Link Components

Figure 11-16 shows the basic components of a simple two-terminal data link. At each end there is the piece of equipment that originates or uses the digital information. This piece of equipment sends or receives the digital data to/from a piece of equipment commonly called a DTE (data terminal equipment) in either parallel or serial form (although parallel is the most popular at this point). The DTE then performs the function of buffering the output data going from the data source to the link or that going from the link to the data receiver, as well as (usually) making the conversion from parallel to serial, or vice versa, via the use of one or more shift registers.

There are many types of DTE equipments, and only a couple of these types will be discussed here. One of the most common for slow-speed data communications is an 8-bit UART (universal asynchronous receiver/transmitter) which has separate 8-bit parallel buses for input and output functions. It has no chip-select decoding and parallel loads or empties its shift registers whenever an enable line is activated (therefore, it is necessary to be sure that the correct data are on the bus going into the UART before those data are transferred into the UART by the activation of this control line, or to be sure that the correct received data are present in the parallel output register of the UART before those data are transferred into the data user's system from the UART). The clocking-in and clocking-out of the serial data are usually done under the command of an external clock signal from the data source. Another type of DTE used in asynchronous data communications is the ACIA (asynchronous communications interface adapter). This chip usually has a bidirectional parallel data bus and chip-select decoding so that it knows when (under the command of the data source or user) to read data from the I/O bus for conversion to serial output or when to put received data onto the parallel I/O bus. Both the UART and the ACIA supply the start and stop pulses needed in the serial asynchronous pulse-train output to the baseband data-link circuit to the receiving DTE or to the MODEM which modulates the baseband signal onto a carrier. The MODEM, fiber-optical modulator, or other device that

DTE = Data Terminal Equipment
DCE = Data Communications Equipment
    (The DCE is Commonly Referred to as a
    Modem, or Modulator/Demodulator)
UART = Universal Asynchronous Receiver/Transmitter
PCI = Programmable Communications Interface
ACIA = Asynchronous Communications Interface Adaptor

Figure 11-16   A simple two-terminal data link.

converts the baseband signal from the DTE into a communications signal is called the DCE (data communications equipment). The interface between the DTE and the DCE may be one of several types which will be discussed shortly.

It should be noted at this point that the terms DTE and DCE may take on slightly different (though somewhat related) meanings when a data communications network is being discussed. In this latter context, the DCE is the device on the network side of the network/user interface, and the DTE is the user's equipment (whatever hardware or software that may include). We will try to minimize the confusion by referring to these latter entities by different names in Section 11.5 on networking.

If synchronous serial data rather than asynchronous are to be transmitted, then the DTE might take the form of a PCI (programmable communications interface) which employs a double-buffering technique where one buffer (shift-register) is loaded or unloaded in parallel while the other register is being unloaded or loaded serially. This permits a continuous flow of uninterrupted bits while still using shift registers of reasonable size (length). In the synchronous serial mode, a block of data (a *frame* of data) is usually transferred at one time on a continuous basis and the sending-end data source must therefore be ready to supply the data to the DTE on an uninterrupted basis (i.e., the DTE register which becomes ready to be loaded in parallel must be loaded before the other register in the DTE which is being clocked out serially is empty and the data in the first register is needed for serial output).

Also, before this synchronous process is begun, the data originator must be certain that the data user at the other end of the link is ready to receive the frame. If this isn't possible, then the receiving-end DTE must have the capability to interrupt the sending-end DTE and prevent it from sending more data out over the link (the sending-end DTE must, in turn, indicate to its data source (or *host* as it is often called) that it cannot accept further data for transmission).

As can be seen from this brief discussion, synchronous data transmission is much more complicated than is asynchronous, and it requires a much higher level of coordination between the data source and user and/or their DTEs than does asynchronous data communications. In return for this inconvenience, synchronous data transfer largely eliminates the overhead of the start/stop pulses of the asynchronous method and therefore provides for more efficient data transfer.

On a synchronous data link there are two levels of synchronization to be achieved: *bit synchronization* and *character synchronization*. Bit synchronization refers to the adjustment of the receiving data communications equipment timing so that it "knows" at what point in time to make the decision as to whether a 1 bit or a 0 bit is currently being received. Character or frame synchronization allows the receiving equipment to determine *which bit* of the received bit stream actually is the first bit in a received character or *which bit* actually begins the data in a data frame. There are two ways in which both of these synchronization problems are simultaneously solved. Which way is actually used depends upon whether a *bit-oriented* or a *character-oriented data-link protocol* is to be used. The operational differences between these two protocols will be discussed in Section 11.3.2.

In both, bit synchronization is obtained by not letting an "idle" or temporarily inactive synchronous data communications line actually be electrically idle for any appreciable length of time. Instead, in many of the bit-oriented protocols, the sequence 01111110 is transmitted on a regular basis when there is no data flow on the line. When this sequence is not actually "on the line," the line is kept at the "1" electrical level for some integer number of bit periods. In some cases the 01111110 sequence is kept on the line on a continuous basis (the presence of two 0's separated by six 1's is sufficient to maintain both the bit and the frame synchronization). For character-oriented protocols, the SYN character is transmitted periodically and in the same way, by the presence of the known 0's in the code used (usually SBT, ASCII, or EBCDIC), maintains both bit and character synchronization.

Frame synchronization for bit-oriented protocols is thus obtained simply by following any of the 01111110 sequences with the stream of data bits in the frame. The end of the frame is indicated by returning to the 01111110 sequences again. This immediately leads us to a problem: What if the sequence 01111110 accidentally occurs in the middle of the stream of true data bits? This would erroneously indicate the end of the data frame. The problem is solved by applying the following rule, called *bit stuffing*:

Whenever five 1 bits occur in sequence anywhere in the data stream to be transmitted, *always* insert a 0 bit after the fifth 1 bit when transmitting the bit stream. At the receiving end, *always* remove from the incoming data stream the first 0 bit which follows five successive 1 bits.

For character-oriented protocols, frame and character synchronization (both are important in this case) are obtained by transmitting two SYN characters in sequence (with no break between them) to define the beginning of the frame. Control characters required for control of the data link (and/or the network in some cases) are imbedded in the bit stream of the frame and the receiving station is always looking for their occurrence (a particular 6-, 7-, or 8-bit sequence).

This presents a problem somewhat similar to that encountered with the possibility of six successive 1's occurring in the actual data stream in the bit-oriented protocol: What if a control character (ETX, ETB, EOT, etc.) or two successive SYN's occur embedded somewhere in the stream of actual data bits in the character-oriented protocol transmitted frame? Again, the solution is to set up some rules as to how the transmitting end modifies the transmitted bit stream whenever this occurs and how the receiving end removes the extra added bits (complete characters in this case) from the received bit stream. Generally, these can be stated for any character-oriented protocol as

1. The transmitter will start the actual block of data with DLE STX and will end it with any appropriate control character (except SYN) preceded by a DLE (DLE is called the *data-link escape* character).
2. If a bit sequence equivalent to the DLE occurs *within* the actual data stream itself, the transmitting end converts this to the sequence DLE DLE. When the

receiver sees two DLEs in sequence, it discards the first and treats the second as part of the actual data stream.
3. If SYN SYN occurs in the actual data stream, the transmitter sends DLE SYN, which the receiver converts back to SYN SYN (since the sequence DLE SYN is not an allowed sequence for control purposes).
(Note: Control characters not preceded by a DLE character are not recognized as control characters.)

This technique is described as *character stuffing* and does not require that the actual data stream itself be organized on a character basis (i.e., the actual data stream can be any arbitrary, up to some maximum, number of bits long).

It should be noted that the bit-stuffing and the character-stuffing operations result in what is termed *transparent mode operation* (or sometimes just *transparency*) on the data link. It is necessary to operate in the transparent mode when the length of actual data within the frame is not a fixed number of bits (either always the same for every data frame or else given in the frame header information). If the length of the actual data stream is always known, then bit or character stuffing is not needed.

11.3.2 Basic Protocols on the Single-Data-Link System

Even with the asynchronous transfer of data, there is still some amount of two-way coordination or checking (often called *handshaking*) that must be done between the two user data devices (shown as the "Data Source/User" blocks in Figure 11-16). One form that this checking procedure (or *protocol*) might take is provided for in the standards for the electrical (or *physical-layer*) interface for the serial baseband connection (as, for example, between the DTE and the DCE at each end of the data communications link). One very popular such standard is the RS-232C standard for serial data interchange circuits. A portion of this standard is outlined in Figure 11-17 and shows the various circuits (wires) that can exist between a DTE and its associated DCE.

Once a full-duplex DCE-DCE carrier-frequency link has been established by either manual or automatic means, there will be two MARK-level carriers present on the link. The presence of these carriers indicates to the DCE that receives one of them that the opposite station is "on line" and *may* be ready to receive data.

In the simplest data-link protocol procedure which makes use of the facilities provided by RS-232C, the presence of the other station's carrier and the fact that the local DCE is turned on (as indicated by the *Carrier Detect* and the *Ready for Sending (RFS)* signals) may be all that is needed to tell the data source that it can begin sending asynchronous data (see Figure 11-18). In more complicated data-link protocols, the data source may first instruct the DTE to put a *Request to Send (RTS)* signal on the link to see if the station at the other end replies with a *Cleared to Send (CTS)* signal before it actually begins transmission of data. It should be noted at this point that the use of these "hardware" RTS and CTS functions requires

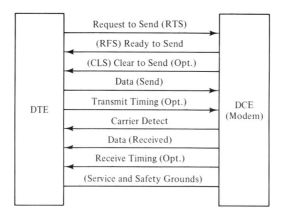

Figure 11-17 Basic RS-232 circuit functions for primary data circuit.

additional virtual data "circuits" on the DCE-DCE link which may or may not be provided for by the types of DCEs (modems) being used.

In the event that these data-link RTS/CTS functions are not provided for by the data-link hardware, then these data-link protocol functions must be handled via software by transmitting special RTS and CTS control frames between the two user devices (we will be discussing this more later). Once cleared to send data, the data source, through its DTE, will begin to output serial data to the DCE on the *DATA (sent)* wire, perhaps along with a transmission timing signal on the *Transmit timing* wire which the DCE may use to synchronize its output bit stream, or which via a separate timing-signal channel it may send to the receiving DCE and DTE. In the receive mode, the RS-232C interface provides for not only the received serial data bit stream from the receiving modem, but may also provide a *Receive timing* signal derived from the incoming data bit stream or from the timing signal sent to it from the other station. The DCE may also feed the DTE with a signal which indicates whether the quality of the received signal is sufficient for dependable low-error-rate operation.

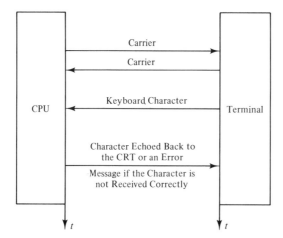

Figure 11-18 Full-duplex time-share terminal link in character mode.

Table 11-3 shows this and some of the other RS-232C signals such as the ringing-indicate signal for dialup purposes and the Baud rate signal (providing the modem is equipped to furnish these types of signals). The RS-232C standard also provides for the simultaneous establishment of two two-way data channels over the same interface (i.e., the secondary channel). In addition to specifying *which* signals may be available for use, the RS-232C standard also specifies the electrical signal levels, impedance levels, grounding procedures, connector pin arrangements, and so on. All in all, the RS-232C is a very complete standard. It is very seldom that all its features are implemented at the same time in any given situation. Also, there are a large number of "unauthorized" modifications of the standard to include

TABLE 11-3 RS-232-C CIRCUITS AS DESIGNATED BY THE ELECTRICAL INDUSTRY ASSOCIATION (EIA) STANDARDS COMMITTEE

Interchange Circuit	C.C.I.T.T. Equivalent	Description	Gnd	Data From DCE	Data To DCE	Control From DCE	Control To DCE	Timing From DCE	Timing To DCE
AA	101	Protective Ground	X						
AB	102	Signal Ground/Common Return	X						
BA	103	Transmitted Data			X				
BB	104	Received Data		X					
CA	105	Request to Send					X		
CB	106	Clear to Send				X			
CC	107	Data Set Ready				X			
CD	108.2	Data Terminal Ready					X		
CE	125	Ring Indicator				X			
CF	109	Received Line Signal Detector				X			
CG	110	Signal Quality Detector				X			
CH	111	Data Signal Rate Selector (DTE)					X		
CI	112	Data Signal Rate Selector (DCE)				X			
DA	113	Transmitter Signal Element Timing (DTE)							X
DB	114	Transmitter Signal Element Timing (DCE)						X	
DD	115	Receiver Signal Element Timing (DCE)						X	
SBA	118	Secondary Transmitted Data			X				
SBB	119	Secondary Received Data		X					
SCA	120	Secondary Request to Send					X		
SCB	121	Secondary Clear to Send				X			
SCF	122	Secondary Rec'd Line Signal Detector				X			

"DTE" = data terminal equipment (terminals, computers, etc.)

"DCE" = data communicatiions equipment (modems, etc.)

other functions not specifically provided for by RS-232C. Some of these modifications are fairly widely accepted, while others are only of local interest.

The RS-232C standard is an example of what is called a *physical layer* (Layer 1) protocol. It establishes the functional wires, electrical signal levels, connector pin numbers, FSK or PSK modulation scheme and frequencies, and so on, for the actual *physical data link* which connects the two ends of the data communications link. The modem standards that were discussed are all parts of the physical layer protocol. On the other hand, the *procedures* by which the physical facilities provided by RS-232C and similar physical layer protocols are actually *used* for end-to-end checking to determine who transmits when, and so on, is part of the next-higher protocol layer called the *data-link layer* (Layer 2) protocols. The function of protocols (or procedures) in both these two layers is to establish the physical and logical data communications link between two points and to insure that the data sent over this link arrives error free at the other end. The error-detection and error-correction techniques discussed in Sections 11.2.2 and 11.2.3 are also part of the data-link layer protocol. As a simple example of the *use* of the error-detecting capabilities which may be provided, Figure 11-18 shows a simple protocol for use by a time-shared computer system to let the human operator at a terminal know that his or her input data have been received by the computer.

Whenever the operator types a character on the terminal keyboard, the code for that character is sent to the computer which stores the character in its memory and also sends back to the CRT screen the code for the character which it *thinks* it has received from the terminal. The operator can then determine if the character which was intended to be sent actually was what arrived at the computer or whether some error has occurred (either operator typing error or communication-link error).

For synchronous data transfer, the more automatic protocol shown in Figure 11-19 might be used wherein after the link has been established and the RTS has been acknowledged by a CLS, an entire block (frame) of data is sent. If the receiving station is not able to receive a block of data at the present time for any reason, it will signal the other end to wait for a while and try again later by sending

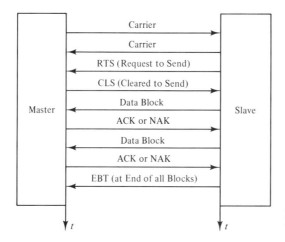

Figure 11-19 Synchronous block transfer with no propagation delay.

back a WACK message.   Once the data frame is sent and received and assuming that everything "goes OK," it is checked for transmission errors by the use of the BCC bits or some other error-detection technique at the receiving end.   If no errors are detected, the receiving end sends back some form of positive acknowledgment to the transmitting end.   If an error *is* detected, the receiving end will send back a negative acknowledgment (or NAK) to the transmitting end or else simply do nothing (i.e., not send back a positive acknowledgment).   If after the data are sent from the transmitting end and no reply of any kind is received from the receiving end for some period of time, the transmitting end will either retransmit the original frame or will send some sort of inquiry message to the other end to see if it is still "alive."

In the preceding example, it is assumed in diagramming the protocol in Figure 11-19 that there is no propagation time between the two stations and that the only delay which occurs is the time required to process the received messages at either end (i.e., the *turnaround* time).   However, in long-distance communications via submarine cable or satellite, the time taken up waiting for signals to propagate between the two ends of the link (anywhere up to 500 msec) may add considerable overhead time to the operation of the system (for example, waiting for the acknowledgment for one data block to come back before transmitting the next block of data).   This problem is shown in Figure 11-20a, while Figure 11-20b shows one possible solution to the problem (keep sending data blocks and then, if the acknowledgment for a particular block does not come back within a reasonable length of time, back up and retransmit it and all the following blocks).   This type of protocol procedure is referred to as *pipelining* and does not allow data frames to be accepted at the receiving end out of the order in which they were sent (i.e., if one frame is lost or damaged, not only it, but all following frames must be retransmitted).   Other, more complicated, protocols do allow the receiving station to accept frames out of order.   We will discuss this further in Section 11.4.3.

11.4 DATA COMMUNICATION PROTOCOLS

Up to this point we have been discussing some of the basic concepts associated with transferring digital data between two "users" without really defining whether the "user" is a piece of hardware, a human being, a process, or whatever.   Also, we have not at all considered what happens when more than two "users" become involved in the situation.   It is apparent that a very complex situation is developing and that some structured approach is needed to subdivide the whole problem into smaller, more manageable, segments.

11.4.1 The ISO's OSI Protocol Model

The complexity of this situation was generally recognized a few years ago, and as a result, a committee was formed within the *International Standards Organization (ISO)* to structure the *manner* in which data communication standards and specifications (i.e., protocols) are to be structured and discussed.   The result was the

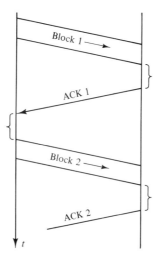

(a) Conventional Method (One Block
    at a Time): Uses Either Half or
    Full Duplex

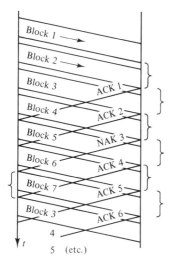

(b) Continuous Sending (Until a NAK
    is Received): Possible With Full
    Duplex Only (Note that There is
    a NAK on Block 3)

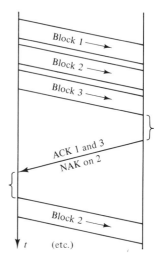

(c) Multiblock Transfer and Acknowledgment:
    Either Half or Full Duplex (Note NAK on
    Block 2)

Figure 11-20 Various types of "ARQ" data-link protocol methods for a single
link ({ or } is the terminal turn-around delay).

Open Systems Interconnection (OSI) Reference Model for Data-Communication Protocols.

Basically, this model divides up the overall problem of a person or process using some type of *digital user device* communicating via a data communications link or network with another person or process using a second user device into seven distinct problem areas.   Each of these subdivided areas is referred to as a *Layer* in the OSI model and is concerned with *what type of things* are to be discussed, what problems are to be solved, and what specifications or standards are to be set for that particular part of the overall problem.   These are summarized in Table 11-4.

The first three of these layers (from the bottom up in Figure 11-21) are concerned with user/network interface protocol problems (i.e., the problems involved with getting data put into the network at one user *port* to come out undamaged at the desired destination "port"). The middle layer (Layer 4 or the *Transport* Layer) has the responsibility of making sure that complete messages get through the network intact and undamaged.   The remaining three "upper" OSI layers are concerned with user/user procedures such as establishing work sessions; agreements on billing for use of each other's facilities; the character set to be used; access security; computer languages, editing procedures, and formats

TABLE 11-4 FUNCTIONS PERFORMED AT EACH LEVEL OF THE OSI REFERENCE MODEL

Layer	Function
1. Physical	This determines what voltage or current levels represent 0's and 1's; what wires and connector pin numbers are assigned to which Layer 2 functions, modem FSK frequencies and phase/amplitude standards, etc.
2. Data Link	Whether transmission is synchronous or asynchronous, the data rate to be used, error detecting and correcting, echo generation, ARQ protocols, link-access procedures, etc.
3. Network	The network layer determines what information is used to route a message or packet through the network and internally how that routing is accomplished and how compensation is made for situations like the loss of a link, etc.
4. Transport	In this layer the conversion is done between the total message and the blocks (packets) into which it is broken for transmission purposes. Also, it makes sure that the whole message is put together at the receiving end with everything in the correct order.
5. Session	This layer is concerned with establishing a virtual *connection* between the users (or between a user and a service). It controls access, cost-of-service charging, security, sign-on/off, etc.
6. Presentation	Basically this layer defines what different characters or computer signals mean (languages, character-code conversions, line length to be used, cursor control, graphic symbols and language and commands, editor operation and commands, control-character functions, etc.).
7. Application	The procedures to be followed in the use or execution of the user software. This level is left almost entirely up to the individual application program that is being run by the user.

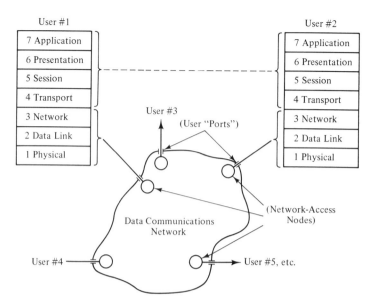

Figure 11-21   The OSI Protocol model as applied to a network.

to be used; and so on.   Since most of the latter are more computer science than data communications oriented, we will not be too interested in them.   Our main concern will be with Layers 1, 2, and 3—the *Physical, Data-Link,* and *Network* layers, respectively.   We have already discussed to some extent the Physical Layer (modem standards, the RS-232C interface standard, etc.)   and some of the Data-Link Layer standards (error-checking codes, basic ARQ protocols, etc.).   In the next two sections (11.4.2 and 11.4.3), we will expand upon the Data-Link Layer protocols; then in Section 11.5 we will examine Layer 3 (the Network Layer) and some of the pertinent parts of Layer 4 (the Transport Layer).

11.4.2 Methods of Protocol Specification

Before actually beginning the discussion of Data-Link Layer protocols in detail, we will first look at some of the techniques used to specify or describe a protocol and for theoretically checking the validity of its operation in order to be reasonably sure that there aren't any unexpected conditions or situations which might occur wherein the protocol operation "breaks down" (i.e., getting stuck in an endless loop or getting into a *trap* where all operation ceases and the system "locks up").

For the purposes of illustrating these various describing and checking techniques, we shall use a simple Data-Link Layer half-duplex *Stop-and-Wait Automatic Repeat Request (ARQ)* protocol as an example.   Basically this protocol governs the operation of two stations (assumed to be identical) at the ends of a single half-duplex data link.   The protocol basically consists of having one station (end "1") send a data frame bearing a *sequence number* (the sequence number is usually represented as *seq*) of 0 or 1 to the other station (end "2"). This data frame will contain in addition to other information a *packet* of actual data.[5] The transmitting end has an input buffer where this data packet obtained from its local *host* computer or terminal is stored prior to being put into the data frame and being transmitted over the data link. Even after transmission of the data that are in end 1's input buffer, those same data are kept in the buffer until an acknowledgment is received from the other end (end 2) that the data frame containing that particular data packet has been received successfully.

The latter condition (successful reception of a data frame number 0 or 1) is indicated by having the end 2 station send the frame sequence number of the last received valid frame back to the end 1 station as part of the next data frame being sent from 2 to 1. This is done in this example by setting the ack number in this frame (the one going from 2 to 1) to the seq number of the last valid frame received from 1. For example, if end 1 sends a frame with seq = 0 to end 2, it is not allowed to send out its frame seq = 1 (containing a new data packet) until it receives a frame from end 2 bearing an ack = 0 acknowledgment number. If 1 doesn't receive this acknowledgment back from 2 within a certain specified *timeout* interval, it will retransmit the original frame (i.e., with seq = 0 and the same input-buffer data packet) to end 2, hoping that this time, it receives a proper acknowledgment from end 2. Figure 11-22 shows how this protocol, as applied to one of the two stations, can be specified in terms of a flow chart.

In this example *sseq* is the sequence number of the frame to be sent out; *expseq* is the sequence number of the other station's frame which is expected to arrive; *sack* is the acknowledgment ack to be sent back to the other station; *sdata* is the input (from the local host) data buffer; and *rseq, rdata,* and *rack* are buffers where the sequence number of the frame just received from the other end, its data, and the ack number it bears are temporarily stored, providing that the received frame is received with no apparent damage or errors. The "Transmit frame" operation is actually the process of constructing and transmitting a complete frame with the sequence number being set equal to *sseq*, including providing synchronization, generation of the BCC bits, and so on, and carrying *sdata* and *sack* as information. Since this is a protocol with a *window size* of 1 (i.e., only *one* outstanding transmitted-but-not-acknowledged frame being permitted to exist at any one time), the only permitted *seq* values are 0 and 1. Thus, whenever any variable referring to a sequence number (i.e., *sseq, expseq, sack, rack*) is "incremented," 0 is changed to 1 or 1 is changed back to 0 using modulo-2 arithmetic, where $(0 + 1) = 1$ and $(1 + 1) = 0$.

Figure 11-23 shows this same protocol specified in terms of a *protocol specification language*. In this case, PASCAL has been chosen as the language, but

---

[5] The term *data frame* as used in this book means a continuous stream of bits, including the synchronization and/or frame-delimiting bits or characters, which is transmitted over a single data link from one station to another. In addition to the synchronization bits, a data frame will normally contain an actual information bit stream obtained from some protocol layer above the data-link layer (we will call this the *data packet* or just simply the "packet"); possibly a *data-link* address; frame sequence numbers (usually denoted as "seq") and acknowledgment numbers (usually denoted as "ack"); error-detecting and/or error-correcting bits; special data-link control codes or characters; and so on.

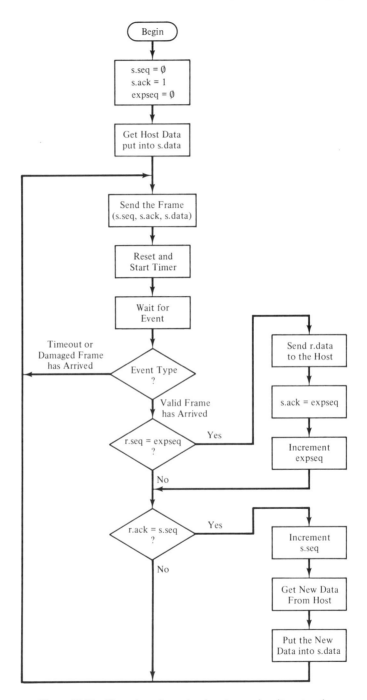

Figure 11-22   Flow chart for a simplex stop-and-wait protocol.

```
type Event = (Valid Frame Arrives,
              error, Timeout);
procedure Sample;
var event: Event
   s, r: Frame
   expseq: Sequence Number
   buffer: Message
begin
   s.seq: = ∅
   expseq: = ∅
   s.ack: = 1
   get from host (buffer);
   s.data: = buffer;
   repeat
      send Frame (s);
      reset and Start Timer;
      wait (event);
      if event = Valid Frame Arrives then
         begin
            get Frame (r);
            if r.seq = expseq then
               begin
                  send to Host (r.data);
                  s.ack: = expseq;
                  increment (expseq);
               end;
            if r.ack = s.seq then
               begin
                  increment (s.seq);
                  get from Host (buffer);
                  s.data: = buffer;
               end;
         end;
      until Forever
end;   Sample
```

Figure 11-23 PASCAL representation of the protocol in Figure 11–22.

most highly structured computer programming languages can be used for this purpose, and a number of special languages (IBM's Format And Protocol Language, FAPL, Schindler's RSPL language, etc.) have also been developed just for the purposes of protocol specification.   Although most of them are based upon or are extensions of some computer programming language, they are not actually themselves programming languages.

In addition to protocol *specification* methods, a number of different techniques have been developed for graphically *verifying* protocol procedures and operations. One such technique is the use of *directed-state diagrams* for the system as a whole. Figure 11-24 is a portion of the directed-state diagram for our example of a simple two-station data-link system.   Each circle specifies one of the possible *states* in which the system may be at any one time.   For one station, the station may have just sent out a seq = 0 or a seq = 1 frame and may be expecting either a 0 or a 1 frame from the other end (four possible combinations or station states).   The line may be bearing eight possible types of messages (four types in either direction) depending upon the seq and the ack number combinations in the messages, or the line may be in the "idle" state (i.e., no messages going in either direction).

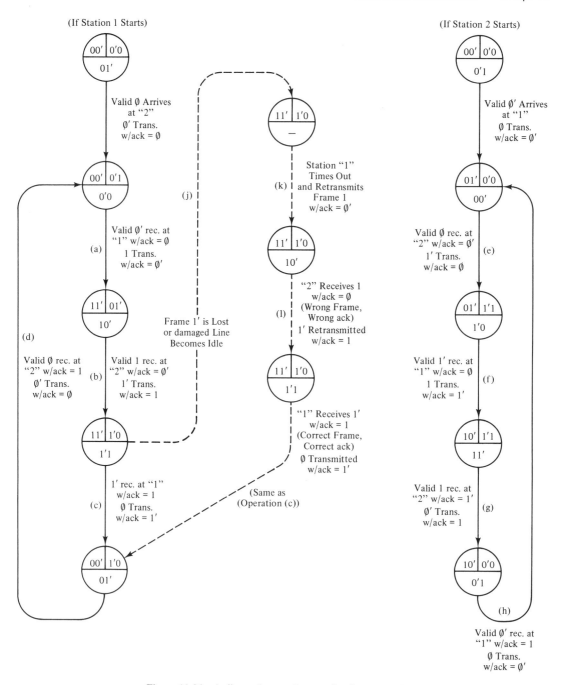

Figure 11-24   A directed state diagram for the protocol of Figure 11-22.

In the state-diagram notation for our example we shall use 0 and 1 to refer to end-1 sequence numbering and 0' and 1' to refer to the end-2 sequence numbering. The upper-left quadrant of each circle will represent the state of station 1 (*sseq, expseq'*), the upper-right quadrant will represent the state of station 2 (*sseq', expseq*). The lower half of the circle will refer to the line state (sequence number of the frame on the line/the ack number it bears). Since there are four states for each station and nine states for the line, there will be $4 \times 4 \times 9 = 144$ possible states, or the complete state diagram *might* have as many as 144 circles on it, providing that it is possible to somehow get into every one of the possible states from some starting or initialization state. The solid lines in Figure 11-24 show what the normal sequence of events is if either station is the first to start operating. The dashed lines show what would happen in *one* (of many) possible cases where a single frame is damaged or lost. In the case shown, after three transition states while the loss of the frame is being corrected, the system operation finally stabilizes to sequencing back around the four normal states on the left-hand side of the diagram. There are seven more similar transitions which may occur as a result of the loss of or damage to just one frame (if two or more frames are damaged in sequence, there are many other possible routes and states that may exist).

Another protocol-mapping approach is the *Petri-net* diagram (see Figure 11-25). In this diagram the individual stations and the line are each assigned to different areas of the diagram, and each circle shows only the state of one of the three entities (end 1, end 2, or the line). The state of the entire system is then denoted by placing a *token* in one of the circles in *each* of the three diagram areas. The transitions between states which are indicated by the lines with arrowheads in the directed state diagram of Figure 11-24 are shown as heavy bars in Figure 11-25 (a one-to-one correspondence between them is indicated by the use of lowercase alphabetic letters). A given transition is allowed to occur (i.e., *enabled*) whenever *all* the circles connected to the input side of a bar possess a token. After the transition takes place the new state is indicated by placing a token in all the circles which are connected to the output side of the bar. Normally some of the tokens do not move during most of the state transitions which take place. The number of inputs to a bar does not have to be equal to the number of outputs (i.e., tokens may be created or lost during some of the state transitions).

As can be seen from both Figures 11-24 and 11-25, both the state diagram and the Petri-net diagram can very quickly become large and complicated, even for very simple protocol examples. Our example protocol has been kept simple by making a number of assumptions: that a local host always has data to send and therefore we can always "piggyback" the ack onto an outgoing data frame; we don't try to speed things up when a damaged frame is received by sending back a *negative* acknowledgment (NAK) frame; we don't provide for the establishment or shutdown of the link; the receiving host is always able to accept data from the link (i.e., no *flow control*); we assume that we never have to interrogate (query) the opposite station about its status or intentions; and so on; and so on. If we get into a more complicated protocol where these assumptions no longer hold, then our flowchart method of specification can also become quite complicated, and for

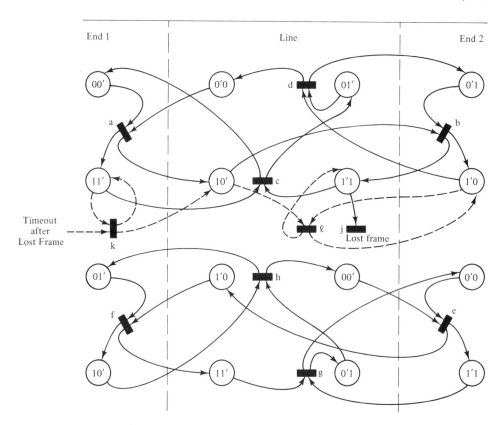

Figure 11-25   The Petri-net diagram for the simple stop-and-wait protocol of Figure 11-22.

the purpose of conveying a *general concept* (but not the operational details) of what a protocol *does,* a somewhat simpler technique of describing the protocol may be desirable. One such method looks at the data frames that are flowing back and forth over the communication channel and maps out what sequences of control and data frames may occur.

An example of one such technique is the *ANSI (American National Standards Institute) protocol map,* an example of which for a half-duplex protocol is shown in Figure 11-26. The protocol shown is a character-oriented protocol which incorporates negative acknowledgments (rather than depending upon the time-out-after-no-positive-acknowledgment method), flow control (through the use of the WACK control frame), and a query (ENQ) control frame. Other simple operations could be easily included into this protocol by adding other routes to the map than those which are shown.

Another type of protocol-mapping technique which is useful when visualizing temporal relationships in the system and when estimating such things as the channel utilization efficiency is the *timeline diagram* wherein time becomes one axis of the

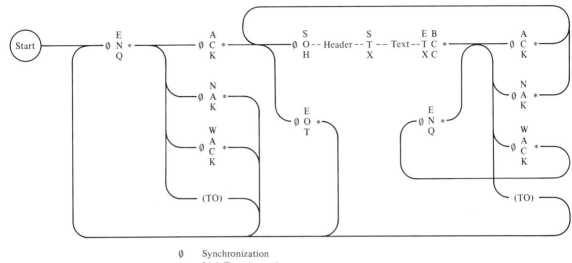

Ø	Synchronization
*	Link Turn Around
ENQ	Query
ACK	Positive Acknowledgment
NAK	Negative Acknowledgment
WACK	Positive Acknowledgment (but "not Ready for More Data")
(TO)	Timeout (Same Station Transmits Again After a TO)
SOH	Start-of-Header
STX	Start-of-Text
ETX	End-of-Text
BCC	Block-Check-Character
EOT	End-of-Transmission

Figure 11-26 The ANSI-chart method of protocol specification.

diagram (usually running toward the right or downward).   Figure 11-20 shows several simple examples of the timeline diagram.

11.4.3 Data-Link Layer Protocols

From the discussion of the OSI protocol model, we saw that it was the responsibility of OSI Layer 1 protocols to assure physically the transfer of a continuous stream of binary digits (i.e., the bits which make up a frame) from one end of the data link to the other.   What this stream of bits means (other than the synchronization bits or characters) is the concern of OSI Layers 2 and 3 and higher.   Layer 2 is concerned primarily with the format of the data frame; the detection and/or correction of errors in the bit stream within the frame, including the retransmission of the frame when needed; frame sequencing when required by the "host"[6] at the

---

[6] In discussing a single data link, the "host" may be an actual user terminal or computer, the network-layer functions of a network access or switching node, or whatever it is that generates and/or receives the *data packets* which are transported within the frame.

receiving end; data-link flow control; getting the frame to the correct piece of hardware on a multidrop link; link arbitration (deciding who transmits when); and so on. The Data-Link Layer protocol is normally not concerned with what is within the data-packet portion of the frame's bit stream (except for the possible bit stuffing or character stuffing that is necessary for transparency). This protocol layer *is* concerned with the format of the data frame (is there a link address and if so, its length; the number of bits in the sequence number and the acknowledgment number; whether there is a frame-type code; whether there is a control field in the data-carrying frame format; permitted lengths of the data packet; length of the BCC; etc.); the type of error detection and/or correction scheme used; the algorithm for retransmitting unacknowledged frames; the data-link flow-control method(s) used; half-duplex versus full-duplex operation; and so on.

In the preceding section, a very simple type of stop-and-wait data-link protocol was used as an example when discussing the various types of protocol specification and analysis techniques. This simple type of protocol can be very effective under the conditions where the link propagation time and the station turnaround time are small compared to the average frame transmission time and where the channel error rate is low. The performance of the link can be improved by the addition of features such as separate control frames (i.e., ones not carrying data packets) that can be sent whenever a damaged frame is received (i.e., a NAK control frame bearing the ack for the last valid data frame received); whenever an ack needs to be sent in the opposite dierection and no returning data frame is available for "piggybacking" (an ACK frame bearing the ack is sent instead); or whenever the station on one end wants to shut off the data flow from the other end (a WACK control frame is sent over).

A flow chart for a stop-and-wait half-duplex protocol employing these latter features is shown in Figure 11-27. In this case the ACK frame is also used when the host computer wishes to indicate that it can begin/continue accepting data from the link. An ENQ frame is provided whenever it is necessary to query the host at the other end regarding its status. The local variables *expseq* and *rwait* are the expected sequence number of the next incoming frame and the flag which tells whether data frames can (*rwait* = 1) or cannot (*rwait* = 0) be sent out. The receive-timeout time, TOR, is set to a value somewhat longer than twice the maximum data-frame-transmit time, plus twice the propagation time, plus the opposite station's TOS timer setting value. This latter, waiting-for-data-from-the-host or TOS, timer is set to a value which is dependent upon how rapidly some sort of an acknowledgment is needed by the other end of the link. Only one frame per turn can be sent out from a station (i.e., a reply must come back for every frame sent out, whether that frame is a data frame or a control frame—the only exception is when the TOR timer "goes off"). All types of frames carry an acknowledgment (i.e., the current ack value) to the receiving station.

The expanded stop-and-wait protocol just described begins to become inefficient whenever the propagation time or the station turnaround time begins to become an appreciable fraction of the time needed to transmit a data frame. Under these conditions, it is best to go to some form of a *pipelined* protocol in which a number of data frames may be transmitted sequentially before the transmitting

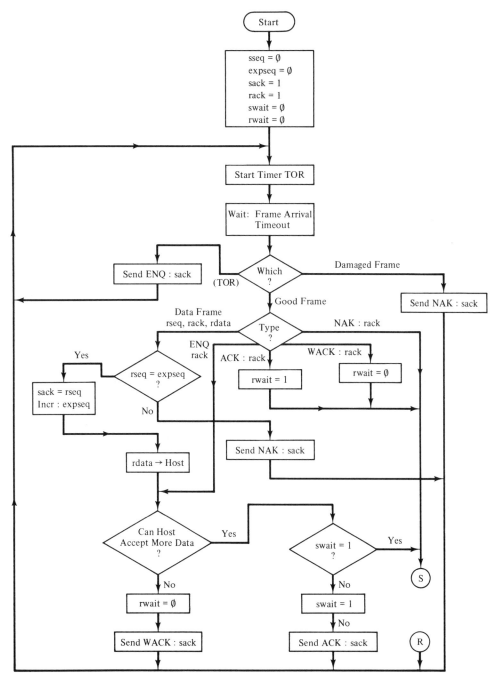

Figure 11-27 Flow chart for a stop-and-wait protocol with flow control and separate ACK.

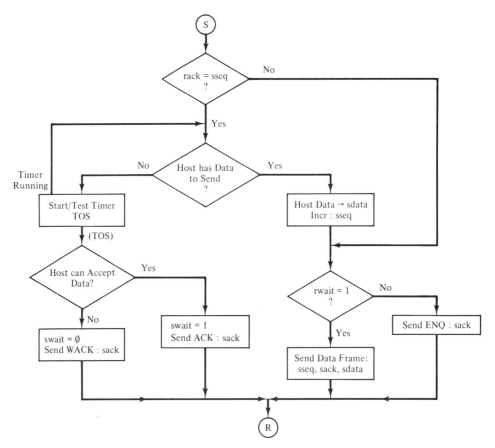

Figure 11-27 Continued

station has to stop and turn the half-duplex link around. If the data link is full-duplex or if it has a slow-speed return-acknowledgment channel, it may be possible for the transmitting station to send data as long as the receiving host is able to accept the data and as long as no transmission errors occur. Whether this is possible or not depends upon the size of the *transmitting window* in comparison to the two-way time delays involved. The maximum window size, $W_{t_{max}}$, is dependent upon the number of bits, n, involved in the sequence number and is equal to $W_{t_{max}} = (2^n - 1)$ which is also the maximum value of the sequence number. In a pipelined data-link protocol which does not accept frames out of sequential order at the receiving end (i.e., if one frame arrives damaged, all subsequently transmitted frames are discarded at the receiving end until the transmitting end stops, backs up, and begins again by retransmitting the damaged and all subsequent frames), the maximum transmitting window size as given above tells how many unacknowledged frames may be outstanding at any one time. A pipelined protocol of this type has a receiving window size of 1 since it can only accept the next frame following the last valid frame received (i.e., all the following ones are discarded). Thus, it only has to buffer one received frame at a time. The transmitter, on the other hand, must buffer all the frames which it has transmitted and for which it has not received an acknowledgment. Thus, the size of its buffer must be equal to $W_{t_{max}}$. It is obvious that a flowchart for this protocol is going to be much more complex than is that of Figure 11-27 since loops must be included in the transmitting portion of the logic for getting data from the host and for transmitting sequential data frames. Also, for a half-duplex link, some type of specific link-turnaround control may be needed.

A still more complicated Data-Link Layer protocol is one in which the receiver will accept frames out of sequential order. In this type of protocol, it can be shown that the receive window size W_r is equal to $W_{t_{max}}$ and that both of these are equal to 2^{n-1} and, therefore, the receiver must be capable of buffering this number of data frames so that when a frame is missing or damaged, it can store the frames following it until the retransmitted damaged frame arrives. The chief advantage of this is that only single unacknowledged damaged frames need to be retransmitted instead of having to retransmit all the frames following the one which was damaged.

11.4.4 The Efficiency of Data-Link Protocols

We have already discussed *encoding efficiency* in Section 11.2.1 (see Equation 11-1) and have mentioned there that we would also later be discussing *transmission efficiency*, *protocol efficiency*, and *channel efficiency*. To discuss these different efficiencies more easily, we will first define some algebraic representations:

D = the number of actual data bits in a frame
H = the number of "overhead" bits in the frame (i.e., the sync bits, header bits, BCC bits, etc.)
F = $D + H$ = the total number of bits in the frame
C = the channel transmission capacity in bits per second
T_{av} = the *average* time needed to get a *valid* (undamaged) frame through the link
T_f = the time required to transmit the data frame
T_p = the one-way transmission propagation time
T_x = the turnaround time (average) at one station
T_c = the time required to transmit a control frame
BER = the bit error rate
PDF = the probability of a frame being damaged
η_{ch} = the channel efficiency in percent
η_{tr} = the transmission efficiency in percent
η_{pr} = the "protocol" efficiency

Using these symbols, what is normally termed the *channel efficiency* may be written as

$$\eta_{ch} = \left(\frac{D}{CT_{av}}\right)(100)\% \qquad (11\text{-}6)$$

On an error-free half-duplex channel which is heavily loaded (i.e., data frames flowing almost continuously in one direction or the other except for the station turnaround time), T_{av} is just the time needed to transmit a frame (F/C), plus the propagation time T_p, plus the station turnaround time T_x. This gives

$$\eta_{ch} = \frac{D}{C(F/C + T_p + T_x)} (100)\% \tag{11-7a}$$

Quite often this is written as

$$\eta_{ch} = \left(\frac{D}{F}\right) \left[\frac{F}{F + C(T_p + T_x)}\right] (100)\% \tag{11-7b}$$

where 100 times the first factor (D/F) is termed the *transmission* efficiency and 100 times the second factor is called the *protocol* efficiency, or

$$\eta_{ch} = \left(\frac{\eta_{tr}}{100}\right) \left(\frac{\eta_{pr}}{100}\right) (100)\% \tag{11-8}$$

For a half-duplex stop-and-wait-with-ACK protocol with no transmission errors and no traffic in the reverse direction, the time required to transmit a frame and get it acknowledged would be

$$T_{av} = \left(\frac{F}{C}\right) + (2T_p) + (2T_x) + TOS + T_c \tag{11-9}$$

where *TOS* is the waiting time before transmitting the ACK control frame from the receiving end. Thus,

$$\eta_{pr} = \frac{F}{F + C(2T_p + 2T_x + TOS + T_c)} (100)\% \tag{11-10}$$

It can be seen that the protocol efficiency (and therefore the channel efficiency) drops considerably when there is no return data flow to piggyback the ack's upon. Going back to the simple stop-and-wait protocol with a timeout-before-retransmitting of TOR seconds, we can find T_{av} for some assumed value of the bit error rate, BER. To begin, the probability of a data frame being damaged during transmission is

$$PDF = \sum_{k=1}^{F} \left(\frac{F!}{k!(F-k)!}\right) BER^k (1 - BER)^{F-k} \tag{11-11}$$
$$\doteq F \times BER \text{ when BER is small}$$

and the probability that the frame will *not* be damaged is

$$1 - PDF = (1 - F\,BER) \tag{11-12}$$

The probability that it will take k tries to get a valid frame through undamaged is given by

$$Pr(k) = (PDF)^{k-1}(1 - PDF) \tag{11-13}$$

The average number of tries to get a valid frame through the data link is then given by

$$k_{av} = \sum_{k=1}^{\infty} k Pr(k)$$

$$= \sum_{k=1}^{\infty} k(PDF)^{k-1}(1 - PDF) \qquad (11\text{-}14)$$

$$= \frac{1}{1 - PDF}$$

The time needed to transmit frames which get damaged and then wait until timeout occurs before trying again is equal to (F/C) plus TOR. The time required to transmit an undamaged frame and get a reply back is equal to (F/C) plus T_p plus T_x (actually twice this value, but in that double amount of time a data frame has also come across from the other end of the link, so we only assign half of the actual required time period to *our* frame's time requirement). Thus,

$$T_{av} = (k_{av} - 1)\left(\frac{F}{C} + TOR\right) + \left(\frac{F}{C} + T_p + T_x\right) \qquad (11\text{-}15)$$

Combining this with the expression for k_{av}, we get

$$T_{av} = \frac{(F/C) + (PDF)(TOR) + (1 - PDF)(T_p + T_x)}{1 - PDF} \qquad (11\text{-}16)$$

which gives

$$\eta_{pr} = \frac{F(1 - PDF)(100)}{C[(F/C) + (PDF)(TOR) + (1 - PDF)(T_p + T_x)]}\% $$
$$= \frac{F(1 - PDF)(100)}{F + C[(PDF)(TOR) + (1 - PDF)(T_p + T_x)]}\% \qquad (11\text{-}17)$$

As we mentioned earlier, the *channel* efficiency is obtained from the *transmission* efficiency and the *protocol* efficiency. If D is the number of bits in the data packet carried by the frame, then we are assuming that all the bits in the data packet are truly data bits. In a practical case involving a network, some of these packet bits may be *network overhead bits* added to the actual data bits by the Network Layer (Layer 3) and/or the Transport Layer (Layer 4) protocols. Thus, to determine an *overall efficiency* for a complete end-to-end data-transmission system, we would have to let D' ($<D$) be the actual number of *true* data bits and let $H' = H + (D - D')$ include the overhead bits in the packet itself, and then use D' and H' in the equations in place of D and H. Also, we would have to include the *encoding efficiency* factor given in Equation 11-1. Thus, the *overall* efficiency would be given by

$$\eta_{overall} = \left(\frac{\log_2 p}{m_{char}}\right)\left(\frac{D'}{F}\right)\left(\frac{F}{CT_{av}}\right) \qquad (11\text{-}18)$$

where m_{char} is the number of bits per character, there are p possible character combinations, and D' is m_{char} times the number of actual data characters in the frame.

In the cases we have been discussing, T_{av} is the average valid frame transmission time for a single link on a one-link system. For the case of a multilink system or network, C times T_{av} would have to be replaced by the summation of similar terms (i.e., $C_i \times T_{av_i}$, where C_i and T_{av_i} are the values for the ith link traversed in the network from one end to the other). Also, a factor would have to be included to account for the overhead time associated with the transmission of Network-Layer and Transport-Layer control packets in the network and for the time consumed in trying to locate and/or resend packets which become damaged or lost in the network. Thus, we see that once we get beyond discussing a single data link, the matter of calculating the overall efficiency gets quite complicated. This will be discussed further, after we first consider some of the other basic concepts dealing with networks.

11.5 DATA COMMUNICATION NETWORKS

Once we get beyond the problem of two data *devices* or *stations* (such as computers, terminals, etc.) directly connected together by a dedicated data link to situations involving two or more stations separated by a large distance and not directly connected to each other, or three or more stations which have to communicate among themselves (although not necessarily all at the same time), we get into the area of data communication *networks*. One of the ways the situation of many stations could be handled, especially if they are all located close physically, would be to connect each device to every other device via a dedicated data link (see Figure 11-28). This particular type of data network *topology* is referred to as a *fully connected mesh* of dedicated links. This is not very practical because of the large number of links, each with its own communications terminal (the circles), involved: $N(N - 1)/2$, where N is the number of devices to be interconnected. Instead, the devices are usually interconnected via a less densely structured (in terms of the number of links involved) network of some type (see Figure 11-29). Usually (but not always) there is only one interface (or network *port*) with the network for each device. This interface is from the device which is located on the outside of the network, through the network port, to a *network access node* which is located just inside the boundary of the network. Each such node may have a single device connected to it, or it may service a number of devices. One of the functions of this type of node is to provide the processing needed for communicating with the external device(s) attached to it.

The physical connection of a device to its access node and the communication which takes place between them are governed by the *network access protocol* for the particular network and node involved. Other *intranetwork nodes* may exist in the interior of the network (i.e., not connected to any external device(s)). Both

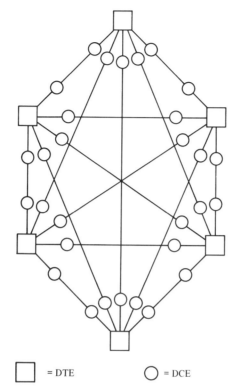

☐ = DTE   ◯ = DCE

Figure 11-28 A fully connected mesh network of dedicated data links.

they and the network access nodes provide the processing needed to route the communications from one external device to another.   This latter processing function is governed by the *intranetwork protocols* for the particular network.   The network-access protocol and the intranetwork protocol layers which are usually involved are OSI-model layers 1, 2, and 3 (the physical, data-link and network layers) and sometimes part of the Layer 4 (the OSI Transport Layer) functions.

Quite often the logical network which interconnects a given set of data devices is not physically a single network at all, but may consist of several interconnected networks of varying types and characteristics.   When this is the case, the connection point between two of the networks is termined a *network gateway*.   In some cases this may be a direct connection from a special access node in one network to a similar node in the other network (see Figure 11-30).   In other cases a special *gateway processor* is installed at point *A* in Figure 11-30 to do the protocol-conversion operations needed to pass data from one of the networks to the other.

The internal structure and operating procedure for a given network may assume several different forms.   The two major divisions of networks are *switched networks* and *broadcast networks*.   In switched networks the transmitted data are routed (or "switched") in some manner by the network to the designated receiving

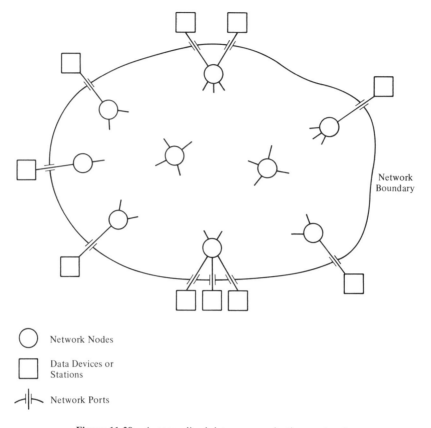

○ Network Nodes

□ Data Devices or
   Stations

⊣⊩ Network Ports

Figure 11-29   A generalized data communications network.

device(s).   Switched networks are subdivided into three classes: *circuit-switched networks, packet-switched networks,* and *message-switched networks.*   Each of these in turn can be subdivided in several ways.   In broadcast networks the data transmitted by one device may be received by all the other devices attached to the network at that time.   It is then the responsibility of each of the receiving devices to make proper use of the information received by it (use it in some way or simply ignore it if the data are not intended for that device).   Again, there are a number of different types of broadcast networks: *packet-radio, satellite-link,* and *bus-type local-area networks,* to name the three most important.

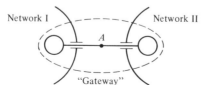

Figure 11-30   A "gateway" between two data-communication networks.

11.5.1 Communication Network Topologies

Logically the function of the data communications network is to provide an apparent, logical one-to-one or one-to-many connection of some type between the various devices attached to the network or, in other words, to emulate to some extent the functions of a fully connected mesh topology. However, the actual internode physical topology of the network itself is usually of another type of topology such as are shown in Figure 11-31 (page 398).

In the *star-connected* network of Figure 11-31a, each access node is directly connected to a master switching node whose function it is to provide a path (either physical or logical) from one node to one or more of the other nodes without going through any nonparticipating node.

In the basic *hierarchical network* topology shown in Figure 11-31b, the switching-node function is divided among three upper-level switching nodes connected by high-data-rate links (the heavy dark lines). These upper-level nodes are often at geographically separated locations. Figure 11-31c shows a hierarchical network modified by the addition of some direct lower-level links between access nodes attached to the same switching node (the "1" links in Figure 31c); between access nodes attached to different switching nodes (the "2" links in the drawing); and between an access node and an alternative switching node (link "3" in the drawing).

Figure 11-31d shows the basic *ring-structured topology*. In this topology each access node also serves the function of a switching node, either delivering incoming data to its own external device or else passing it on to the next node in the ring. The direction of data flow around the ring can be either unidirectional (in either a clockwise or a counterclockwise direction), or bidirectional. In the latter case the links between each pair of nodes are usually full-duplex links.

Figure 11-31e shows a *partially connected mesh* network. This topology bears some resemblance to portions of several of the other topologies: some access nodes serve also as switching nodes, while others operate only as access nodes. There may also be interior nodes that serve only as switching nodes.

Figure 11-31f shows the basic *bus topology*. This topology is usually associated with broadcast-type data networks. The logical bus may take a number of different physical forms, ranging from baseband hard-wired parallel buses (the *IEEE-488 instrumentation bus*, for example); coaxial cables carrying audio, HF or VHF carriers modulated with serial data; optical fiber "buses"; the transponder of a communications-relay satellite; and an HF, VHF, or UHF radio frequency (or a pair of frequencies with a radio repeater station used to "relay" the signal received on one frequency from a transmitting access node to all the other access nodes by retransmitting the signal on the other frequency).

11.5.2 Characteristics of Switched Data Networks

As mentioned previously, a switched data network may have the characteristics of circuit switching, packet switching, or message switching. These terms may be applied *either* to the external characteristics of the data network *or* to its internal

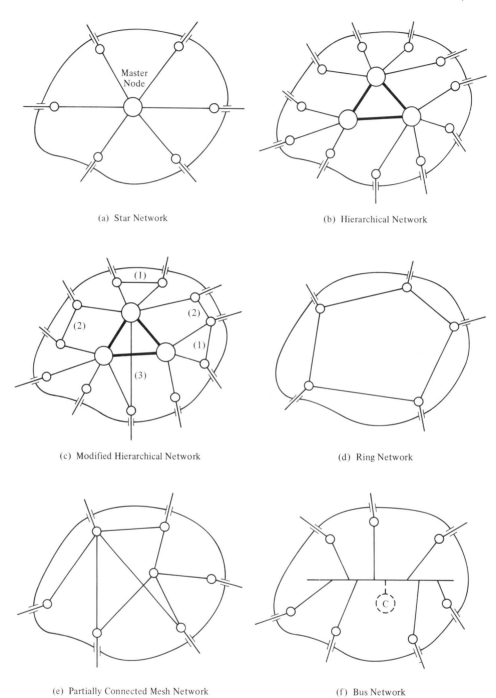

(a) Star Network

(b) Hierarchical Network

(c) Modified Hierarchical Network

(d) Ring Network

(e) Partially Connected Mesh Network

(f) Bus Network

Figure 11-31 Different network topologies.

organization and method of operation.   For the present, we will consider the network's *external* characteristics.   Circuit switching implies setting up a *virtual* communications *link* between two devices for the period of time that the two devices wish to carry on a communications session.   Externally, it *appears* as if a dedicated continuous (in time) physical data link exists between the two devices, even though internal to the network no such physical path actually exists (i.e., the network might employ packet switching *internally*, for example).   In providing the services of a virtual data *link,* the communications network leaves it up to the two user devices to provide most of the OSI layer 2 (i.e., the Data-Link Layer) protocol functions such as error detection, sequencing, and so an between themselves (i.e., the network only provides what "looks like" a dedicated physical data link over which individual data frames—sometimes of some specified maximum length— can be transmitted).

Message switching is the opposite extreme of the service range and involves putting the complete text of a message (such as a letter from one person to another) into the message-switched data network and having it *apparently* handled by the network as one complete unit.This implies that the user need only provide the destination address once (in the *message header* at the beginning of the message) and that the message that follows the header may be of any reasonable length.   A computerized mail system for receiving (from the sender) and delivering (usually upon request at the receiving end) an electronic "letter" to the designated "ad- dressee" device or station would be a good example of one of the functions of a message-switched network external characteristic.   In some cases the network may even be able to provide the sender of the "letter" with a confirmation of when the letter was delivered to its final destination.

In between these two extremes of network service, a network which externally provides a *packet-switched* service requires that the user sending device pass to it *packets* of data of limited length (usually a few dozen to several thousand bytes) with the address information present in the header of each individual packet.   It then becomes the responsibility of the sender to divide its message to be sent up into packets, attach the appropriate sequencing and address information to each packet, and deliver each packet to the network node.   At the receiving end, the attached user device must receive the packets from the network and reassemble them into the complete message.   How difficult a job this latter is depends upon the level of service provided by the packet-switched network to its users.   The two most common levels of service are represented by the *datagram model* and the *virtual-circuit model* (don't confuse the latter with the *virtual data-link* service provided by the *circuit-swtiched external network characteristic*).

In the datagram model each packet is treated by the data communications network as a separate entity not related to any other packet.   There is no guarantee that packets will arrive at the receiving end in the same order in which they were put into the network at the sending end or even that they will arrive at all if the datagram service used does not provide for confirming delivery.   This means that the transport-layer protocol of the user devices must provide for receiving and storing packets which arrive out of order, and also must provide for requesting the retransmission of lost packets when this becomes necessary.

In the virtual circuit model a two-way link is first initiated between the caller and the callee. The calling station must initiate the process by sending a *Call Request* control packet to the called station. If the called station is then able to participate in a dialogue with the caller, it sends back a *Call Accept* control packet. The network, which has been monitoring this process, then sets up a logical path or route called a *virtual circuit* (vc) over which packets given to it by the sending end can be routed to the receiving station *in order as received* until one of the stations transmits a *Clear Request* control packet to indicate that it wants to "hang up" and the other station replies with a *Clear Accept* (or Call Disconnect) control packet to indicate that it agrees with terminating the "connection." At this point the network clears the virtual *circuit*. Unlike the virtual *link* provided by the *circuit-switched* network external characteristic, the virtual circuit model does provide the Data-Link Layer protocol functions of error detection, retransmission of damaged or missing packets, sequencing, and so on.

11.5.3 Examples of Circuit-Switched Networks

Very often a time-shared computing system, especially one that is located within a few miles or so of the user, may be accessed via the public dialup telephone network using a voice-grade audio AFSK modem and a computer terminal which transmits one character at a time. The full-duplex signals from the terminal and the modem will usually travel to their respective telephone company local switching offices via a baseband *local subscriber loop* such as was described in Chapter 8. At that point, if both the terminal and the computer are on the same local switching exchange, their baseband loops may be directly connected together if the local switch is one of the older crossbar or reed-relay "space" switches as were described in Chapter 8, thus giving a direct audio-frequency connection or data link between the two. In the case of a more modern switching office, the "direct" audio baseband connection may be emulated by analog sampling and TDM multiplexing the baseband audio AFSK signals in each direction onto the common bus of a modern electronic "time-switched" local telephone switching office. If a longer distance is to be covered, the AFSK signals may be SSB modulated upon a carrier in an FDM cable or microwave trunk system; or sampled and digitized and transmitted as PCM signals over a time-multiplexed digital link such as the T-1 digital link described in Chapter 8; or put onto an optical-fiber link such as is discussed in Chapter 10. Whatever the technique, the *apparent* result is a physical link from one end to the other, and once the telephone connection is established, the fact that the signals may be transversing a very complex electronic network is transparent to the user. Thus, the user need then only be concerned with the Physical Layer and the Data-Link Layer protocols and not the Network Layer protocol.

The establishment of the telephone connection by the user may be a manual process of listening for the dial tone, dialing the telephone number of the computer, listening for the carrier tone from the computer's modem, making the electrical or acoustical connection to the terminal's modem, and then typing in the specified access character(s) or code. This manually performed procedure constitutes the Network Layer (Layer 3) protocol that must be performed. In some cases where a "dialup" modem with the equipment for generating the dialing pulses or tones and for recognizing the "dial tone" is used, the ringing signal and the various types of "busy" signals are provided as part of the Layer 1 protocol, the dialing/connecting procedure may be made semiautomatic if a "dumb" terminal is used or fully automatic if the "terminal" is actually a personal computer with the proper "automatic dialup" (Layer 3) software.

Another type of virtual data *link* through a network which is not circuit-switched internally but is instead packet-switched has been mentioned in the previous section.   In this case the virtual link which is provided is set up whenever it is needed by following the setup (access) procedures prescribed by the network operator.   Again, once this protocol has been executed and the virtual link established, the user may forget about further Network Layer protocols until it comes time to "hang up" the link.   The fact that the signals are passing through a complex packet-switched network are again transparent to the user.   The only noticeable difference may be that the end-to-end delay time might be larger than on a network which is actually circuit-switched (either time or space switching) internally because the latter handles the data link signals passing through it in "real time" whereas the data passing through the *virtual* data link of the former (packet-switched) network will experience some storage-time and processing delay at each of the various nodes in the network.

11.5.4 Packet-Switched Networks

The concept of the *external* packet-switching *characteristic* for a network has already been mentioned, as well as the actual *internal* packet-switching *modes* of operation. At this point we shall just very briefly consider the internal *virtual circuit* mode of operation and how it is implemented, and then move on to consider the *external* packet-switched network *characteristics* as defined by the 1984 CCITT Recommendation X.25 for access protocols to such a network.

11.5.4.1 The virtual circuit (vc) mode of internal operation for a packet-switched network.   Figure 11-32 shows a typical packet-switching network.   It consists of a partial-mesh network topology with various-speed data links between the different nodes (in some cases there may be multiple links between the same pair of nodes, but to keep things simple, we will show only one link between each pair).   There are two functions performed by a node: network access for a user (host) and internal network packet switching.   Each node has two or more data links coming into it.   These may be links across the network boundary (access ports) or links to another node in the network.   The internal links are governed by the internetwork protocols (Layers 1, 2, and 3) adopted by the network operator for that particular network, whereas the access-port links are governed by the X.25 protocol if this is a public packet-switched data network.   At each node there is a physical connection to each data link coming into that node.   This physical connection is specified by the Layer 1 protocol for either the network-access type (X.25) links or the internal-type links.   The data frames that pass over these

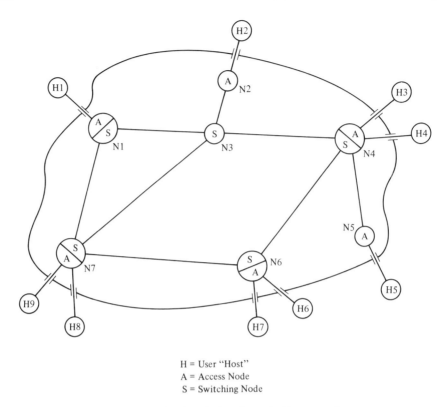

H = User "Host"
A = Access Node
S = Switching Node

Figure 11-32   A typical packet-switching network.

physical links are governed by the appropriate Data-Link Layer protocol for the pair of stations at the two ends of the link (either user-host/access-node, governed by X.25, or network-node/network-node, governed by the internetwork Data-Link Layer protocol). The frames which pass over each data link may be control frames for that one data link, or they may be data-type frames which carry "data" in the form of *packets* which are passed from host-to-host via the network.

At each node the Data-Link Layer protocol for one of the attached links (the No. 1 data link in Figure 11-33, for example) removes the packet from an incoming frame from the opposite end of that link and passes the packet to the Network Layer (Layer 3) protocol procedures of the node's computer. In a sense, this network-node computer which handles the packet according to the Layer 3 protocol procedures fulfills the function of the "host" in the discussions of the Data-Link Layer protocols in Section 11.4.3. The node "host" computer then examines the header information in the packet passed to it by the Data-Link Layer protocol for the incoming data link and, based upon this packet-header information, gives the packet to the Data-Link Layer protocol for the appropriate outgoing link to be inserted into a data-link transmission frame and sent on to the next node along its route to the destination host computer (or directly to the user host itself if the outgoing link is a network-access link).

Now that we have looked at how a node actually handles a packet, we will see how we go about getting information to the node as to how that node is to determine where to send the packet. We have already mentioned that one way of doing this is to set up a route through the network from one user process (or program) to a specified process at the destination host computer by setting up a vc between the two user processes. This virtual circuit is set up or established via the interchange through the network of two special control packets called the Call Request and the Call Accept packets. These special packets contain complete destination addresses for the host computer and process as well as the vc identifier number assigned by the computer which originates the Call Request packet. The network uses the address information in the Call Request packet to set up a route through the network, and at the destination end, the network notifies the destination process that a virtual circuit is being set up and gives the destination process a vc number for that incoming virtual circuit (not the same number which was assigned by the originating computer for reasons to be discussed later). Internally, the network may set up routes for the vc by establishing at each node a table which identifies each vc by a pair of *logical channel numbers* (LCNs).

One such LCN is associated with the data link on which the vc "grows" toward the node as it is established (this LCN is determined by the station at the opposite end of the link), and the other LCN is a number assigned by the node itself as it "grows" the vc out on another (outgoing) link to the next station.

Figure 11-34 shows a set of such routing tables for the nodes of the network shown in Figure 11-32. The left-hand pair of columns represent the station from which the virtual circuit grew toward the node and the LCN assigned by that station to the vc, while the right-hand pair of columns represent the station toward which the node continued to grow the vc and the LCN which the node assigned to the vc for the outgoing link. The entries in these tables represent the establishment

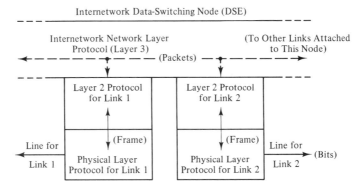

Figure 11-33 The OSI protocol structure at a network data switching exchange (DSE).

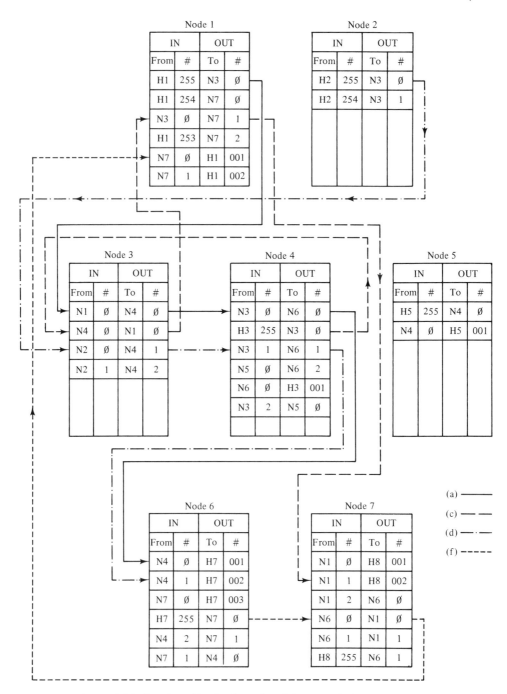

Figure 11-34   Routing tables for the nodes of the network shown in Figure 11-32.

of the following nine virtual circuits (designated by the letters a through i) one-at-a-time in the order indicated:

(a) H1–N1–N3–N4–N6–H7
(b) H1–N1–N7–H8
(c) H3–N4–N3–N1–N7–H8
(d) H2–N2–N3–N4–N6–H7
(e) H1–N1–N7–N6–H7
(f) H7–N6–N7–N1–H1
(g) H5–N5–N4–N6–N7–N1–H1
(h) H8–N7–N6–N4–H3
(i) H2–N2–N3–N4–N5–H5

with the routes for the first, third, fourth, and sixth circuits established (a, c, d, and f) shown as various types of lines on the drawing. Note that for the host/node links that three-digit LCN numbers have been used while for the node/node links one-digit numbers have been used. We will discuss the reason for this in the next section. It should be obvious from the table that several of the data links are carrying more than one vc. In practice there may be hundreds of virtual circuits using the same data link concurrently. Also, in our example we only "grew" one vc at a time. In the practical network case, there may be a large number of virtual circuits in the process of being established or torn down at any one time. The latter operation is performed by the passage of a Clear Confirm control packet along the route of the vc through the network. Once the use of a virtual circuit number or an LCN number is discontinued, it may then be reused again in establishing a new vc.

How a particular network internally handles the routing of packets for what externally appears to be an X.25 virtual circuit depends upon how the network operator wants to do it. The method just described establishes a fixed route through the network via the use of routing tables maintained at each node. This requires some memory to be provided at each node to store the table, but allows the packets to be handled by reference only to the vc number or LCN associated with each packet. On the other hand, the network designer may choose to establish a network that can dynamically route each packet for a given vc via whatever route is available or has the least time delay at any given time and condition of the network. This technique, sometimes called the "internal datagram" approach, requires that the network pass along with each packet the same complete address information that went with the Call Request type of packet, thus using up more physical channel capacity on each data link which the packet transverses. However, no routing tables have to be maintained, and by allowing dynamic routing, the vc can be maintained in most cases where a data link "crashes" by simply routing the packets normally using that data link via some "detour" route instead. In the routing table method, when a given data link goes down, so do all the virtual circuits being routed over that particular link.

11.5.4.2 The CCITT recommendation X.25 for the access port to a public packet data network.

This is a recommendation (or "specification") for the protocols for the access-port data link into a public packet-switched data network. It specifies *only* the external network characteristics and *is not concerned with* the internal operation of the network which was discussed in the previous section. The X.25 Recommendation is divided into a number of sections. Section X.25.1 discusses the possible physical (OSI layer 1) protocols which might be employed. Basically, these involve using CCITT X.21 or X.21 bis (the latter interim version of X.21 is similar to RS-232-C) as the standard (see Section 11.3.2).

The brief discussion of the physical layer protocol is followed by a long discussion of what X.25 calls the "link access procedures" in section X.25.2, which essentially amounts to specifying the OSI data-link (Layer 2) protocol standards. The ISO's High-Level Data-Link Control (HDLC), with procedures LAP-B (Link Access Protocol-Balanced) is taken to be the standard at this level. The data-link frame format associated with this protocol is shown in Figure 11-35.

The frame check sequence (FCS)—or block check code (BCC) as it is often called—is generated by using the CRC-CCITT shift register shown in Figure 11-8c, presetting the register stages all to 1 before clocking the data stream through the register. The *1's complement* of the remainder left in the CRC at the end of the stream of actual data bits is what is transmitted as the FCS (or BCC). At the receiving end, the CRC is similarly preset to all 1 bits before clocking through the entire transmitted data stream (the data bits plus the 16 FCS bits). If no errors have occurred, the final result remaining in the receiver's shift register should be $(0001\ 1101\ 0000\ 1111)_2$. Appendix G shows the CRC-CCITT tables for this for

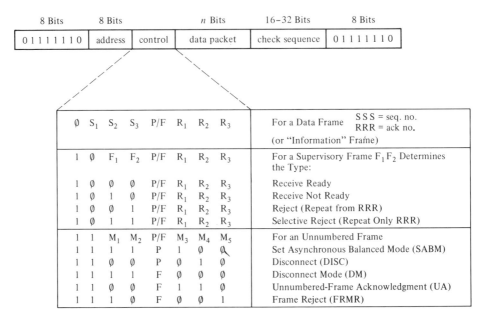

Figure 11-35 Format for the HDLC/LAP-B data and control frames used in X.25.

the case of a typical short bit stream.   The next portion of X.25.2 (i.e., X.25.2.3 through X.25.2.7—about 52 single-spaced typed pages in the 1984 edition of X.25) specifies the LAP-B Data-Link Layer protocol procedures for the network-access data link.   Obviously we won't discuss this protocol in detail here.   We will merely observe that it is a full-duplex pipelined link protocol using 3-bit or 7-bit sequence numbers with provisions for operating several parallel data links logically as a single link between the user host (which X.25 refers to as the DTE) and the network access node (the DCE in X.25 terminology).

Sections X.25.3 through X.25.7 and Annexes X.25.A through X.25.G discuss the "packet level" specifications and procedures (this part of the Recommendation is about 114 typed pages long).   These protocols cover all of OSI Layer 3 (the Network Layer) and some aspects of Layer 4 (the Transport Layer).   We will briefly discuss some of the more important aspects of these packet-level formats and procedures.   First, it is important to realize that when we discuss *packets*, much of what we observe in the way of *packet formats* and talk about concerning *packet-level protocols* will bear some resemblance to our previous discussions about data frames and Data-Link Layer protocols.   It is important that we don't confuse the two.   We can somewhat keep the two concepts distinct if we always keep in mind the fact that *packets* are carried by *frames* (see Figure 11-35) over physical links governed by Data-Link Layer protocols and are passed by these Layer 2 protocols to the Network-Layer protocol operations at the DTE and the DCE. These Layer 3 protocols in turn make use of the information in the *packet* header and the information conveyed by the reception of *control packets*.

Figure 11-36 shows the types of control packets used by the X.25 packet-level procedures, and Figure 11-37 shows additional details for the Call Request and the Call Accept types of control packets which are used to establish the virtual circuit through the network between the two DTEs.   These two packet types contain the complete addresses of both the "calling" user process at the originating DTE and the "called" user process at the destination DTE.   In addition, they contain what is called a *facilities field* which can designate such things as packet redirection (to another alternate destination DTE, cost-charging information (a "collect" call, etc.), a packet which belongs to a "closed" group of users, nonstandard packet length, network throughput negotiations, and other special network features that may be provided).   From the names of some of the types of control packets, one can surmise some of the packet-level procedures which may be carried on (a DTE may indicate that it is ready or not ready to receive additional packets of data; a receiving DTE may interrupt a sending DTE (or the *network*, for *network flow control* purposes, may choose to have a particular DCE interrupt its DTE); the Reject packet is sent when for some reason a receiving DTE wants to reject an arriving data packet (similar to a NAK frame at the data link level); plus packets to conduct diagnostic operations or to back up and start over again at some certain point).   As can be seen, some types of control packets piggyback *packet* ack numbers and some don't.   In X.25 the ack number represents the sequence number of the *next data packet expected* at the destination DTE (note that none of the control-type packets are sequence numbered).

The format of the data-type packet is shown in Figure 11-38.   Bits 3 through

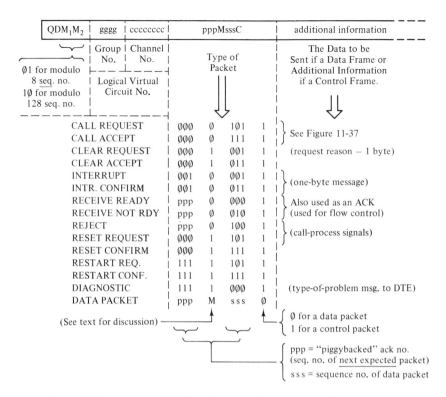

Figure 11-36 Format of X.25 packets with a listing of control-packet types.

16 have the same meaning as in the control-type packets (i.e., $M_1M_2 = 01$ means that 3-bit sequences and ack numbers will be used, whereas 10 means that 7-bit sequence numbers will be used; the following 12 bits give the virtual circuit number). The Q bit identifies the data within the packet as *qualified* ($Q = 1$) or nonqualified ($Q = 0$) data (this depends upon user definition of what "qualified" means), and the D bit determines whether the arrival of the transmitted packet is to be ac-

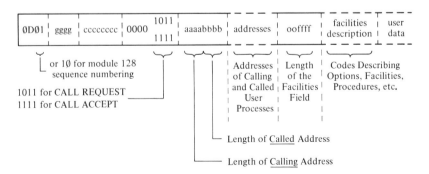

Figure 11-37 Format of an X.25 call-request and call-accept control packets.

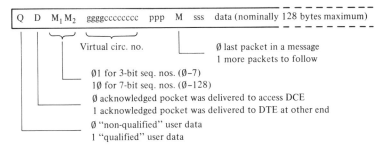

Figure 11-38   Format of the X.25 data packet.

knowledged by the destination user host (D = 1) or by the network (D = 0). The bits sss and ppp represent 3-bit packet sequence and ack numbers, respectively (if 7-bit sequencing is used they become sssssss and ppppppp). The M bit following the ack is used to indicate whether this packet of data is the last packet of a message which has been split into packets (M = 0), or whether there will be additional packets to follow to complete the message (M = 1). Following the 3- or 4-byte data-packet header will be 128 bytes of actual data (if M = 1) or possibly less (if M = 0). This maximum length of 128 bytes can be modified via the *facilities* field of the Call Request packet.

The X.25 network-access protocol provides for three basic types of service: *permanent virtual circuits* which are set up on a semipermanent basis and are always available to the host for sending packets of data to one other host; the *called virtual circuit* which must be established each time a session is initiated by the exchange of the Call Request and the Call Accept packets and "disconnected" after the session is over by the exchange of *Clear Request* and *Clear Confirm* packets between the DTE and its DEC; and the *Fast Select Facility* (FSF) option which is the 1984 X.25 replacement for what was previously known as *Datagram Service*. The latter class of packet-network service provides for a one-packet-each-way exchange of data such as might occur between an automatic bank-teller terminal and the banking system's main computer. This is done by the terminal transmitting a Call Request packet with a special code in the *facilities* field and up to 128 bytes of user data. This sets up a vc through the network in just the same way as does a regular Call Request packet. The receiving-end DCE sends a Call Indication packet containing the user data to its DTE (the bank's computer). If the computer is able to complete the transaction by sending a simple reply back to the terminal, it gives its DCE a special type of Clear Request packet containing up to 128 bytes of reply data to the send to the teller terminal. The teller terminal's DCE converts this to a Clear Indication packet which is similar to a Clear Confirm packet except that it has the reply data tacked on. The virtual circuit is disconnected as the special Clear Request packet propagates back through the network to the teller terminal. On the other hand, if the bank's computer wishes to solicit an additional response from the teller terminal, it can essentially convert the temporary vc established by the special Call Request packet from the terminal into a regular *called virtual circuit* by sending back an ordinary Call Accept packet. In addition to providing these services specified under X.25, many public network operators may elect to provide additional services such as virtual data *links*, message-handling features (i.e., message-switched facilities), and so on. The X.25 recommendation does not cover these additional types of offerings, but it doesn't prohibit them either.

In setting up the routing tables of Figure 11-34, we used a three-digit LCN number for the host/access–node (i.e., the DTE/DCE links) and one-digit numbers for the internetwork links. We did this to emphasize that the DTE/DCE logical circuit number was determined by the X.25 network-access protocol, while the internetwork LCN numbering is determined by whatever layer 3 protocol the network is using for its internal operation. The X.25 Recommendation says that when Call Request packets go from the DTE to its DCE, the DTE assigns LCN numbers starting at some maximum value (often 255 or 4095) and working downward, while at the destination end where the vc "grows" from the DCE to the DTE, the DCE assigns LCN numbers starting at 001 and working upward (LCN 000 is reserved for special network-originated messages to the DTE).

There are, of course, many other details and features specified for just the *external* characteristics of a packet network by the X.25 protocol. In addition, there is much we could study about the internal operation of the various types of packet networks which can provide the X.25 service. We could look at topics such as flow control, routing algorithms, congestion and delay-time problems, data-link failures, data-link and network testing and analysis methods and test sets, and so on. However, these topics would provide enough material to fill several textbooks (remember, the 1984 version of the X.25 Recommendation itself is 180 pages long!), so we will have to move on to the consideration of some other basic types of data networks.

11.5.5 Broadcast-Type Data Networks

We have already mentioned broadcast-type networks in Section 11.5.1 when we were discussing network topologies. The bus topology and the ring topology are the most commonly used broadcast networks, but we will limit our discussion mostly to two types of bus networks: a communications satellite transponder accessible by a number of ground stations and an example of a local area network (LAN) that uses a baseband coaxial cable as the bus. In either case we have a situation wherein any station may put a signal on the bus and all stations are able to hear that signal. The fact that any station can put a signal on the bus leads to an immediate problem: How is it decided who is permitted to transmit? This is one of the major issues to be dealt with in broadcast-type networks, although as we shall see, it is actually dealt with at the Data-Link Layer level rather than at the Network Layer level.

11.5.5.1 Communications satellite "bus" data networks. The communications satellite transponder was discussed in Chapter 9. Such a transponder may, of course, be used as a single end-to-end data link between two stations.

This type of operation was mentioned in Section 11.4. The only major concern there was the long two-way delay time due to propagation (approximately 240 to 300 milliseconds) which necessitated using large sequence-number windows on this type of data link. The use we are now going to put the satellite transponder to is somewhat different. We are essentially going to use it as a broadcast station to get the signal from any *one* of a large number of possible senders delivered simultaneously (or nearly so) to an equally large number of receivers (even though the contents of the signal may be intended for the use of only one of the receivers). The propagation time delay which is involved becomes a major factor in the operation of the system. The receiving stations don't know who started transmitting until about a quarter-second after one of them begins transmitting. This can cause some problems in deciding who transmits when.

To begin considering the problem of *bus arbitration* (or control of who is transmitting on the data bus), we will assume that a standard-length data frame has been agreed upon by all users. One possible solution to the arbitration problem might be to assign each user a certain time slot in which to transmit if it has anything to send. This becomes quite inefficient when there are a large number of *possible* senders, but only a relatively few that are interested in sending a data frame at any one time, so this technique isn't much used. Another scheme might be some type of reservation system wherein when a station wants to transmit, it tries, during a designated time slot, to send a short reservation-request message to the link controller. Because of *collisions* with the requests from other stations, it may take several tries to get the reservation request through to the controller and have a time slot assigned in which to transmit a data frame. Despite such problems, there are a number of practical systems employing some variation of this type of access control.

Another group of interesting protocols for the satellite transponder bus fall into the general category of "try-and-try-again-until-you-don't-get-clobbered." The initial work with these types of protocols began at the University of Hawaii, and they have thus acquired the name of "Aloha" protocols. Several different variations of the Aloha protocol have been developed. The first that we shall discuss is usually referred to as *Pure Aloha* (see Figure 11-39). In this version a station which wants to send a data frame simply begins transmitting the frame whenever it feels like doing so, in the hope that it won't get "clobbered" or suffer a *collision* with a frame which some other station already has begun to transmit or which some other station will begin to transmit before "our" station completes its transmission. If a collision *does* occur (our station can determine this by listening to what comes back from the satellite about $\frac{1}{4}$ second later), we simply wait a random length of time and try to retransmit the same frame again. If we let the time required to transmit the standard-length data frame be T, and assume that there are a large number of stations, N, using the system, where the probability of any station wanting to begin sending (or resending) a frame during a period of time equal to T is p, then the *average* number of frames beginning during any length of time equal to T will be equal to $G = N$ times p. The probability that exactly k frames will be initiated during any length of time equal to T is given by the Poisson

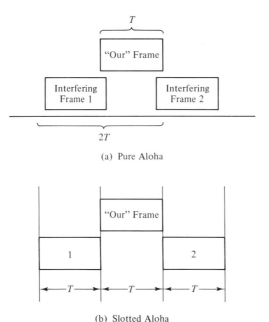

(a) Pure Aloha

(b) Slotted Aloha

Figure 11-39 "Pure" and "Slotted" Aloha.

probability function

$$Pr(k) = \frac{G^k e^{-G}}{k!} \tag{11-19}$$

So as not to have another frame collide with our frame, no other frame may start within the time period of our frame or in the length of time T seconds before the start of our frame. Since our frame is also T seconds in length, this gives a time period of $2T$ during which $k = 0$ other frames may be initiated. Since the average number of frames which will start in this length of time is equal to $G' = 2G$, the probability that *no other* frame will start during this time interval (and hence also the probability that our frame will therefore not be "clobbered") is given by

$$Pr(k = 0) = \frac{(G')^0 e^{-G'}}{0!} = \frac{(1)e^{-2G}}{(1)} = e^{-2G} \tag{11-20}$$

where $G' = 2G$ and $k = 0$.

The *average* number of frames which get through without a collision per length of time, T, is given by the average number generated, G, times the probability of a generated frame getting through, or

$$S = G\,Pr(k = 0) = G\,e^{-2G} \tag{11-21}$$

A plot of this is shown as the line for Pure Aloha in Figure 11-40. It can be seen that the *throughput*, S, "peaks out" at $S = 0.1839$, or optimum use is being made of the satellite channel when $G = 0.5$ or when $p = 0.5/N$. This says that as the average number of active stations, N, becomes large, the probability of any one station sending a frame should be decreased to maintain optimum throughput on the satellite channel. This can be done via having each station wait a longer average random time before retransmission of a collided-with frame before retransmitting the second, third, . . . , and so on time. A number of algorithms have been developed to accomplish this general effect.

Another area of concern is the average number of tries needed to get a frame through when the G of the system is a certain value. This is obtained by summing the probability of a frame getting through on try n times the value of n for all n. The probability that a frame gets through on try n is the probability that it *doesn't* get through on the first $(n - 1)$ try times the probability that it *does* get through on the nth try, or

$$\left. \begin{array}{c} \text{probability of success} \\ \text{on try } n \end{array} \right\} = (1 - Pr(k = 0))^{n-1}(Pr(k = 0)) \qquad (11\text{-}22)$$

Thus, the average number of tries to get through is

$$N_{av} = \sum_{n=1}^{\infty} n(1 - e^{-2G})^{n-1}(e^{-2G})$$

$$= e^{-2G} \sum_{n=1}^{\infty} n(1 - e^{-2G})^{n-1} \qquad (11\text{-}23)$$

$$= e^{-2G}(e^{-2G})^{-2} = e^{2G}$$

At the optimum value of $G = 0.5$ for Pure Aloha, $n_{av} = 2.72$ tries. It can be seen that the average number of tries goes up quite rapidly as the system becomes

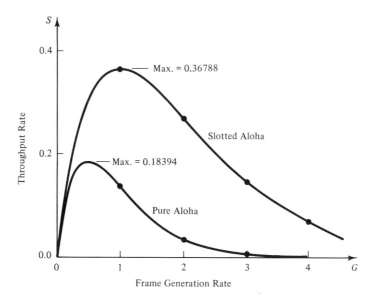

Figure 11-40   Throughput curves for Pure and Slotted Aloha.

more heavily loaded. If the probability of a new packet of data being presented to the sender per unit of time T is given by p_{new} and is fixed, then if n_{av} starts becoming much larger than 1.0, the frame presentation rate per unit of time for each station, p, begins to become much larger than p_{new} due to all the "backlogged" packets which begin to pile up waiting for transmission. When this begins happening at all N stations, G begins to become large very rapidly, and the throughput, S, of the system drops off. This, in turn, causes a further increase in n_{av} and G and the system very rapidly becomes unstable (unless some type of "slowdown" or "back-off" algorithm is used to reduce the input, G, to the system).

As can be seen from Figure 11-40, the throughput of Pure Aloha does not make efficient use of the satellite channel (only about 18.39% of its transmission capacity is used for frames which don't suffer collisions). One way to improve this situation is via the use of *Slotted Aloha*. This protocol differs from Pure Aloha in that any frame which is transmitted must be started at some fixed point in time (these "official" start points are spaced slightly more than T seconds apart). Thus, if at some arbitrary time a station wants to begin transmitting a frame, it is constrained to wait until the next designated starting point. By doing this we only have to worry about colliding with frames from stations which "felt the urge to start" during the last time slot (of approximately T seconds), and we don't have to worry about frames from stations which want to begin transmitting during the T seconds after we begin our transmission (they have to wait until the next time slot). This increases the probability of *no* collision from e^{-2G} to e^{-G}. This causes the throughput equation then to become

$$S = Ge^{-G} \qquad\qquad (11\text{--}24)$$

This is plotted as the Slotted Aloha line in Figure 11-40 and has a maximum of about 0.3679 when $G = 1.0$, which means that the Slotted Aloha can handle about twice the data flow per unit time as can Pure Aloha. However, at this optimum rate of channel utilization, we still have $n_{av} = e^G = e^{1.0} = 2.72$, which is the same as for the Pure Aloha case. Since Slotted Aloha is not much more complicated than is Pure Aloha, it is usually what is implemented in practical systems.

Another type of system which is very closely related to the satellite link is the use of a ground-based VHF or UHF radio transponder (or radio *repeater* station). There are two obvious differences: the coverage area is much smaller in terms of distances, and the propagation delay time is much smaller. The latter characteristic may make it possible to use some other more efficient protocol than Slotted Aloha. Very often some type of LAN protocol such as will be discussed in the next section can be used (in fact, there is one LAN system designed for a limited (several hundred feet in radius) coverage area which uses a very-low-power VHF radio repeater for the purpose of eliminating the coax or fiber-optic cables of the more traditional LAN systems which we will be discussing).

11.5.5.2 local area networks. A LAN network differs from long-distance broadcast-type data networks in several aspects: it covers a relatively small physical area (such as an industrial or office complex, university campus, etc.); it has a relatively high channel capacity, $C;$ because of the close proximity of the stations to each other, each station can more readily determine if anyone else is transmitting; and the characteristics of the data flow among the various stations on the LAN may bear a greater resemblance to the communications among the devices in a large computing center than to the types of messages communicated over long distances. LANs are often used in a commercial office environment to interconnect computers with word-processing terminals, terminals used to enter accounting data, copy machines, printers, and "executive" terminals. The LAN may also have a connection through an X.25/LAN *gateway* to a public or private packet data communications network.

The first type of LAN we shall examine is a common-bus system such as is shown in Figure 11-41. In this type of system some type of electrical or optical transmission line serves as the signal-transmission medium. A number of stations are "tapped" into this medium. Most of these stations have the capability to put a signal into the bus medium (some, however, may be *receive-only* stations). The signal put in by a transmitting station then propagates in both directions along the bus to all the other stations on the bus. It is important that no signal reflections occur on the bus, or else confusing, error-producing multipath interference may result. This means that the ends of the bus must be properly terminated (in resistive loads equal to the characteristic impedance of the transmission medium of the bus), as shown by T_a and T_b in Figure 11-41. Also, care must be taken that no reflections are produced by any of the station tap-ins, regardless of whether any equipment is connected to the tap-in or not.

Once we have decided upon the Layer 1 considerations such as the type of medium to use for the bus (twisted-wire pair, baseband coaxial cable, AF or RF carrier coax, etc.), we must next decide upon how we are going to manage or control the operation of the stations on the bus and the types of data frames that are going to be exchanged among the stations (i.e., the Data-Link Layer) and how these data frames are utilized to exchange data among the various users on the system (i.e., the Network Layer protocol). It can be seen from this that an additional function has been added to the Data-Link Layer: that of controlling the access to the bus. Thus, this layer now can be subdivided into two logical sublayers: a media-access-control (MAC) sublayer (on the bottom next to the physical layer) and a logical-link-control (LLC) sublayer (on the top next to the Network Layer).

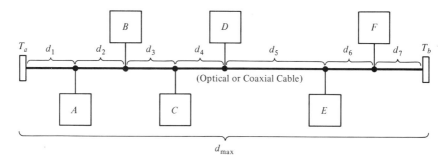

Figure 11-41   A bus-type local area network.

The operation of the LLC sublayer is very similar to that of any of the Data-Link Layer protocols we have already discussed. The MAC sublayer will be discussed in the next paragraph. One of the Institute of Electrical and Electronic Engineers (IEEE) Standards Committees (Committee 802) has set up standards for both of these sublayers and how they relate to the OSI model:

IEEE 802.1   How the remainder of the 802 standard relates to the OSI Physical Layer.

IEEE 802.2   Describes the LLC sublayer (this is very similar to the X.25 HDLC/LAP-B protocol procedures).

IEEE 802.3   These are the MAC sublayer standards for the "CSMA/CD"
803.4   bus, "token" bus and "token" ring LAN media-access pro-
802.5   cedures, respectively.

We have already discussed HDLC/LAP-B in connection with the X.25 standard, so we won't elaborate on that aspect of the LAN 802 standard at this point. Instead, we shall discuss in some detail the *Carrier Sense Multiple Access with Collision Detection (CSMA/CD)* mode of operation for a LAN bus network such as has already been shown in Figure 11-41. The particular LAN system we shall look at is the *Ethernet* system pioneered by the Xerox, Digital Equipment (DEC), and Intel Corporations. Its operation is very similar to IEEE 802.3, and it uses a 10-Mbps baseband coaxial cable as the medium.

The basic operational problem faced by the LAN bus system is very similar to that of a number of stations wanting to access the satellite transponder channel discussed in the previous section. Thus, the Pure Aloha or Slotted Aloha protocols would be a possible solution to determine who is to transmit. However, the LAN but has one big advantage over the satellite transponder channel: every station on the bus hears every other station *almost* simultaneously. Therefore, if "our" station "listens" before it begins to transmit into the bus, its data frame will not collide with that of another station unless that other station begins transmitting less than T_p seconds before or after our station begins its transmission, where T_p is the propagation time between the two stations. This is given by d/v_p, where d is the bus distance between the stations and v_p is the propagation velocity on the cable.

In Ethernet, d is limited to 2500 meters, which gives $T_p = 23$ microseconds for the type of coax cable specified for the Ethernet system. This *listen before transmitting* technique is the CSMA (carrier sense multiple-access) part of the Ethernet MAC protocol. Although Ethernet uses a baseband signal, it employs Manchester coding of the baseband signal (see Section 11.2.5), which produces a strong 10-MHz component which substitutes for the "carrier" to be sensed. Once a station (at one extreme end of the bus) is able to transmit for 46.4 microseconds (two times $T_{p\max}$ plus 0.4 microseconds for the process of "detecting" a collision) without interference from another station, the transmitting station can assume that no collision will occur during the frame. If two stations *do* begin transmitting within 23 microseconds of each other, they are both required to stop transmitting as soon as each of them detects that a collision has occurred. They then must wait a random length of time before trying to resend their respective data frames. The collision detection (the "CD" portion of CSMA/CD) is done by each station listening to the signal that is "on the bus" and comparing that signal with the signal that the station is putting into the bus. If the two signals are not the same, then another signal is present and a collision has occurred.

When a collision (or collisions) is (are) occurring on the Ethernet bus, the system is said to be in the *contention* state. If a single station is successfully transmitting its data frame on the bus, the bus is in the *transmission* state, and if no signal is present on the bus at all, the system is, of course, in the *idle* state.

The CSMA method used by Ethernet is termed *p-persistent CSMA*. It involves dividing the random waiting period into discrete intervals (or "slots") of time having a width T_w and assigning a probability, p, of beginning retransmission in any one particular slot following a collision. If each station were to wait an *analog* random time length before beginning retransmission, we would have what is known as *nonpersistent* CSMA, and if each station tried to retransmit as soon as possible after detecting that a collision had occurred, we would have *1-persistent* CSMA.

For *p*-persistent CSMA, if there are *n* stations wanting to transmit or retransmit a frame at any give time, the probability that one and only one will begin is given by

$$Pr(\text{success}) = n\, p^1 (1 - p)^{n-1} \tag{11-25}$$

It can be shown[7] that the average time taken before *one* of the stations gets a message through is given by

$$T_{\text{wait,av}} = T_w \sum_{j=0}^{\infty} j[Pr(\text{success})][1 - Pr(\text{success})]^{j-1} \tag{11-26}$$

$$= \frac{T_w}{Pr(\text{success})}$$

where *j* is the number of the slot where the wait ends (with reference to the slot where the collision occurred). The probability of one and only one station acquiring a slot is maximized when $p = 1/n$, where $Pr(\text{success}) = 1/e$. Thus, under this condition $T_{\text{wait,av}} = eT_w$ (where T_w is twice the maximum propagation time on the bus). To maintain this condition, each station on the bus must somehow estimate how many stations are waiting to transmit and adjust its value of *p* accordingly. The throughput efficiency of the Ethernet bus can then be estimated by

$$\eta = \frac{(\text{time to transmit a data frame, } T_f)(100)}{(T_f) + (\text{average waiting time, } T_{\text{wait,av}})} \tag{11-27}$$

$$= \frac{(T_f)(100)}{(T_f + (eT_w))}$$

[7] See Andrew S. Tanenbaum, *Computer Networks* (Englewood Cliffs, N.J.: Prentice-Hall, 1981), p. 294.

The IEEE 802.2 standard varies from the Ethernet standard in that it permits bit rates of 1, 5, 10, and 20 Mbps instead of just the 10-Mbps rate of Ethernet. It also limits the length of a cable segment to 500 meters (instead of the 2500 meters of Ethernet), but permits five such segments to be interconnected by means of duplex repeater stations.

IEEE 802.4 and 802.5 are concerned with "token-passing" protocols on bus and ring networks. In this case the problem of deciding who transmits when is solved by sequentially passing a "permit to transmit" or "token" among the stations according to some algorithm. The algorithm must take into account what to do when a station isn't "on line." Also, more important stations or those with more generated data traffic must get a chance to transmit more often than lower-priority stations. This must all be worked into the token protocol and must be altered every time the network configuration or operating conditions change.

11.5.6 *The Integrated Services Digital Network (ISDN)*

Since much of the present public telephone network (except for the subscriber local loop and the older space or analog-TDM switching offices) is now digital, it seems reasonable to extend the advantages of digital technology right out to the subscriber's premises. The major advantage of doing this is that many extra services such as *electronic mail*—the sending and receiving of printed messages via a CRT terminal and/or teleprinter; *voice mail*—a "one-way" telephone conversation that is put into the network whenever the sender desires and is taken out of the network whenever and wherever the receiver desires to "check the mail"; *teletext service*—public or consumer-oriented messages displayed on a CRT or TV screen (similar to the *videotext* service now provided by some TV cable systems); test and data communications of a number of different types (packet-switched would probably be the most popular); *telefax*—"printed" pictures, and so on, produced by a facsimile machine; voice telephone service; and other communication services.

ISDN is visualized as being an international service and therefore the CCITT is the group most active in developing the standards for this type of communications service. For the services provided to the subscriber, four different types of channels are visualized:

      A channel,   4 KHz BW analog

      B channel, 64 Kbps digital

      C channel,   8 or 16 Kbps digital

      D channel, 16 Kbps digital

with the A channel being a transitional provision while the changeover is being made from the analog local telephone loop to the all-digital system. Upon completion of the transition to digital, CCITT envisions the *Basic Subscriber Service* (for residential service) as consisting of one D channel and two B channels. The D channel is used as the signaling and control channel and to support the lower-speed data communication features such as CRT or teleprinter terminals, videotext, home energy management (power system load control, etc.), emergency signaling services, and so on.   The B channels will support digitized voice, high-speed packet-switched data communications, facsimile, and slow-scan video, among others.   Since there are two B channels, more than one of these services can be provided simultaneously at any given time.   All this (two B plus one D channel) can be provided over one 192-Kbps digital circuit.   For business applications CCITT's *Primary Service* standard specifies having either 23 or 30 B channels and one 64 Kbps signaling channel, utilizing a 1.544- or a 2.048-Mbps subscriber digital link. Other services in between these, and some with higher-data-rate logical channels, are also envisioned as a part of ISDN.

   ISDN is only one of a number of very rapidly developing types of new communication systems largely or entirely based upon digital technology.   Many others (such as the cellullar radio of Chapter 8) will also be a part of the very-near-future public communications picture.   It will be a great challenge to the communications engineer to keep up with all the new hardware technology and operating methodology of these new systems as they become available.

PROBLEMS

11-1. An 8-bit code is used to encode a graphics "character" set having 107 possibilities. What is the coding efficiency (in percent)?

11-2. Data are sent at a rate of 50 Kbps.   What is the Baud rate and the minimum channel BW (in Hertz) for
   a. NRZ coding.
   b. Manchester coding.

11-3. A QAM signal with 16 possible amplitude/phase combinations is to be sent over a channel with a 26-dB S/N ratio and a 6-MHz bandwidth.   What is the maximum *bit* rate which can be used?

11-4. What is the *Hamming* distance for the set of codewords:
   101010
   110100
   111111
   000000
   001101

11-5. A system employing a cyclical redundancy checking scheme with the generating function $G(x) = X^4 + X^3 + X + 1$ has the following sequence appear at the receiver: 110110101010011.   Is this sequence correct?   Why or why not?

11-6. Draw the block diagram of the CRC register for the $G(x)$ of Problem 11-5.

11-7. Block-check characters are to be generated by the use of a CRC-type shift register. The generating function $G(x)$ is [10 110 11].
                                                MSB   LSB
   a. Draw the CRC register circuit and explain how it operates.
   b. The message $M(x) = [1110101101]$ is to be sent with the above block-check code. What is the codeword $T(x)$ which will be sent (assume all zeros in the shift register initially).

11-8. A communications channel has a bandwidth of 4 KHz and an *S/N ratio* of 1000.

  a. What is the *maximum possible* rate at which data could be sent over this channel?

  b. If NRZ binary baseband encoding of digital data is used, what is the maximum data rate on this channel?

  c. If QAM modulation of a carrier is used with 16 possible phase/amplitude combinations, what is the maximum data rate?

11-9. Consider the convolutional encoder shown.

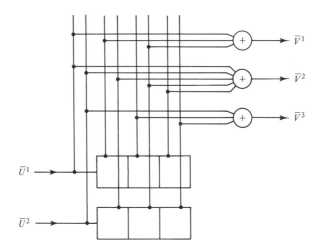

  a. Find the six generating functions $\bar{g}_p^j$ ($j = 1, 2, 3$ and $p = 1, 2$).

  b. If $\bar{U} = (11, 01, 10, 00, 10, 11, 10, 01, 10)$, find the output vector $\bar{V}$. Assume the registers are initially both set to zero.

11-10. What is meant by a stop-and-wait protocol? What window size does it have? Compare it with a Pipelined protocol and with a protocol that accepts frames out of order. Discuss how acknowledgments are handled, the sizes of the transmitting and receiving data buffers, and the window sizes as a function of the number of bits used for the sequence numbers.

11-11. Discuss what is meant by a sliding window protocol. What determines the sizes of the transmitting and receiving windows at any one time (use a 4-bit sequence number in your example(s)).

11-12. Discuss transparent versus nontransparent data-link protocols. What is meant by "bit stuffing" and "character stuffing," and why are they used (give examples)?

11-13. Assume a data-link channel with $C = 1$ Mbps, a one-way delay of 270 milliseconds, and an error rate of 0.0001 per data frame. All acknowledgments are "piggybacked" on data frames returning on a similar channel in the opposite direction and are inserted into the data stream in the frame whenever it is time to send them. There are 50 overhead bits and 4950 data bits in each frame. An 8-bit sequence number is used with a pipelined protocol. What is the overall efficiency of the use of the channel? (There are no separate ACK or NAK frames; transmit timeout is 1.5 seconds; turn-around time is negligible).

The following applies to Problems 11-14, 11-15, 11-16, and 11-17.

  The following specification for a half-duplex two-way data link with identical stations at either end use a stop-and-wait protocol with $W = 1$ and 0 and 1 is the valid sequence numbers (1-bit modulo 2 arithmetic).

Each host always has a message needing to be sent and each frame consists of:

The channel capacity is 4 Kbps, the one-way propagation time is 100 milliseconds, there is negligible processing time at either end, and if a frame arrives damaged or with the wrong sequence number, it is ignored completely. The timeout timer goes off in 500 milliseconds.

The protocol for one end of the link is

type	*Event* =	(*Valid* {an undamaged frame with the correct sequence number arrives}, *NoGood* {a damaged or incorrect frame arrives}, *TimeOut* {timer goes off})
var	*Sent*	{sequence number of last frame sent out}
	Exp	{sequence number of next expected frame}
	Msg	{buffer area for outgoing message}
	r	{received frame}
	s	{sent frame}

```
begin
    Sent := 0;
    Exp := 0;
    FromHost (Msg);
    s.Data := Msg;
    s.FrNo := Sent;
    s.Ack := 1;
    transmit s;
    reset timer;
repeat
    Wait (Event);
    if Event = Valid then
        begin
            get incoming frame r;
            ToHost (r.Data);
            increment Exp;
            if r.Ack = Sent then
                begin
                    increment Sent;
                    From Host (Msg);
                end
        end
    S.FrNo.  := Sent;
```

$S.Data := Msg;$
$S.Ack := Exp - 1;$
transmit frame;
reset timer;

until *forever*
end

11-14. Assume that a frame is never lost or damaged, what is the channel efficiency and the transmission efficiency for (a) <u>Data</u> 100 bits long, (b) <u>Data</u> 800 bits long, and (c) <u>Data</u> 8000 bits long?

	Channel efficiency		*Transmission efficiency*	
a.	_____	%	_____	%
b.	_____	%	_____	%
c.	_____	%	_____	%

11-15. Draw a flowchart for the protocol for one station.

11-16. In setting up a Petri net or a State diagram you need to know what possible states each machine can assume and what possible states the line can assume. Define the possible states for one of the machines and for the line.

11-17. Suppose the BER (bit error rate) is given (i.e., you have a number for BER). How do you calculate the probability of a frame arriving undamaged when <u>Data</u> (a) is 100 bits long, (b) 800 bits long, and (c) 8000 bits long? How do you use this information to modify the channel efficiency results of Problem 11-14?

11-18. Compare the following (describe each one and/or tell how they differ from each other) for X.25:
    a. A circuit-switched virtual data link characteristic.
    b. A permanent virtual circuit packet-switching characteristic.
    c. A called virtual circuit packet-switching characteristic.
    d. A "Fast-Select-Facility" packet-switching[8] characteristic.

11-19. Explain the difference in what is done with "overhead" information (e.g., address fields, sequence number fields, acknowledgments, control fields, CRC fields) in data frames versus what is done with the "overhead" information in packets. Where (physically) in the system does each apply? And how (or for what purpose) is each used? Relate how each applies to the OSI protocol model.

11-20.

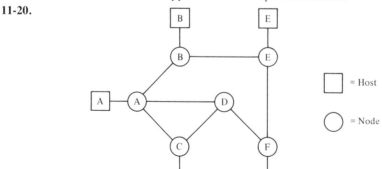

---

[8] This replaced Datagrams in the 1984 version of X.25.

IN	OUT
STA CIR	STA CIR

IN	OUT
STA CIR	STA CIR

IN	OUT
STA CIR	STA CIR

IN	OUT
STA CIR	STA CIR

IN	OUT
STA CIR	STA CIR

IN	OUT
STA CIR	STA CIR

Called virtual circuits in the above X.25 packet-switched network are established in
the following order: (A →D →F →E), (A → B → E → F), (E → F → D → C →
A), (C → D → A → B → E), (F → D → A), (E → B → A → D → C). Fill in
the virtual circuit reference tables for each node.

11-21. Virtual switched circuits are established in the following order:

A → B → S → D
A → B → C
E → F → S → C → B → A
D → S → B → A
C → S → D
D → S → B
D → E → F → S → B → A
C → S → B → A

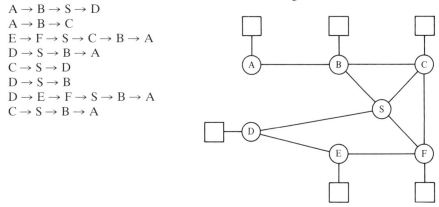

Give the virtual circuit routing maps for the seven nodes. Use "H" for the host
associated with the node.

11-22. Two stations are located at a *propagation time* of 18 μsec apart on an Ethernet bus.
It takes a station 1.3 μsec to detect a collision after the colliding signal has begun
to arrive at the station and 1.0 μs to detect the presence of another carrier before
beginning to transmit (i.e., a station may *begin* transmitting just as another carrier
is arriving or for 1 μs afterwards, and it takes 1.3 μs to detect the collision after *both*
signals are present). (a) What is the minimum length of time that one of the stations

may transmit?   (b) What is the maximum length of time one of the stations may transmit if a collision is going to occur?   (c) How long does the opposite-end station transmit in cases (a) and (b)?   (d) Under what conditions do both transmit for the same length of time and how long is this time?

11-23. On an Ethernet bus the propagation velocity is 1.1×10^8 meters per second.   If two stations are 1700 meters apart on the bus,

   a. How long does it take either station to detect that a collision has occurred if both begin to transmit at the same time?

   b. And if station B begins to transmit 14 μsec after A starts to transmit, how long does A transmit before *it* detects the collision?   How long does B transmit before *it* detects the collision?

11-24. A Slotted ALOHA satellite packet network uses the Poisson distribution ($Pr(k) = (G^k e^{-G})/k!$ to define the probability that exactly k packets will be presented to the network in any one lot when the average no. per slot is G.   The network is operating with a throughput of $S = 0.25$.   Under what two possible conditions could this occur?   On the average, how many times must a new packet be transmitted before it gets through?   (Hint: A reasonably accurate graphical solution is acceptable.)

11-25. Consider the case of Slotted Aloha with an infinite number of stations.   (a) If 80% of the available slots are never contended for,* what is the average throughput rate and, on the average, how many times must each packet be presented before getting through successfully?   Is this a stable condition?   (b) Repeat (a) if only 30% of the slots are never contended for.

11-26.

   a. What "overhead" information is contained in a packet and where in the system is it used and for what purposes?

   b. What "overhead" information is present in a frame (carrying any type of data) and where in the system is it used and for what purposes?

---

* Not used or no collision occurs in the slot.

12

TELEMETRY SYSTEMS

Telemetry systems are communication systems designed to communicate physical data of some type to a remote location in either *real time* or by means of recording the data upon some medium, transporting that medium to the remote location, and "playing back" the recorded information. From the discussions of Chapter 11, it should be obvious that one type of telemetry system would be any digital data communications system *once the physical data are put into some type of digital format.* Thus, one of the topics that will be discussed in this short chapter is the conversion of physical data into digitally encoded representations that then may be immediately transmitted or else stored in a digital memory or digitally recorded for later transmission or for use directly from the storage medium. The other major topic area will be the direct transmission or recording/playback of the data in an analog form.

12.1 THE PREPROCESSING OF PHYSICAL DATA

Physical data come in many different forms, but most of these forms are analog. The only exceptions to this are on/off (or true/false or go/no-go or whatever you want to call two-state types of data) such as the operation of limit switches, the condition of a two-state signal line, and the opening of a burglar-detection window foil, and data that are of a counting-type nature such as event counters and nuclear-particle detectors. Also, with the advent of modern digital measuring instruments, physical data that are inherently analog in nature may be more readily available in digital measured form than in analog measured form. The digital frequency counter is a good example of the latter type of situation. In the following paragraphs we will be discussing physical data that are either in nonelectrical analog physical form or that have already been converted into an analog electrical signal. We shall reserve the handling of information which is already in digital form and the conversion of analog information into a digital electrical signal for later in the chapter.

Although some of the analog data in which we may be interested for transmitting or recording may already be in the form of an electrical signal, oftentimes we will have to convert some other (than electrical) physical quantity into an electrical signal of the proper voltage or current range. This is done via the use of what is known as a *transducer*. Analog transducers may take on many different forms, some of which directly convert the physical quantity in question to an electrical voltage or current signal, while other types convert the physical quantity into a change in resistance, inductance, or capacitance. In the second type, the change in R, L, or C must then be converted into a voltage or current of the proper range. A short and incomplete tabulation of some analog electrical transducers would be:

> Piezoelectric devices: These are usually composed of special ceramic materials which produce a small voltage when placed under mechanical strain and can therefore be used to measure small mechanical displacements or used to produce an electrical output from some type of transducer which converts some nonmechanical physical quantity into a small mechanical displacement—the "crystal" phonograph cartridge is a good example of this type of transducer.
>
> Mechanical Strain Gauges: These elements change resistance with small mechanical displacements, but now we must have some means of converting the resistance variations into variations in voltage or current—this will require the addition of a stable battery or well-regulated power supply to our transducer system.
>
> Variable-reluctance pickups: These devices consist of an electrical inductance coil with a magnetic circuit—the reluctance of the magnetic circuit is varied via a movable piece of iron (or sometimes brass) in such a way that the inductance of the coil changes—this change in inductance may then be detected by means of an a.c. bridge or similar type of circuit.
>
> Variable-capacitance pickups: These operate in a similar manner to the variable-reluctance type of pickup, except that the capacitance is varied proportional to a mechanical displacement—very often this type of pickup is "biased" with a voltage in series with the capacitance so that when the

capacitance is varied an a.c. current is produced through a load resistance
in series with the battery and the variable capacitance.

"Magnetic" pickups: This type of device is most commonly represented by
the ordinary *dynamic* audio microphone or phonograph pickup—it consists
of a moving coil attached to the microphone diaphragm, the phonograph
needle, or any other type of mechanical input—a permanent magnet supplies a magnetic field through which the coil is moved, thereby producing
an induced electrical voltage proportional to the velocity of the mechanical
movement.

Photovoltaic devices: These devices produce an electrical voltage when radiation flux falls upon their sensitive surfaces—the silicon "solar cell" is a
good example of this type of device.

Photoconductive devices: These are also used to measure radiation flux, but
instead of producing an electrical voltage directly, their resistance changes
as a function of the amount of impinging flux—thus, they must be biased
with a battery or power supply.

Thermocouples: Thermocouples are devices which produce a *contact potential*
which varies with temperature—when two such devices are connected back-
to-back and are held at different temperatures, a net detectable voltage is
produced which is roughly proportional to the temperature difference between the two thermocouple *junctions*.

Thermistors: These are pieces of doped semiconductor material whose resistance varies greatly with their temperature—used with a biasing source they
can produce a large voltage change for a fairly small temperature change—
it should be noted that semiconductor *p-n* junctions can also be used for
this purpose.

The foregoing is a very short summary of some of the more important types of
physical/electrical transducers. The main characteristic of all of them is that they
are capable of (either by themselves or in conjunction with some other electrical/
electronic device(s)) producing an analog electrical voltage signal of some polarity
and amplitude. This may either be a d.c. voltage that is related to the quantity
being measured or an ac voltage that is either directly related to the quantity or
is the time derivative of it (such as in the dynamic microphone where the induced
voltage is proportional to the derivative of the mechanical displacement). Once
this voltage has been obtained, the next problem is one of *conditioning* that voltage
so that it may be used as the input to a standard analog telemetry system or analog
data recorder.

Analog signal conditioning involves (1) getting the signal's d.c. average and
its variational amplitude scaled to the correct values and limits, (2) limiting the
bandwidth of the analog signal to that which it is desired to transmit (very often
this involves limiting the bandwidth so as to exclude excess thermal noise, 60-hertz
"hum" pickup, ignition, or other industrial electrical noises, etc.), and (3) *equalizing* the frequency/phase characteristics of the signal by passing it through an

equalizer which has an amplitude/frequency and/or a phase/frequency response characteristic that compensates for any undesirable such characteristics in the transducer or other portion of the measuring system. Also, often we are not interested in all the possible information that may be contained in the analog signal, and by thus first bandwidth-limiting or otherwise conditioning the signal, we may conserve the transmission or recording bandwidth that is available to us for the recording of an ensemble of analog data signals (i.e., we "throw away" the information in which we are not interested, even though it may be perfectly good data).

The analog signal conditioning is typically accomplished by adding or subtracting a stable d.c. bias signal to the transducer output, amplifying or attenuating the signal to get it into the correct (desired or "standard") voltage range, filtering with either passive (usually R/C or L/R/C) or active filters, and so on. This is done to each data signal in the ensemble of data signals to be transmitted or recorded. At this point the signals are then ready to be applied to the analog telemetry transmission system, the analog data recorder, or the digitizing system that will convert them into equivalent digital data signals to be handled via digital communication techniques. The analog techniques of transmission and recording will be discussed in Section 12.2, and the processing of telemetry data for digital transmission and recording will be discussed in Section 12.3.

12.2 ANALOG TELEMETRY AND RECORDING METHODS

There are quite a number of techniques by which an ensemble of analog data signals may be accurately transmitted or recorded. One of the techniques which was developed after World War II when the Americans began test-firing captured V-2 missiles at White Sands, New Mexico, was the IRIG FDM/FM (inter-range instrumentation group frequency-division-multiplexed frequency-modulated) telemetry system.

Basically, this system consists of a number of standardized carrrier frequencies in the range from 400 Hz to 165 KHz. Each of these carriers may be FM modulated up to a peak frequency deviation of $\pm 7.5\%$ of the carrier frequency for a "standard" channel or $\pm 15\%$ for a "wide-band" channel. These peak FM deviations are for standardized data input voltage of ± 2.5 volts (or 0 to $+5$ volts if a standard -2.5-volt bias voltage is applied to the input data signal). There are 21 standard ($\pm 7.5\%$ deviation) channels whose carrier frequencies are listed in Table 12-1. Also listed are the equivalent peak deviation and the maximum data signal frequency which may be used on each channel. The last column is the maximum commutation rate for each channel—we will say more about that later. Table 12-2 lists the eight possible wide-band channels that can be selected (although not all eight at the same time). For each wide-band channel that is used three standard channels must be eliminated. Also, no two adjacent wide-band channels can be used (i.e., if E is used, then D and F cannot be used). The maximum data-frequency specification is given to insure that no one channel interferes with any adjacent channel. This means that the data input to each particular channel must be low-pass-filter limited to the maximum data frequency specified.

TABLE 12-1   THE IRIG "STANDARD" (7.5% DEVIATION) CHANNELS

IRIG Band Designation	Center Frequency	Peak Deviation	Maximum Data Frequency	Maximum Commutation Rate
1	400	± 30	6.0	1.52
2	560	42	8.4	2.00
3	730	55	11.0	2.76
4	960	72	14	3.38
5	1,300	98	20	5.00
6	1,700	128	25	6.25
7	2,300	173	35	8.75
8	3,000	225	45	11.20
9	3,900	293	59	15.0
10	5,400	405	81	20.0
11	7,350	551	110	27.5
12	10,500	788	160	40.0
13	14,500	1,088	220	55.0
14	22,000	1,650	330	82.0
15	30,000	2,250	450	112.0
16	40,000	3,000	600	152.0
17	52,500	3,940	790	196.0
18	70,000	5,250	1050	263.0
19	93,000	6,975	1400	—
20	124,000	9,300	1900	—
21	165,000	12,375	2500	—

From examination of Tables 12-1 and 12-2, it can be seen that a wide range of data bandwidths is available. Very often when telemetering a complex system this is an advantage because of the different types of physical data available and the wide range of inherent baseband bandwidths of these different types. However, in some cases there may be more data channels to be telemetered than the 21 IRIG channels would permit if each channel were to be used directly. One way of solving this problem if many of the extra channels are low-bandwidth channels is to time-division-multiplex (TDM) several of these low-data-bandwidth channels onto one of the higher-numbered IRIG channels. For example, IRIG wide-band channel D can be TDM commutated at switching rates up to 400 steps per second. This means that two data channels could be sampled 200 times per second (allowing 100-Hz data bandwidths on each when the Nyquist criterion is applied), or four channels could be sampled 100 times per second (50-Hz maximum bandwidth on each), and so on. Such an arrangement is shown in Figure 12-1b where there are three non-TDM (or "direct") channels and three TDM channels feeding into the six FM subcarrier oscillators. One TDM multiplexor has four inputs and the other two have three each, making a total of 10 TDM data channels and three direct channels or a grand total of 13 data channels.

Another possibility for transmitting a number of data channels is not to use FDM at all, but simply, as shown in Figure 12-2, to have one FM oscillator and

TABLE 12-2   THE IRIG WIDE-BANDWIDTH (15% DEVIATION)
CHANNELS

Band	Center Frequency	Peak Deviation	Maximum Data Frequency	Maximum Commutation Rate
A	22,000	3,300	600	165
B	30,000	4,500	900	222
C	40,000	6,000	1200	303
D	52,500	7,880	1600	400
E	70,000	10,500	2100	526
F	93,000	13,950	2800	—
G	124,000	18,600	3700	—
H	165,000	24,750	5000	—

(a) The IRIG "wide-band" FM channels

When using channel A, omit channels 13, 14, and 15
When using channel B, omit channels 14, 15, and 16
When using channel C, omit channels 15, 16, and 17
When using channel D, omit channels 16, 17, and 18
When using channel E, omit channels 17, 18, and 19
When using channel F, omit channels 18, 19, and 20
When using channel G, omit channels 19, 20, and 21
When using channel H, omit channels 20 and 21
When using a 100-KHz reference signal, do not use channels 19, F, or G.

(b) Correlation of the wide-band and narrow-band channels

TDM all of the data channels, using a strapped multiplexor (such as was described earlier in Section 4.8) if different bandwidth input data channels are to be serviced. In this case it is usually desirable to keep one of the input channels at a constant accurately known voltage for use as an FM-detection reference level upon detection of the received signal.

Once the IRIG subcarrier signals have been generated (covering a maximum frequency range of from about 350 Hz to 200 KHz), this composite signal may be transmitted directly over a narrow-band video link (transmission bandwidth up to 200 KHz), recorded onto magnetic tape using a wide-band "audio" instrumentation recorder, or else transmitted on an RF carrier using the composite IRIG signal to AM, FM, or PM modulate the RF signal. If the IRIG signal is to be recorded on tape, then an accurate subcarrier-frequency signal is usually recorded. Usually the frequency which is used is 100 KHz, and the use of this requires that IRIG channels 19, F, and G must not be assigned. Figure 12-3 shows such a record/playback system. The d.c. voltage derived from the 100-KHz discriminator at the playback end can be used for two purposes: first, it possibly can be used to control the speed of the playback tape "deck" if the mechanism is equipped with such a speed compensation system, and second, whether the speed-control servoing is used or not, the d.c. signal from the 100-KHz discriminator is added to the output of each of the other discriminators. This corrects for any playback speed error (even that which remains after the speed correction) and compensates for errors introduced by "wow" and "flutter" in the tape-drive mechanism.

Other types of multichannel analog modulation methods can also be used, such as TDM PAM or TDM PDM. However, the IRIG system has been one of the more accurate and more popular up until the present time. Although it is largely being replaced with various types of digital telemetry systems such as will be described in the next section, it still retains some degree of usage because of its relative simplicity when the digital format is not desired at the receiving end, and therefore equipment for this standard is still being manufactured yet today.

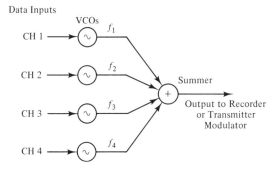

(a) IRIG System with Four Direct Channels

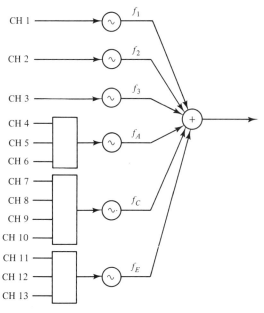

(b) IRIG System with TDM on Three IRIG Channels

Figure 12-1 IRIG FM analog telemetry.

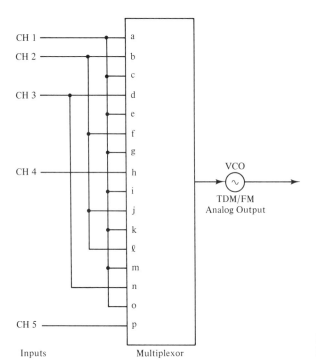

CH 1
CH 2
CH 3
CH 4
CH 5

Inputs

Multiplexor

a
b
c
d
e
f
g
h
i
j
k
ℓ
m
n
o
p

VCO

TDM/FM
Analog Output

Figure 12-2   Using TDM instead of
FDM on an analog telemetry system.

12.3 DIGITAL TELEMETRY METHODS

In sending data about physical variables via digital communication techniques, there are several different combinations of approaches which may be used. The physical data itself may first be converted into analog electrical signals and then these analog signals *digitized* into binary digital information, or the physical data may be directly converted into digital form (usually in parallel format) by means of a *digital encoder*. Once the digital signal is available, it may be transmitted immediately, usually as a serial binary signal, or it may be stored and a group or *frame* of digital data values transmitted in one burst, sometimes employing some of the digital data communications methods discussed in the last chapter.   When the data are stored before being transmitted they are sometimes digitally processed according to some algorithm for the reduction of noise and/or errors (i.e., *digital filtering*; the elimination of unneeded or redundant information (sometimes referred to as *digital bandwidth-compression techniques*); or to reduce the amount of data processing needed at the receiving end of the system.   In the following section we very briefly discuss the means of obtaining the physical data in digital form and then conclude with a very brief discussion of digital processing methods.

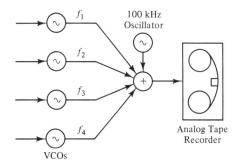

(a) Recording System

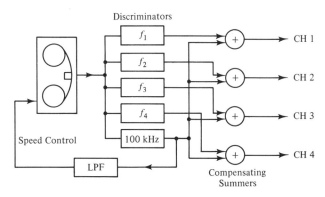

(b) Playback System

Figure 12-3   The IRIG FM data recording system.

12.3.1 *Digital Encoders and A/D Converters:*

One of the more common digital encoders is the mechanical "gray-code" encoder. These come in either a linear (straight-line mechanical motion) or a rotary form. In the latter form they are often referred to as *shaft encoders*. Figure 12-4 schematically shows a linear digital encoder. We could form this into a shaft encoder by "wrapping" the scale around in a complete circle so that the left and right ends join together. One of the problems with which we would be faced if we were to use a common binary-numbering sequence such as 0000, 0001, 00010, 0011, 0100, 0101, and so on is that if the analog mechanical displacement happened to be crossing a boundary such as that between 0011 and 0100, we would have three digits that must change state *simultaneously*. This is very difficult to make happen with most electromechanical arrangements that can be contrived, and what may occur is perhaps a sequence of one-bit changes so that instead of 0011 to 0100, we might get a sequence such as 0011, 0111, 0101, 0100. Thus, if we happen to look

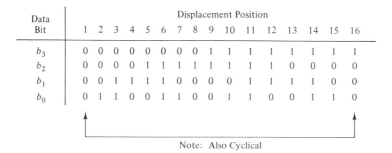

Data Bit	Displacement Position															
	1	2	3	4	5	6	7	8	9	10	11	12	13	14	15	16
b_3	0	0	0	0	0	0	0	0	1	1	1	1	1	1	1	1
b_2	0	0	0	0	1	1	1	1	1	1	1	1	0	0	0	0
b_1	0	0	1	1	1	1	0	0	0	0	1	1	1	1	0	0
b_0	0	1	1	0	0	1	1	0	0	1	1	0	0	1	1	0

Note: Also Cyclical

Figure 12-4   A "gray code" digital electromechanical data encoder.

at the output of the encoder while it is in the transition state, we might read 0111 or 0101, neither of which are values between 0011 or 0100. To get around this problem, we use a binary code wherein only one digit changes at a time as we progress from one increment to another.

One type of such code is the one shown in Figure 12-4. Note that the one-bit-per-transition change rule not only applies to all the transitions 1-2, 2-3, and so on up to 15-16, but also applies to the 16-1 "wraparound" transition. This makes the encoder *cyclical* and means that we can use this code for a continuous shaft encoder (i.e., we don't have to limit the rotation in either direction).

Just as we were able to produce analog voltage signals from a variety of physical data "signals" by first converting the physical variable we wanted to telemeter into a mechanical displacement and then converting the mechanical displacement into an analog electrical signal via the use of some type of analog electromechanical transducer, we can also "digitize" many types of physical data by converting that data into a mechanical displacement that actuates a linear or rotary digital encoder such as has just been described.

The other technique which may be employed is to first obtain an analog voltage corresponding to the physical variable value and then to "digitize" this electrical signal by the use of some type of analog-to-digital converter. This method has the advantage of being able to handle data that are more rapidly changing than those which can be handled by the mechanical digital encoder and also to permit some analog preprocessing (filtering, biasing, etc.) prior to doing the digitizing operation.

Figure 12-5 shows a number of different types of electronic A/D converter techniques. Figure 12-5a shows the use of a voltage-controlled oscillator whose frequency is dependent upon the value of the analog input. By determining the number of oscillator cycles in a known time period the oscillator frequency can be determined , and if a f/v calibration curve is known for the oscillator, the value of the analog signal can be found. In this case some standard counting period is used, and the digital data which is sent is just the output of the binary counter. Knowing the count period and the oscillator calibration curve, the receiving end of the telemetry system can obtain the value of the analog input signal.

Figure 12-5b shows the *ramp-type A/D converter*. In this case, rather than counting the cycles of a variable-frequency oscillator for some fixed period of time,

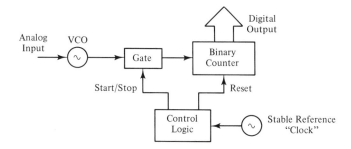

(a) VCO-Type A/D Converter

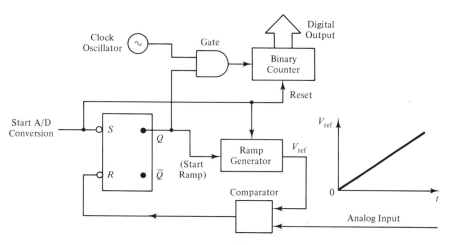

(b) Ramp-Type A/D Converter

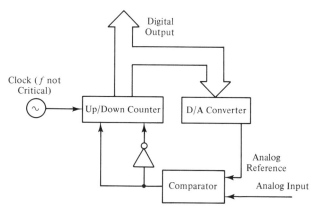

(c) Tracking-Type A/D Converter (Tends to "Dither" — not
Good to use with TDM Systems)

Figure 12-5   Different types of A/D converter circuits.

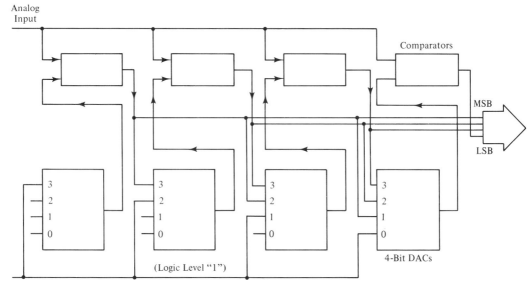

(d) Successive-Approximation-Type A/D Converter

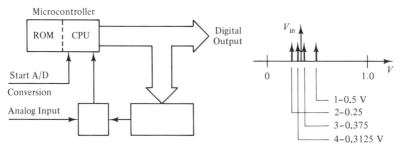

(e) Half-Interval-Search-Type A/D Converter

Figure 12-5   (*Continued*)

the cycles of a stable oscillator of some known frequency are counted for a length of time which is dependent upon the value of the analog input. This length of time is determined by comparing the analog input with a generated ramp function which starts at zero and increases linearly at some fixed known rate. When the ramp-function voltage is equal to the analog input voltage the counting period is ended and the number of cycles counted will then be proportional to the value of the analog input voltage.

Another type of A/D converter which gives a continuous digital output is the tracking type shown in Figure 12-5c. In this A/D circuit a digital-to-analog converter is used to continually convert the digital output value back into an analog voltage signal which is then compared with the analog input voltage. The 0 or 1 output of the comparator is then used to control an up/down counter in such a way that if the analog input decreases, the counter will *count down* (lower its digital value by counting oscillator cycles) until its digital value reflected through the D/A converter is equivalent to the new value of the analog input voltage.

Figure 12-5d shows a *successive approximation* type of A/D converter wherein a change in the analog input voltage will "ripple" back and forth through the chain until the best approximation to the analog input voltage is obtained.

The last type shown (in Figure 12-5e) is what is known as the *half-interval-search* type. It uses the classical half-interval-search algorithm often employed in software in digital data searches conducted on a computer. This algorithm is stored in the microcontroller and causes progressively closer digital values to be sent to the D/A converter which then produces an analog voltage which is compared with the analog input. The comparator tells the microcontroller whether the analog input signal is larger or smaller than the analog reference signal. The program in the controller uses this information to control the direction of the search algorithm and hence the digital value that is put out and applied to the D/A converter.

An example of a four-step algorithm that digitizes a 0- to 1.0-volt analog signal into a four-bit value is shown. The four steps would be start at 0.5 volt; since the input is smaller, subtract 0.25 volt to get 0.25 volt; since the input is now larger, add 0.125 volt to get 0.375; and finally, since the input is smaller than this, subtract 0.0625 to get 0.3125 volts. Additional steps could be added to get closer and closer to the actual value of the analog input voltage.

Most of the types of A/D converters just discussed are available as single silicon chips with varying number of significant bits, digitizing accuracy, and cost. It should also be remembered that the digital sample value of the analog variable must be extracted often enough to satisfy the Nyquist criterion or else *aliasing* of the data will result. In the case of the electromechanical encoders, this merely means that the parallel digital output must be read often enough. In the case of the analog electrical signal being converted to digital data, it means that if the A/D converter's analog input is mutiplexed among a number of analog data signals, each such signal must be "digitized" often enough to satisfy the Nyquist criterion.

12.3.2 *Digital Processing of Telemetry Data:*

Once the physical data value has been converted to a digital form, it may be processed either at the sending end or at the receiving end to enhance the usefulness of the data and/or to reduce the amount of data which must be sent. Transform-type digital filtering algorithms can be employed if a number of digital samples of the same physical data are available in memory at the same time. In this case the usual procedure is to take the Fast-Fourier Transform (FFT) of the data to obtain the transform (frequency-domain) functions. These are then multipled by frequency-dependent factors which represent the frequency and phase characteristics of the filter transfer function and the inverse FFT taken of this modified "spectrum" to give the filtered digital "time function." The other technique is a continuous time-domain convolution operation using the current digital value and a fixed number of preceding digital data values of the same physical variable. These are each multiplied successively by the different values of the convolving function; the sequence of sums which result from each cycle of such multiplications becomes the new data-function sequence that is then stored or transmitted.

Sometimes the digital filtering process described may result in an output data sequence wherein the "sampling" rate is excessive for the filtered data. In this event it is possible to "throw away" some of the data. This can be done by simply discarding every other data point, or two out of three points, for example. Or it can be done by averaging two, three, or more points into one. This is one method of reducing the "bandwidth" of the digital signal to be transmitted. Other techniques such as the *run-length* coding, as described in Chapter 7 for video BW compression, can also be applied to reduce the amount of digital information which actually needs to be transmitted. The many methods of digital processing of data signals derived from all different types of physical sources is too large a subject to discuss in detail here, and the reader is referred to the many texts and papers on digital filtering, digital bandwidth-compression techniques, and image enhancement methods for further information in these areas.

PROBLEMS

12-1. Design a time-division multiplexing system using a strapped sequential multiplexor which will sample the following five channels at a regular rate and produce a composite TDM/natural-PAM output. Design the system so that the multiplexor clock (stepping) rate is as low as possible while still keeping a reasonable frame length. Include a synchronizing signal consisting of three full-amplitude pulses transmitted at the start of each frame. The data on the analog input channels have the following information bandwidths:

> Channel A: 10 Hz
> Channel B: 20 Hz
> Channel C: 30 Hz
> Channel D: 50 Hz
> Channel E: 70 Hz

12-2. Design an IRIG FM telemetry system to handle the analog input channels listed below (19 channels total) within the smallest video (the FDM output signal from all of the IRIG VCOs) bandwidth possible. You may TDM as many of the IRIG VCOs as you wish, but you are to keep the total cost of the system as low as possible (VCOs cost $200 each and each TDM multiplexor costs ($400 + 35$N$) each, where N is the number of inputs on the TDM multiplexor).

number of channels	(having an)	information bandwidth of
6		10 Hz
5		15 Hz
4		30 Hz
2		50 Hz
2		100 Hz

12-3. An FM/FM satellite telemetry system uses voltage-controlled subcarrier oscillators on IRIG Channels 7 through 15 inclusive, and uses a maximum subcarrier modulation index on each channel of 5. The maximum peak (+ or −) analog signal input voltage for each channel (i.e., that which gives the specified peak deviation for the VCO for that channel) is 10 volts.

a. What will be the lowest and the highest IRIG output signal frequencies present?

b. If the data signal on channel 10 is a 5-volt rms cosine wave, what will be the peak deviation on Channel 10, and what is the highest frequency that the input data signal can have?

c. If the IRIG signal from the outputs of all nine oscillators is used to frequency modulate an RF transmitter with a carrier of 150 MHz, and if the peak FM modulation index for the RF signal is also set at 5, what will be the maximum frequency deviation of the RF signal? Approximately what RF bandwidth will it occupy?

d. For Channel 12, complete the following table for the various data-signal examples given:

Data Signal	f Deviation (Peak)	m_f of the VCO
$10.0 \cos(2\pi \times 160t)$		
$4.0 \cos(2\pi \times 160t)$		
$10.0 \cos(2\pi \times 40t)$		
$4.0 \cos(2\pi \times 40t)$		
$10.0 \cos(2\pi \times 1.0t)$		
$0.1 \cos(2\pi \times 160t)$		

e. At the data processing center on the ground, the magnetic tape on which the composite IRIG signal (i.e., the demodulated output of the FM receiver itself) was recorded is played back into an array of IRIG filters and discriminators which recover the original data signals. Sketch a block diagram of this system, showing, for each channel:

(1) The passband of the video filter ahead of the discriminator.

(2) The discriminator center frequency.

(3) The passband of the low-pass filter following the discriminator.

f. What advantage is there in using an FM subcarrier modulation index of 5 instead of the standard 7? What disadvantage?

Appendix A
THE THERMAL RADIATION LAW

From physics we have Planck's law of thermal radiation which gives the power density radiated *per unit of solid angle per unit of radiating area* in a direction perpendicular to the radiating surface *in the frequency passband* Δf centered about the frequency $f = c/\lambda$. Using the relationship that $\Delta f = (c/\lambda^2)(\Delta\lambda)$, this law becomes

$$p(\lambda)\,\Delta\lambda = \left(\frac{2c^2 h}{\lambda^5}\right)\left(\frac{\Delta\lambda}{(e^{ch/kT\lambda} - 1)}\right) \text{ Watts/m}^2\text{-steradian} \qquad \text{(A-1)}$$

where $c = 3 \times 10^8$ meters per second (the speed of light), h is Planck's constant $(6.625 \times 10^{-34}$ Joule-seconds), and k is Boltzman's constant $(1.381 \times 10^{-23}$ Joule/°K). If the product λT is large, then the factor $e^{ch/kT\lambda}$ can be approximated by a series

$$e^{ch/kT\lambda} \doteq 1 + \left(\frac{ch}{\lambda kT}\right) + \frac{(ch/\lambda kT)^2}{2!} + \cdots \qquad \text{(A-2)}$$

Also, since $\lambda = c/f$, $\Delta\lambda = (c/f^2)(\Delta f)$ as given. Making these substitutions into Equation A-1 gives

$$p(f)\Delta f = 2\left(\frac{k\,T\,\Delta f}{2}\right) \text{ Watts/m}^2\text{-steradian} \qquad \text{(A-3)}$$

in the bandwidth $B = \Delta f$ centered about the frequency $f = c/\lambda$. The power density per square meter at the receiver located a distance ℓ away from the radiating surface of area dA_p is then given by[1]

$$dp_{NR} = p(f)\,B\,(\ell^{-2})\,(dA_p) = \frac{2\,k\,B\,T\,dA_p}{\lambda^2\ell^2} \text{ Watts per meter}^2 \qquad \text{(A-4)}$$

which is the same as Equation 6-41 if T_s is substituted for T. Here the bandwidth Δf has been replaced by the notation B for the bandwidth of the receiving system to which the antenna or receiving area is attached.

If the black body is an isotropic radiator, its effective aperture area is $\lambda^2/4\pi$. The total radiated noise from this isotropic source which can be picked up and used is then given by

$$P_{NR} = \tfrac{1}{2}\left[p(f)\left(\frac{\lambda^2}{4\pi}\right) 4\pi B\right] = \tfrac{1}{2}\left[\frac{2\,k\,T\,B}{\lambda^2}\right]\lambda^2$$
$$= (k\,T\,B) \qquad \text{(A-5)}$$

where the factor of $\tfrac{1}{2}$ accounts for the fact that only the power in one polarization can be intercepted (received) by a normal antenna. If the radiator is considered to be a resistor at ambient temperature T attached to an isotropic radiator antenna matched to its environment, then Equation A-5 gives the noise power output of the resistor and leads to Equations 3-5 and 3-6. The value of the resistor R is the radiation resistance of the antenna being fed by the resistor.

[1] Where the solid angle at the source due to an increment of area at the receiver is given by $\Delta\Omega = \Delta A_r/\ell^2$. This is the reason for the ℓ^{-2} factor in Equation A-4.

DERIVATION OF THE S/N RATIO AT THE OUTPUT OF A T-PAD PASSIVE ATTENUATOR ELEMENT

This appendix is a development of the proof that the excess noise power added by the attenuator to its output signal is given by the relationship $N_o = kT_DL\Delta f$. The T-pad attenuator attached to a signal-and-noise source and to a matched load is shown in Figure B-1. For an attenuator we can define a factor α where α is the reciprocal of the voltage gain of the attenuator. Since the output voltage will be less than the input voltage, α will be greater than unity. Also, the power gain of the attenuator is then given by α^{-2} if the input and output impedances are equal. We will assume that the system is one in which both the input and output impedances are matched and at the impedance level R_o. In this case the load resistor will have the value R_o and the Thevenin impedance of the source will also be R_o. We now give the formulas for the three resistors making up the T-pad attenuator itself:

$$R_1' = R_1 = R_o \left(\frac{\alpha - 1}{\alpha + 1} \right) \tag{B-1}$$

$$R_2 = \frac{R_o 2\alpha}{\alpha^2 - 1} \tag{B-2}$$

As a first step, let us prove that given Equations B-1 and B-2 the input impedance to the attenuator at terminals $a - a'$ is equal to the characteristic impedance level R_o.

$$Z_{in} = R_1 + \frac{R_2(R_1 + R_o)}{R_2 + (R_1 + R_o)}$$

$$= \left(\frac{\alpha - 1}{\alpha + 1} \right) + \frac{[2\alpha/(\alpha^2 - 1)][1 + (\alpha - 1)/(\alpha + 1)]R_o}{\{[2\alpha/(\alpha^2 - 1)] + (\alpha - 1)/(\alpha + 1) + 1\}}$$

$$= R_o \left\{ \left(\frac{\alpha - 1}{\alpha + 1} \right) + \frac{[2\alpha(\alpha + 1 + \alpha - 1)/(\alpha + 1)]}{[2\alpha + (\alpha - 1)(\alpha - 1) + (\alpha^2 - 1)]} \right\} \tag{B-3}$$

$$= R_o \left\{ \frac{(\alpha - 1) + (2\alpha)(2\alpha)/(2\alpha + (\alpha^2 - 2\alpha + 1) + (\alpha^2 - 1))}{(\alpha + 1)} \right\}$$

$$= R_o \left[\frac{(\alpha - 1) + (4\alpha^2/2\alpha^2)}{(\alpha + 1)} \right] = R_o \left[\frac{(\alpha + 1)}{(\alpha + 1)} \right] = R_o = Z_{in}$$

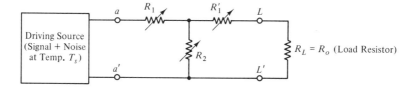

Figure B-1   System employing a T-pad attenuator.

Now, let each resistor in the attenuator be represented by an equivalent Thevenin noise generator (see Figure B-2).   Then the system shown in Figure B-1 becomes as shown in Figure B-3.   Next we find the mean-squared noise current through R_o by the use of the power superposition method (where noise *powers* from different sources add rather than the noise voltages or currents).   Define the following:

$$\overline{i_{A_1}^2} = \text{ the component of } \overline{i_N^2} \text{ due to the noise source}$$
$$\text{in resistor } R_1$$
$$\overline{i_{A_{1'}}^2} = \text{ the component of } \overline{i_N^2} \text{ due to the noise source}$$
$$\text{in resistor } R_{1'}$$
$$\overline{i_{A_2}^2} = \text{ the component of } \overline{i_N^2} \text{ due to the noise source}$$
$$\text{in resistor } R_2$$
$$\overline{i_s^2} = \text{ the component of } \overline{i_N^2} \text{ due to the noise source in}$$
$$\text{the signal generator}$$

Adding these together gives

$$\overline{i_N^2} = \overline{i_{A_1}^2} + \overline{i_{A_{1'}}^2} + \overline{i_{A_2}^2} + \overline{i_s^2} \tag{B-4}$$

where each individual term is given by the following equations:
For $\overline{i_{A_1}^2}$ (see Figure B-4a)

$$\overline{i_{A_1}^2} = \left(\frac{1}{\alpha^2}\right) \frac{\overline{v_{A_1}^2}}{4R_o^2} \tag{B-5a}$$

for $\overline{i_{A_{1'}}^2}$ (see Figure B-4b)

$$\overline{i_{A_{1'}}^2} = \frac{\overline{v_{A_{1'}}^2}}{\left[R_1 + R_o + \dfrac{R_2(R_1 + R_o)}{(R_1 + R_2 + R_o)}\right]^2} \tag{B-5b}$$

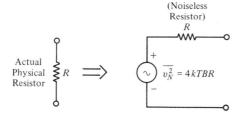

Figure B-2   Equivalent Thevenin circuit of a noisy resistor.

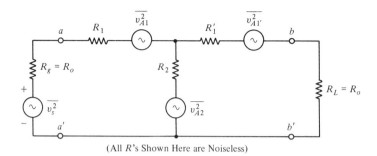

(All R's Shown Here are Noiseless)

Figure B-3 Equivalent circuit of the T-pad, source, and load for determining the noise power to the load.

for $\overline{i_{A_2}^2}$ (see Figure B-4c)

$$\overline{i_{A_2}^2} = \left(\frac{1}{2}\right)^2 \frac{\overline{v_{A_2}^2}}{\left[R_2 + \dfrac{1}{2}(R_1 + R_o)\right]^2} \tag{B-5c}$$

and for $\overline{i_s^2}$ (see Figure B-4d)

$$\overline{i_s^2} = \left(\frac{1}{\alpha^2}\right) \frac{\overline{v_s^2}}{4R_o^2} \tag{B-5d}$$

Now, substituting for R_1, R_1', and R_2 in Equations B-5b and B-5c

$$\overline{i_{A_{1'}}^2} = \frac{\overline{v_{A_{1'}}^2}}{R_o^2\left[\dfrac{(\alpha-1)}{(\alpha+1)} + 1 + \dfrac{(2\alpha/(\alpha^2-1))/((\alpha-1)/(\alpha+1)+1)}{(\alpha-1/\alpha+1+2\alpha/(\alpha^2-1)+1)}\right]^2} = \frac{\overline{v_{A_{1'}}^2}}{4R_o^2} \tag{B-6}$$

$$\overline{i_{A_2}^2} = \frac{\overline{v_{A_2}^2}}{4R_o^2\left[(2\alpha/(\alpha^2-1)) + \dfrac{1}{2}((\alpha-1)/(\alpha+1)+1)\right]^2}$$

$$= \frac{\overline{v_{A_2}^2}}{4R_o^2\alpha^2[(\alpha+1)^2/(\alpha-1)^2]} \tag{B-7}$$

Equation B-4 then becomes

$$\overline{i_N^2} = \left(\frac{1}{\alpha^2}\right) \times \left(\frac{kT_sB}{R_o}\right) + \frac{kT_DB}{R_o}\left(\frac{1}{\alpha^2}+1\right)\frac{(\alpha-1)}{(\alpha+1)} + \frac{2kT_DB(\alpha-1)}{\alpha R_o(\alpha+1)}$$

$$= \frac{kB}{R_o}\left[\frac{T_s}{\alpha^2} + \frac{(\alpha-1)}{(\alpha+1)}T_D\left(\frac{1}{\alpha^2}+1+\frac{2}{\alpha}\right)\right] \tag{B-8}$$

$$= \frac{kB}{R_o}\left[GT_s + \left(1-\frac{1}{\alpha^2}\right)T_D\right] = \frac{kB}{R_o}[GT_s + LT_D]$$

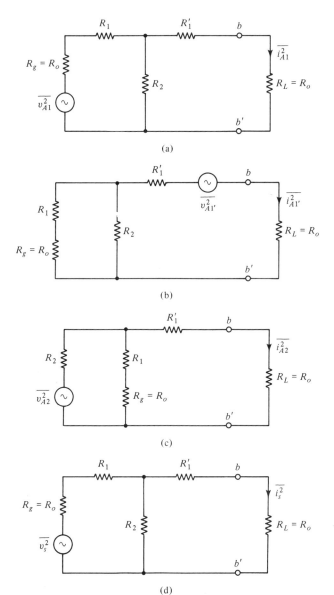

Figure B-4 Partial equivalent noise circuits.

and the output noise power delivered by the attenuator to the load resistor $R_L = R_o$ is then given by

$$P_{N,\text{out}} = R_o \overline{i_N^2} = kB(GT_s + LT_D) \tag{B-9}$$

where the first term in the brackets is the contribution of noise from the signal source passed through the attenuator and the second term represents the excess noise added by the resistive nature of the attenuator itself. T_D is the actual ambient temperature of the attenuator, and T_s is the *noise* temperature of the noise appearing on the signal going into the attenuator. Once again, remember that the attenuator loss factor L is related to the attenuator G by the relationship $L = G - 1$ where G is less than unity for an attenuator.

Appendix C
THE BALANCED MODULATOR

One way of thinking of the balanced modulator is as two separate RF amplifiers whose inputs are in phase with each other and whose outputs are 180 degrees out of phase. Let the gain of each amplifier be controlled by an applied voltage $v_g(t)$ (an audio voltage which varies much more slowly than the RF signal does). Then an equivalent circuit can be drawn as shown in Figure C-2, where the two gain-control voltages $v_{g1}(t)$ and $v_{g2}(t)$ are derived from a split-secondary audio transformer or phase splitter and are thus 180 degrees out of phase with each other. Analyzing the circuit of Figure C-2, we get

$$v_1(t) = A_1 v_c(t) \tag{C-1}$$

$$v_2'(t) = -A_2 v_c(t) \tag{C-2}$$

but since the gains A_1 and A_2 are themselves time functions given by

$$A_1 = K v_{g1}(t) \tag{C-3}$$

$$A_2 = K v_{g2}(t) \tag{C-4}$$

the net output signal becomes

$$
\begin{aligned}
v_o(t) &= v_1(t) + v_2'(t) \\
&= [K v_{g1}(t)] v_c(t) - [K v_{g2}(t)] v_c(t)
\end{aligned}
\tag{C-5}
$$

If the time dependence of the two control signals is given by

$$v_{g1}(t) = C s_m(t) \tag{C-6}$$

$$v_{g2}(t) = -C s_m(t) \tag{C-7}$$

the output signal then becomes

$$
\begin{aligned}
v_o(t) &= [K C s_m(t)] v_c(t) - \{K[-C s_m(t)]\} v_c(t) \\
&= 2KC[s_m(t)][v_c(t)]
\end{aligned}
\tag{C-8}
$$

which is what we wanted to accomplish (i.e., the multiplication of the RF carrier signal by the modulating signal). The K and C used here are simply multiplying dimensionless constants.

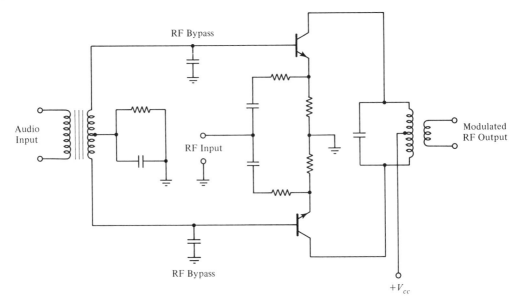

Figure C-1 One way of thinking of the balanced modulator is as two separate RF amplifiers whose inputs are in phase with each other and whose outputs are 180° out of phase. Let the gain of each amplifier be controlled by an applied voltage $v_g(t)$ (an audio voltage which varies much more slowly than the RF signal does). Then an equivalent circuit can be drawn as shown in Figure C-2 where the two gain-control voltages $v_{g1}(t)$ and $v_{g2}(t)$ are derived from a split secondary audio transformer or phase splitter and are thus 180° out of phase with each other. Analyzing the circuit of Figure C-2 we get:

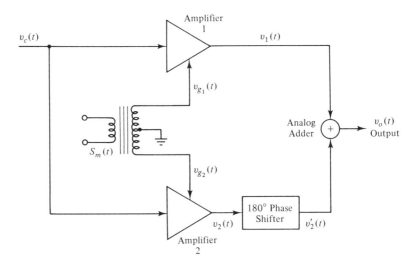

Figure C-2

Appendix D
THE FOSTER-SEELEY DISCRIMINATOR AND THE RATIO DETECTOR CIRCUITS FOR FM

Figure D-1 is a sketch of an equivalent RF circuit for either the Foster-Seeley discriminator or the ratio detector. These circuits employ a transformer whose secondary is tapped at the exact center. In this equivalent circuit loose primary-secondary coupling is assumed and the voltage induced into the secondary from the primary is modeled by two controlled-voltage generators each having a root-mean-square value of $\hat{V}_{in}/2$. Self-induced secondary voltage is accounted for in the usual manner by the inductances L_p and L_s. The effects of secondary-to-primary coupling are ignored. Thus, the circuit impedance and the voltage at point A are given by

$$\hat{Y}_p \doteq \frac{1}{R_p} + j\left(\omega C_p - \frac{1}{\omega L_p}\right) \tag{D-1}$$

$$\hat{V}_A \doteq \frac{\hat{I}_g}{\hat{Y}_p} \tag{D-2}$$

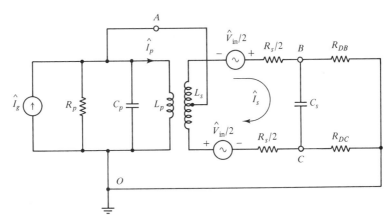

Figure D-1

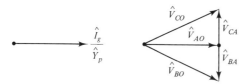

Figure D-2

Accounting for the coupling from the primary to the secondary by the controlled voltage generators we obtain

$$\hat{V}_{in} = j\omega M \hat{I}_{Lp} = j\omega M \frac{\hat{V}_A}{j\omega L_p} = \frac{M}{L_p} \frac{\hat{I}_g}{\hat{Y}_p} \tag{D-3}$$

and solving the secondary loop equation letting I_s be the current circulating in the secondary resonant circuit and assuming R_{DB} and R_{DC} to be large

$$\hat{I}_s = \frac{\hat{V}_{in}}{(R_s + j(\omega L_s - 1/\omega C_s))} = \frac{M\hat{I}_g}{L_p \hat{Y}_p \hat{Z}_s} \tag{D-4}$$

Then the voltage from points B and C to ground is given by

$$\hat{V}_B = \hat{V}_A + \frac{\hat{V}_{in}}{2} - \frac{1}{2}(j\omega L_s)\hat{I}_s - \frac{1}{2}R_s \hat{I}_s \tag{D-5}$$

$$\hat{V}_C = \hat{V}_A + \frac{\hat{V}_{in}}{2} + \frac{1}{2}(j\omega L_s)\hat{I}_s - \frac{1}{2}R_s \hat{I}_s \tag{D-6}$$

which becomes when substitution for $\hat{V}_A$ is made

$$\hat{V}_B = \frac{\hat{I}_g}{\hat{Y}_p} + \frac{M\hat{I}_g}{L_p \hat{Y}_p}\left(1 - \frac{R_s}{2\hat{Z}_s} - j\frac{\omega L_s}{2\hat{Z}_s}\right) \tag{D-7}$$

$$\hat{V}_C = \frac{\hat{I}_g}{\hat{Y}_p} - \frac{M\hat{I}_g}{L_p \hat{Y}_p}\left(1 - \frac{R_s}{2\hat{Z}_s} - j\frac{\omega L_s}{2\hat{Z}_s}\right) \tag{D-8}$$

When the quantity $j\omega L_s$ is much greater than R_s and $\hat{Z}_s$ (ie., for R_s small and the circuit at resonance) these equations become

$$\hat{V}_B = \left(1 - j\omega\frac{ML_s}{2L_p\hat{Z}_s}\right)\frac{\hat{I}_g}{\hat{Y}_p} \tag{D-9}$$

$$\hat{V}_C = \left(1 - j\omega\frac{ML_s}{2L_p\hat{Z}_s}\right)\frac{\hat{I}_g}{\hat{Y}_p} \tag{D-10}$$

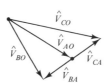

Figure D-3

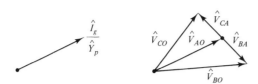

Figure D-4

Now first consider the case where no deviation is present and $f = f_c$. Then $\hat{Z}_s$ is real and $\hat{Y}_p$ is real and the phasor diagrams are as shown in Figure D-2. $\hat{V}_B$ and $\hat{V}_c$ are the rms values of the RF voltages to ground at points B and C, respectively. When the instantaneous frequency is above the the carrier frequency, f is larger than f_c and $\hat{Z}_s$ has a positive impedance angle resulting in $|\hat{V}_C|$ being greater than $|\hat{V}_B|$ as is shown in Figure D-3. When the instantaneous frequency is less than the carrier frequency, f is less than f_c and $\hat{Z}_s$ has a negative impedance angle, giving $|\hat{V}_C|$ less than $|\hat{V}_B|$ as is shown on Figure D-4. Remember that $|\hat{V}_B|$ and $|\hat{V}_C|$ are the rms values of the RF voltages prior to diode rectification.

The Foster-Seeley discriminator is shown in Figure D-5. Note that both diodes are facing in the same direction. The value of the quasi-d.c. (audio) voltages across the RF bypass capacitors C_1 and C_2 at any point in time are proportional to $|\hat{V}_B|$ and $|\hat{V}_C|$, respectively, and the value of the output voltage is

$$v_{\text{out}}(t) = v_{C_1}(t) - v_{C_2}(t) \qquad (\text{D-11})$$

or

$$v_{\text{out}}(t) \propto \{|\hat{V}_B| - |\hat{V}_C|\} \qquad (\text{D-12})$$

Thus, we see that at any point in time the audio output voltage is proportional to the RF signal strength *and* to the deviation of the signal frequency from f_c at that time. To eliminate the dependence of the audio output voltage upon the RF signal strength the Foster-Seeley discriminator FM detector must be preceded by a limiter amplifier. Figure D-6 shows how $|\hat{V}_B|$, $|\hat{V}_C|$, and the differences ($|\hat{V}_B| - |\hat{V}_C|$) vary with the frequency of the detector input signal.

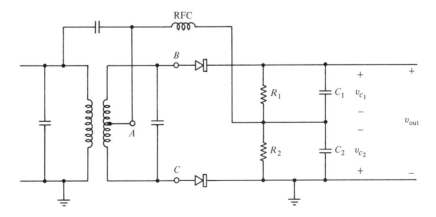

Figure D-5

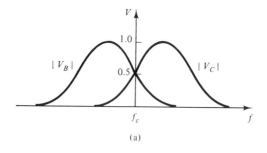

(a)

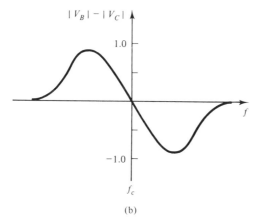

(b)

Figure D-6

Figure D-7 shows the connection of the diodes and the audio circuit for the ratio detector. Unlike the discriminator circuit, the two diodes face in opposite directions giving d.c. voltages across C_1 and C_2 which add rather than cancel. Since C_3 is much larger than C_1 or C_2 the voltage across C_3 is the long-term average of the sum of $|\hat{V}_B| + |\hat{V}_C|$ (see Figure D-8). Represent this long-term average by

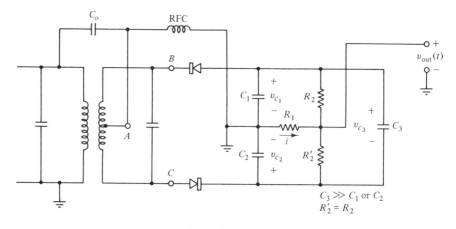

Figure D-7

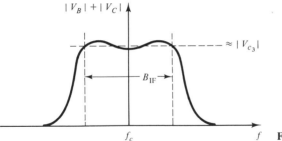

Figure D-8

$LT(|\hat{V}_B| + |\hat{V}_C|)$. The time-varying d.c. (audio) voltages are proportional to the current rms values of $|\hat{V}_B|$ and $|\hat{V}_C|$ or

$$v_{C_1}(t) = K \times |\hat{V}_B| \qquad (D\text{-}13)$$

$$v_{C_2}(t) = K \times |\hat{V}_C| \qquad (D\text{-}14)$$

However, the sum of these two voltages must equal the long-term average or

$$v_{C_1} + v_{C_2} = v_{C_3} \qquad (D\text{-}15)$$

which gives

$$K(|\hat{V}_B| + |\hat{V}_C|) = LT(|\hat{V}_B| + |\hat{V}_C|) \qquad (D\text{-}16)$$

where "LT" represents "the long-term-average-of" function. From this we then get

$$K = \frac{LT(|\hat{V}_B| + |\hat{V}_C|)}{(|\hat{V}_B| + |\hat{V}_C|)} \qquad (D\text{-}17)$$

K can be considered as a type of "gain" function which eliminates rapid variations in $(|\hat{V}_B| + |\hat{V}_C|)$ due to amplitude modulation (noise, etc.) on the RF signal. It does this by changing the bias on the diodes as the signal level changes. Also, C_3 may be thought of as a swamping capacitor which absorbs the energy in any noise spikes which are present on the RF signal. The detector output then becomes

$$v_{\text{out}}(t) = (v_{C_2}(t) + v_{C_1}(t)) \frac{R_1}{(2R_1 + R_2)} = (|\hat{V}_C| - |\hat{V}_B|) \frac{K \times R_1}{(2R_1 + R_2)} \qquad (D\text{-}18)$$

$$= \frac{R_1}{(2R_1 + R_2)} + \frac{(|\hat{V}_C| - |\hat{V}_B|)}{(|\hat{V}_C| + |\hat{V}_B|)} \times LT(|\hat{V}_C| + |\hat{V}_B|)$$

Thus we see that the ratio detector output voltage is proportional to the frequency deviation and the long-term average of the signal level.

Appendix E
IMPORTANT TRIGONOMETRIC IDENTITIES USED IN THE ANALYSIS OF COMMUNICATION SYSTEMS

$$\sin (A \pm B) = (\sin A)(\cos B) \pm (\cos A)(\sin B)$$

$$\cos (A \pm B) = (\cos A)(\cos B) \mp (\sin A)(\sin B)$$

$$(\sin A)(\sin B) = \tfrac{1}{2}[\cos (A - B) - \cos (A + B)]$$

$$(\cos A)(\cos B) = \tfrac{1}{2}[\cos (A - B) + \cos (A + B)]$$

$$(\sin A)(\cos B) = \tfrac{1}{2}[\sin (A - B) + \sin (A + B)]$$

$$\cos (z \sin \theta) = J_0(z) + 2 \sum_{k=1}^{\infty} J_{2k}(z) \cos (2k\theta)$$

$$\sin (z \sin \theta) = 2 \sum_{k=0}^{\infty} J_{2k+1}(z) \sin \{(2k + 1)\theta\}$$

$$\cos (z \cos \theta) = J_0(z) + 2 \sum_{k=1}^{\infty} (-1)^k J_{2k}(z) \cos (2k\theta)$$

$$\sin (z \cos \theta) = 2 \sum_{k=0}^{\infty} (-1)^k J_{2k+1}(z) \cos \{(2k + 1)\theta\}$$

Fourier Series of a Pulse Train

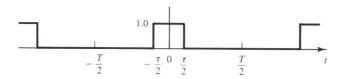

τ = pulse length T = pulse period

$$p(t) = \frac{\tau}{T} + \frac{2}{\pi} \sum_{n=1}^{\infty} \frac{\sin\left(\dfrac{n\pi\tau}{T}\right)}{n} \cos\left(\frac{2\pi nt}{T}\right) = \frac{\tau}{T} + \left(\frac{2\tau}{T} \sum_{n=1}^{\infty}\right) \frac{\sin\left(\dfrac{n\pi\tau}{T}\right)}{\left(\dfrac{n\pi\tau}{T}\right)} \cos\left(\frac{2\pi nt}{T}\right)$$

454

Appendix F
THE GAUSSIAN DISTRIBUTION

In working with practical communication systems it is often found that random noise signal voltages or currents have a Gaussian probability distribution concerning the occurrence of any particular value of voltage or current. Also, many of the information signals encountered in communication systems have amplitude probability distributions which approach the Gaussian distribution. For this purpose the Gaussian distribution is of great importance to the study of communication systems.

Consider a noise or information signal voltage at some point in the system. If it is not possible to write a deterministic function for this voltage we can still learn much about the system effect upon this signal if we know the *statistics* of the signal. One set of statistics is associated with the probability that if we look at the value of the voltage at any instant of time there is some probability of finding the voltage value within the small range of voltage Δv centered about some particular voltage value v_i. As we change the value of v_i, we of course expect to find different probabilities of the voltage falling in the vicinity of each value of v_i. If we plot these probabilities as a function of the values of v_i, we get a plot that looks like Figure F-1, and if we let the size of Δf shrink to zero, we obtain a smooth curve of probability *density* as is shown in Figure F-2. The particular *probability-density-function* curve $p(v)$ shown in F-2 is called the Gaussian distribution function. It is symmetrical about some point v_μ and the width of the curve is dependent upon a parameter called the *standard deviation*. As it turns out, v_μ (the mean or average voltage) corresponds to the d.c. value of the voltage time function and the standard deviation (represented by the symbol v_σ) corresponds to the rms (root-mean-square) value of the a.c. component of the voltage time function. If we want to find the probability that the voltage at any particular instant of time will fall between v_1 and v_2 and call it P_{1-2}, then

$$P_{1-2} = \int_{v_1}^{v_2} p(v) \, dv \tag{F-1}$$

If we let $v_1 = v_\mu - v_\sigma$ and $v_2 = v_\mu + v_\sigma$, we find that P_{1-2} is equal to 0.6825 or that 68.25% of the area under the distribution curve falls within one v_σ

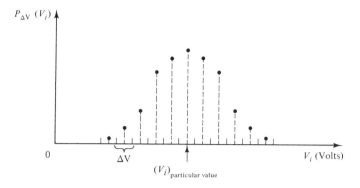

Figure F-1

of the mean value.   If we let $v_1 = v_\mu - 3v_\sigma$ and $v_2 = v_\mu + 3v_\sigma$, then we find that $P_{1-2} = 0.9975$ or that 99.75% of the area under the total curve falls within $3v_\sigma$ of the mean voltage value.   Very often we let the $\pm 3v_\sigma$ points represent the maximum and minimum voltage values (i.e., positive and negative peak voltages) of the signal.

The equation for $p(v)$ is given by

$$p(v) = \frac{1}{\sqrt{2\pi} \, v_\sigma} \, e^{-\left[\frac{(v - v_\mu)^2}{2v_\sigma^2}\right]} \tag{F-2}$$

where this function has been normalized so that its integration from $-\infty$ to $+\infty$ gives 1.0 for the value of the integral.   Further information about the Gaussian distribution function and its use may be found in any good book on statistical distribution functions.

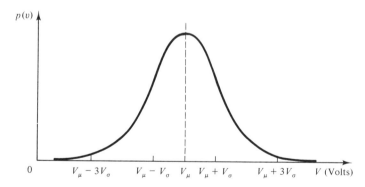

Figure F-2

Appendix G
THE CRC-CCITT REGISTER WITH THE DIGITAL MESSAGE
$(1100\ 0000\ 1111\ 1100)_2$

For the transmitter we first preload the register with 1 bits and then clock the message sequence through the register:

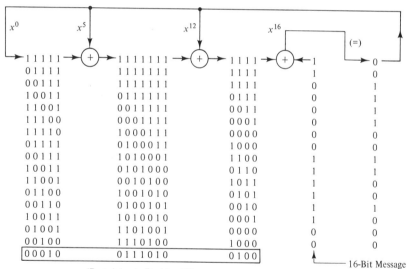

(Remaining in Register After
"Clocking In" Bit 16 of the Message)

The 1's complement of the bottom row of the shift-register portion is what is sent as the FCS character for the X.25 data frame, with the x^{16} bit being sent the first. Thus, the FCS for this case is 1101 1010 0011 0111 (time running from left to right).

When the FCS derived on the preceding page is added to the data bits, T is then 1100 0000 1111 1100 1101 1010 0011 0111. This is clocked through the shift register at the receiver (this shift register is also preloaded with 1 bits):

```
  x⁰            x⁵              x¹²            x¹⁶
                                                      (=)
     1 1 1 1 1 →(+)→ 1 1 1 1 1 1 1 →(+)→ 1 1 1 1 →(+)← 1        0
     0 1 1 1 1        1 1 1 1 1 1 1        1 1 1 1      1        0
     0 0 1 1 1        1 1 1 1 1 1 1        1 1 1 1      0        1
     1 0 0 1 1        0 1 1 1 1 1 1        0 1 1 1      0        1
     1 1 0 0 1        0 0 1 1 1 1 1        0 0 1 1      0        1
     1 1 1 0 0        0 0 0 1 1 1 1        0 0 0 1      0        1
     1 1 1 1 0        1 0 0 0 1 1 1        0 0 0 0      0        0
     0 1 1 1 1        0 1 0 0 0 1 1        1 0 0 0      0        0
     0 0 1 1 1        1 0 1 0 0 0 1        1 1 0 0      1        1
     1 0 0 1 1        0 1 0 1 0 0 0        0 1 1 0      1        1
     1 1 0 0 1        0 0 1 0 1 0 0        1 0 1 1      1        0
     0 1 1 0 0        1 0 0 1 0 1 0        0 1 0 1      1        0
     0 0 1 1 0        0 1 0 0 1 0 1        0 0 1 0      1        1
     1 0 0 1 1        1 0 1 0 0 1 0        0 0 0 1      1        0
     0 1 0 0 1        1 1 0 1 0 0 1        0 0 0 0      0        0
     0 0 1 0 0        1 1 1 0 1 0 0        1 0 0 0      0        0
     0 0 0 1 0        0 1 1 1 0 1 0        0 1 0 0      1        1
     1 0 0 0 1        1 0 1 1 1 0 1        1 0 1 0      1        1
     1 1 0 0 0        0 1 0 1 1 1 0        0 1 0 1      0        1
     1 1 1 0 0        1 0 1 0 1 1 1        1 0 1 0      1        1
     1 1 1 1 0        1 1 0 1 0 1 1        0 1 0 1      1        0
     0 1 1 1 1        0 1 1 0 1 0 1        1 0 1 0      0        0
     0 0 1 1 1        1 0 1 1 0 1 0        1 1 0 1      1        0
     0 0 0 1 1        1 1 0 1 1 0 1        0 1 1 0      0        0
     0 0 0 0 1        1 1 1 0 1 1 0        1 0 1 1      0        1
     1 0 0 0 0        0 1 1 1 0 1 1        1 1 0 1      0        1
     1 1 0 0 0        1 0 1 1 1 0 1        0 1 1 0      1        1
     1 1 1 0 0        1 1 0 1 1 1 0        0 0 1 1      1        0
     0 1 1 1 0        0 1 1 0 1 1 1        0 0 0 1      0        1
     1 0 1 1 1        1 0 1 1 0 1 1        1 0 0 0      1        1
     1 1 0 1 1        0 1 0 1 1 0 1        0 0 0 0      1        1
     1 1 1 0 1        0 0 1 0 1 1 0        0 0 0 0      1        1
(Remainder) 1 1 1 1 0   0 0 0 1 0 1 1      1 0 0 0
                                                  └── Message Plus FCS
```

Reading this register remainder from right to left, we get $(0001\ 1101\ 0000\ 1111)_2$, which is the specified value if no bit errors have occurred during the transmission process.

It is interesting to look at these operations in algebraic notation, letting $M(x)$ be the message polynomial, $G(x) = x^{16} + x^{12} + x^5 + 1$ be the generating polynomial, $C(x)$ the remainder, $T(x)$ the codeword and letting $L(x) = x^{15} + x^{14} + \cdots + x^1 + x^0$ (i.e., 16 1's in binary form). At the transmitting end, the preloading of the CRC register with all 1's amounts to adding $x^k L(x)$ to $x^{16} M(x)$, where k is the length of $M(x)$ (16 bits in the example given here). Division by $G(x)$ gives:

$$\frac{x^{16} M(x) + x^k L(x)}{G(x)} = Q(x) + \frac{C(x)}{G(x)}$$

where $C(x)$ is what remains in the CRC register after 16 shift operations. The codeword $T(x)$ then consists of $T(x) = x^{16}M(x) + [C(x) + L(X)]$, where adding $L(x)$ to $C(x)$ is equivalent to forming the 1's complement of $C(x)$. At the receiving end the presetting of the register to all 1's and clocking $T(x)$ through it for 32 cycles corresponds to dividing $[T(x) + x^kL(x)]$ by $G(x)$:

$$\frac{(x^{16}M(x) + [C(x) + L(x)] + x^kL(x)}{G(x)} = Q(x) + \frac{C(x)}{G(x)} + \frac{[C(x) + L(x)]}{G(x)}$$

where the contents of the CRC register after *32* cycles* corresponds to

$$\frac{C(x) + C(x) + L(x)}{G(x)} = \frac{L(x)}{G(x)}$$

This is represented by the modulo-2 division of $(1111\ 1111\ 1111\ 1111)_2$ by $(1\ 0001\ 0000\ 0010\ 0001)_2$. The contents remaining in the CRC register after cycle 32 represent the *remainder* of this division.

---

* If shifting of the *transmit* CRC register had been continued for 32 cycles instead of only 16, we would have obtained the remainder of $C(x)$ divided by $G(x)$ instead of just $C(x)$ itself as the contents remaining in the CRC register.

BIBLIOGRAPHY

This bibliography lists what this author believes to be the most valuable reference sources in a number of different areas. The lists are by no means complete, and a large number of other fine textbooks exist for each of the topic areas listed. In addition to the titles listed, most of which are text or reference books, the reader is referred to the various periodicals covering each of the subject areas. The publications of the Institute of Electrical and Electronic Engineers (IEEE) (its *Proceedings*, the *Transactions* of its various technical and professional groups, and the published results of its standards committees and various annual conferences) are quite valuable. In addition, the monthly and weekly magazine and newspaper-type publications (some of which are "freebies" to qualified subscribers) and the "house organs" of various manufacturers should not be overlooked as a means for keeping current with the latest developments in a given technical area.

This bibliography is divided into several of the technical subareas involved in communication systems, but some of the materials listed under a particular area may also contain information pertaining to one or more of the other areas listed. Some of the books listed may now be out of print. However, most of these were "old standards" in the field at one time and should be available at most of the larger technical libraries. They are included here because of their value as historical or tutorial references.

COMMUNICATION SYSTEMS (GENERAL)

Proceedings of the National Communications Forum, The National Engineering Consortium, Chicago, 1985.

RODDY, DENNIS, and JOHN COOLEN. *Electronic Communications*, 3rd ed. Reston, VA: Reston, 1984.

COMMUNICATION SIGNAL AND NOISE THEORY AND MODULATION THEORY

GREGG, W. DAVID. *Analog and Digital Communication*. New York: John Wiley, 1977.

HAYKIN, SIMON. *Communication Systems*, 2nd ed. New York: John Wiley, 1983.

RODEN, MARTIN S. *Analog and Digital Communication Systems*, 2nd ed. Englewood Cliffs, NJ: Prentice-Hall, 1985.

SWARTZ, MISCHA. *Information, Transmission, Modulation and Noise*, 3rd ed. New York: McGraw-Hill, 1980.

ZIEMER, RODGER E., and ROGER L. PETERSON. *Digital Communications and Spread Spectrum Systems*. New York: Macmillan Publishing Company, 1985.

COMMUNICATION SYSTEM CIRCUITS AND HARDWARE

CANNON, DON L., and GERALD LUECKE. *Understanding Communication Systems*, 2nd ed. Dallas, TX: Texas Instruments, 1984.

FIKE, JOHN L. and GEORGE E. FRIEND. *Understanding Telephone Electronics*, 2nd ed. Dallas, TX: Texas Instruments, 1984.

MANDL, MATTHEW. *Principles of Electronic Communications*. Englewood Cliffs, NJ: Prentice-Hall, 1973.

SMITH, JACK. *Modern Communication Circuits*. New York: McGraw-Hill, 1986.

TERMAN, FREDERICK E. *Radio Engineer's Handbook*. New York: McGraw-Hill, 1943.

YOUNG, PAUL H. *Electronic Communication Techniques*. Columbus, OH: Charles E. Merrill, 1985.

ANTENNAS AND PROPAGATION

American Radio Relay League (ARRL), *Antenna Anthology*, Newington, CT: ARRL, 1978.

American Radio Relay League (ARRL), *Antenna Compendium*, Newington, CT: ARRL, 1985.

American Radio Relay League (ARRL), *The ARRL Antenna Handbook*, 14th ed., Newington, CT: ARRL, 1982.

KRAUS, JOHN E. *Radio Astronomy*. New York: McGraw-Hill, 1966.

KRAUS, JOHN E. *Antennas*. New York: McGraw-Hill, 1950.

KUECKEN, JOHN A. *Antennas and Transmission Lines*. Indianapolis, IN: Howard M. Sams, 1969.

WALTER, CARLTON H. *Traveling Wave Antennas*. New York: McGraw-Hill, 1965.

COMMUNICATION AND NAVIGATIONAL SATELLITE SYSTEMS

BALAKRISHNAN, A. V., ed. *Space Communications*. New York: McGraw-Hill, 1963.

GAGLIARDI, ROBERT M. *Satellite Communications*. Belmont, CA: Lifetime Learning, 1984.

GOULD, R. G., and Y. F. LUM. *Communication Satellite Systems*. New York: Institute of Electrical and Electronic Engineers, 1982.

OPTICAL COMMUNICATION SYSTEMS

LACY, EDWARD A. *Fiber Optics*. Englewood Cliffs, NJ: Prentice-Hall, 1982.

MILLER, STEWART E., and ALAN G. CHYNOWETH. *Optical Fiber Telecommunications*. New York: Academic Press, 1979.

TELEVISION

Committee on Colorimetry of the Optical Society of America. *The Science of Color*. Washington, DC: Optical Society of America, 1963.

Federal Communications Commission. *Rules and Regulations of the FCC*. Washington, DC: Title 47, Section 73 of the United States National Code, published annually.

DATA COMMUNICATIONS

FIKE, JOHN L., GEORGE E. FRIEND, JOHN C. BELLAMY, and CHARLES BAKER. *Understanding Data Communications*, 2nd ed. Dallas, TX: Texas Instruments, 1984.

LIN, SHU, and DANIEL J. COSTELLO. *Error Control Coding*. Englewood Cliffs, NJ: Prentice-Hall, 1983.

MCNAMARA, JOHN E. *Technical Aspects of Data Communications*, 2nd ed. Bedford, MA: Digital Equipment Corp., 1982.

POOCH, U. W., W. H. GREENE, and G. G. MOSS. *Telecommunications and Networking*. Boston: Little, Brown, 1983.

STALLINGS, WILLIAM. *Data and Computer Communications*. New York: Macmillan Publishing Company, 1985.

STALLINGS, WILLIAM. *Local Networks*. New York: Macmillan Publishing Company, 1984.

TANENBAUM, ANDREW S. *Computer Networks*. Englewood Cliffs, NJ: Prentice-Hall, 1981.

HANDBOOKS AND REFERENCES

American Radio Relay League (ARRL), *The Radio Amateur's Handbook*. Newington, CT: ARRL, published annually.

FINK, DONALD G., ed. *Electronic Engineers' Handbook*. New York: McGraw-Hill, 1975.

FREEMAN, ROGER L. *Telecommunication Transmission Handbook*, 2nd ed. New York: John Wiley, 1981.

JORDAN, EDWARD, ed. *Reference Data for Radio Engineers*, 7th ed. Indianapolis, IN: Howard W. Sams, 1985.

INDEX

464

466

468

470